Synthesis, Properties and Mineralogy of Important Inorganic Materials

Synthesis, Properties and Mineralogy of Important Inorganic Materials

Terence E. Warner
Associate Professor of Materials Chemistry
University of Southern Denmark

A John Wiley and Sons, Ltd., Publication

This edition first published 2011
© 2011 John Wiley and Sons Ltd

Registered office
John Wiley & Sons Ltd, The Atrium, Southern Gate, Chichester, West Sussex, PO19 8SQ, United Kingdom

For details of our global editorial offices, for customer services and for information about how to apply for permission to reuse the copyright material in this book please see our website at www.wiley.com.

Library of Congress Cataloging-in-Publication Data

Warner, Terence E., 1960–
  Synthesis, properties and mineralogy of important inorganic materials /
Terence E. Warner.
      p. cm.
  Includes bibliographical references and index.
  ISBN 978-0-470-74612-7 (cloth) – ISBN 978-0-470-74611-0 (pbk.)
  1. Inorganic compounds.–Synthesis. 2. Inorganic compounds.–Properties.
I. Title.
  QD156.W37 2011
  541'.39–dc22
                                                    2010034769

A catalogue record for this book is available from the British Library.
ISBN HB: 9780470746127
ISBN PB: 9780470746110
e-Book: 9780470976029
e-pub: 9780470976234
Set in 10/12pt. Sabon by Thomson Digital, Noida, India.
Printed and bound in Singapore by Markono Print Media Pte Ltd.

For Kasper and Lars,
and in memory of the late Professor
Hans Toftlund

Contents

Foreword

This book is both timely and essential as, at present, there is a decline in the interest in inorganic chemistry as more and more bio related courses are introduced into University curricula. It is important to rectify this situation as the synthesis of inorganic materials with important physical and chemical properties will underpin many of the innovative industries of the future, and these industries will depend on the skills taught in this book. This applies especially to start-up companies where these specialist skills can be used to create novel products without an enormous capital investment. Given the present content of many courses, there is a dearth of trained scientists in this area. A significant number of the novel materials for the future will be synthesised by the techniques described in this book that meets a pressing need as chemists, physicists and materials scientists lacking in these essential skills from their undergraduate studies, can quickly learn the necessary techniques in order to make high-purity materials with the required properties.

Terence Warner has created a scholarly work that describes the synthesis of a selection of modern and ancient inorganic materials within the context of their mineralogy, properties and applications, which support many of our industries. It is informative to read how many of our advanced materials, or their closely related analogues, occur in nature. The text covers a wide range of materials from minerals that have been known to mankind for several generations, up to the latest synthetic electroceramic phases.

For graduate students and others who require a source of these materials for the study of their physical properties, it is always better and, perhaps, more satisfying, to prepare their own materials rather than rely on a supply from third parties. Furthermore, it is always preferable to synthesise the materials in-house so that a much greater understanding of the materials is obtained and the purity is assured. Although this book is primarily directed to final-year undergraduates and postgraduates, it also gives an excellent resume of the skills that had previously been taught in Universities for those working in industry. There is a continuing link between university research and industrial aspirations throughout this book.

Attention is also paid to the health and safety aspects of the preparations. Again, with a much greater emphasis on health and safety in Universities, some Departments are reluctant to undertake experiments involving high-temperature conditions and molten phases. This book very clearly demonstrates that it is possible to make these materials, without undue risk, using chemicals and equipment that are relatively cheap. For example, small pottery kilns are much cheaper than furnaces, yet deliver the same performance.

Overall, materials chemistry is an exciting area and this book ensures that the skills necessary to synthesise the relevant materials have been documented in a single volume. This book includes the science associated with the materials and their history of development and their applications, interspersed with fascinating facts, which makes a joy to read and, at the same time, for the reader to be stimulated.

Professor Derek J. Fray, *FREng FRS*
Cambridge

Preface

The synthesis of high-quality material is an essential step in the process of obtaining meaningful information about the material's properties, and therefore, is an important link between physics and chemistry. Semiconductors; superconductors; solid-electrolytes; glasses; pigments; dielectric, ferroelectric, thermoelectric, luminescent, photochromic and magnetic materials; are technologically important classes of material, that are represented by numerous inorganic phases. Yet how many of us are aware of their precise chemical compositions, and have sufficient knowledge to actually make them?

This book attempts to address this problem by offering the reader clear and detailed descriptions on how to prepare a selection of fifteen inorganic materials that exhibit important optical, magnetic, electrical and thermal properties; on a laboratory scale. The materials and chemical syntheses have been chosen so as to illustrate the large variety of physicochemical properties encountered in inorganic materials, and to provide practical experience covering a wide range of preparative methods, with an emphasis on high-temperature techniques. The majority of the materials described in the book relate to the macroscopic state and are prepared as polycrystalline materials. Several of these are fashioned into ceramic monoliths, whilst a few are retained in powder form or as single crystals.

The reader is given guidance, for example, to the quantities of material to prepare; sizes and types of crucibles and ampoules; use of furnaces; reaction times; heating and cooling rates; materials handling, such as, milling, mixing and separation procedures; multistage preparations; and the fabrication of ceramic monoliths. This book therefore aims at removing the rather abstract mysteries of chemical formulation and preparative methodology, whilst at the same time, focusing on the technologically important business of synthesising electroceramic materials.

The book is based on fifteen case studies, which illustrate how the properties of materials and the methods adopted for their syntheses, are dependent upon the chemical nature of the elements from which they are composed. Five of these materials contain the element copper situated in different chemical environments, which creates a convenient forum to discuss various topics, including: chemical affinity; electronegativity; acid–base chemistry; oxidation states; coordination chemistry; chemical bonding; electronic structure; crystal structure; and defect chemistry. These discussions extend throughout the book to incorporate the other ten materials.

Many useful ideas concerning this subject can be gained from a knowledge of mineralogy; which involves the study of naturally occurring inorganic crystalline phases. Therefore, references to mineralogy are made throughout the book. However, it is not essential for the reader to have a prior knowledge of mineralogy in order to understand the text since there are plenty of footnotes offering explicit information regarding mineralogical terminology. Likewise, certain historical details are included, so as to bring some of the curious tales surrounding the discovery of these materials, and the development of their syntheses, to the reader's attention.

The book has been written primarily for senior undergraduate and postgraduate students, and is intended to serve as a textbook for laboratory courses in chemistry, ceramics, materials science, and solid state physics degree programmes. It is written in a style that should encourage students to make the transition from *reading* to *researching* for their degree, and thereby help bridge the gap between conventional textbooks on the subject, and the research literature as published in scientific journals.

Many researchers working today in the field of materials science and technology have a background that may not have equipped them with sufficient chemical knowledge in order for them to prepare their own materials. This is often the case for newly graduated physicists, geologists and engineers who are engaged with postgraduate or postdoctoral research, where they are obliged to prepare inorganic materials as part of their research activities. This book should be of benefit to them. For instance, after attempting the syntheses described in this book, the reader should have acquired sufficient skills and knowledge so as to be able to prepare many similar and, unusual, inorganic materials.

The book is of appeal to university lecturers and school teachers who are interested in extending their repertoire of teaching material into the realms of high-temperature synthesis. It is also of interest to professional chemists, physicists, materials scientists and technologists, ceramicists, mineralogists, geologists, geochemists, archaeologists, metallurgists, engineers, and non-specialists, who are interested in learning more about how inorganic materials and artificial minerals are made.

Finally, the author assumes that the reader is familiar with the basic principles and concepts of materials chemistry (or at least has access to such knowledge), such as: thermodynamic equilibrium; interpreting phase diagrams; chemical bonding; crystal structure; powder X-ray diffractometry;

optical and electron microscopy; crystal defects and Kröger–Vink notation; and the properties of materials. These subjects are covered extensively in undergraduate and postgraduate textbooks on physical chemistry; inorganic chemistry; solid-state chemistry; solid-state physics; materials chemistry; and mineralogy; and so they will not be repeated here. In this respect, the various footnotes refer the reader to a selection of relevant texts.

Acknowledgements

I wish to begin by thanking Dr Adam N. Bowett and Professor Keith Murray for encouraging me to put pen to paper and write this book. The nature of the subject brought me into dialogue and correspondence with numerous persons from a diverse range of professional backgrounds. I very much appreciate their willingness to share their knowledge and to provide literary and illustrative material, translations and experimental data. Consequently, I am grateful to Professor Andrew Bond, Professor John R. Cooper, Professor Judith L. Driscoll, Dr Mark Dörr, Professor Peter P. Edwards, Professor Dr Mohammed Es-Souni, Kristine Gable, Associate Professor Per Lyngs Hansen, Dr Yvonne Harpur, Associate Professor Ole Johnsen, Dr Dedo von Kerssenbrock-Krosigk, Associate Professor Venkatesan Manivannan, Associate Professor Per Morgen, Dr Boujemaa Moubaraki, Assistant Professor Ulla Gro Nielsen, Dr Ngo Van Nong, Associate Professor Ole V. Petersen, Associate Professor Victor V. Petrunin, Associate Professor Birte Poulsen, Dr Nini Pryds, Associate Professor Kaare L. Rasmussen, Professor Thilo Rehren, Dr Chris J. Stanley, Associate Professor Paul C. Stein, Mrs Sylvia Swindells, Professor Emeritus Michael S. Tite, Mary E. Vickers, Associate Professor Stefan Vogel, Nicola Webb, Professor Mark T. Weller, Associate Professor Emeritus Ole Wernberg, Professor Emeritus Jaime Wisniak, Dr Chris J. Wright and Dr Jianxiao Xu.

I am also grateful to my students who performed the syntheses, and in certain instances, some of the research described in the book. I also convey my thanks and appreciation for the technical assistance given by Mrs Susanne J. Hansen, Mr Poul Bjerner Hansen, Mr Torben H. Jensen and Mr Tommy Nørnberg.

I thank the referees (who remain anonymous to me) for their comments and suggestions regarding my original proposal. I also express my deep gratitude to Professor Derek J. Fray, dr. techn. Finn Willy Poulsen, Dr Nevill M. Rice and Professor Eivind M. Skou for reading the typescript and offering their constructive criticisms and comments. Their efforts have helped enormously to improve the book. It is, however, only fair to add that I remain entirely responsible for all errors and inconsistencies in the published text.

I also wish to thank Associate Professor Ole Johnsen, for kindly allowing the reproduction of four of his photographs of mineralogical specimens in this book; three of these were published originally in *Mineralernes Verden* (GadsForlag, 2000). I am also sincerely grateful to Professor Derek J. Fray for writing the Foreword.

I thank my wife, Tine, for her unremitting support that ensured the completion of this work. Sadly, whilst I was writing this book, my colleague and friend Professor Hans Toftlund passed away. I wish to express my appreciation for the knowledge and inspiration that Hans so generously gave me and that created the foundation for this book. His company is greatly missed.

Finally, I wish to express my thanks to my publishers for their enthusiastic and helpful co-operation throughout the production of this book, and particularly to Alexandra Carrick, Richard Davies, Jon Peacock and Emma Strickland for their editorial advice and guidance.

1

Introduction

There is an island off the west coast of Scotland whose name is given to a mineral that was first discovered there by Bowen *et al.* in 1924 [1]. The mineral occurs within an argillaceous rock[1] that has undergone thermal metamorphism[2] as a consequence of becoming entrapped within a hot magma during the Tertiary period. The mineral is quite remarkable in that the synthetic equivalent is produced within terra-cotta[3] (baked clay) and thereby must surely represent one of the earliest crystalline inorganic ternary phases prepared artificially by mankind. Furthermore, terra-cotta artefacts have been fabricated since antiquity; thus marking the advent of ceramics, both as a material and a subject.

The mineral in question is *mullite*, as from the Isle of Mull, Scotland. Mullite has a chemical composition corresponding to the solid-solution series, $Al_{4+2x}Si_{2-2x}O_{10-x}$ ($0.17 \leq x \leq 0.59$), and occurs as tiny colourless acicular crystals that are normally invisible to the naked eye (see Figure 1.1). It has a crystal structure closely related to sillimanite, Al_2SiO_5 (the mineral that it was originally mistaken for); but it is commonly reported as simply, $3Al_2O_3 \cdot 2SiO_2$ [2]. Mullite is formed in nature under the unusual geological conditions of high temperature and low pressure (corresponding to the sanidinite facies: $T = 800 - 1000\,^\circ\text{C}$ and $p \leq 2\,\text{kbar}$) and is found in certain metamorphic glassy rocks, such as buchite [3].

---

[1] *Argillaceous rocks* are a group of sedimentary rocks (e.g., clays, mudstones, shales and marls) that have a high content of clay minerals.

[2] *Thermal metamorphism* is the process by which rocks within the earth's crust are altered under the influence of heat alone; cf. contact metamorphism.

[3] *Terra-cotta* is a crude earthenware that is usually brownish-red due to the presence of hematite, Fe_2O_3 cf. the more refined potteries e.g., faience, stoneware and porcelain.

Synthesis, Properties and Mineralogy of Important Inorganic Materials   By Terence E. Warner
© 2011 John Wiley & Sons, Ltd

Figure 1.1 Scanning electron micrograph of mullite needles formed hydrothermally in small druses of volcanic rocks of the Eifel mountain, Germany (Reproduced with permission from Mullite by H. Schneider, Fig. 1.3, xvii, Edts. H. Schneider and S. Komarneni, Copyright (2005) Wiley-VCH)

When ball clay (a natural earthy mass of kaolinite, $Al_2Si_2O_5(OH)_4$ and other clay minerals) is heated artificially, it first undergoes dehydration, followed by an exothermic reaction at $\sim$950 °C, where upon the dehydrated material transforms into an impure composite of mullite, β-quartz and γ-alumina; commonly known as terra-cotta. This is a complex metastable system in which the fraction of mullite increases with annealing time and temperature, at the expense of β-quartz and γ-alumina. Hence, ordinary pottery is a more intricate material than what first meets the eye. Given the primitive conditions under which terra-cotta was prepared by prehistoric man, it is not too surprising that this material comprises the three most abundant elements in the earth's crust, namely; oxygen, silicon and aluminium. This also reflects the ubiquitous nature of the clay minerals, which can be collected quite readily with bare hands, from surface deposits.[4]

Terra-cotta artefacts have been manufactured in Mesopotamia since at least 6000 BC [6].[5] This predates the smelting of copper and the synthesis of

---

[4] The vast majority of clay minerals are essentially hydrous aluminium silicates. The reader is referred to Schneider and Komarneni [4] for further details regarding the subject of mullite; and to Holdridge and Vaughan [5] for details concerning the thermal treatment of kaolin.

[5] *TheTerracotta Army* of Qin Shi Huang the First Emperor of China, discovered near Xi'an Shaanxi province, China, is estimated to comprise over 8000 terra-cotta figures (mostly depicting soldiers as life size), and is an impressive example of the use of terra-cotta dating from *ca.* 210 BC.

Figure 1.2 Nest of triangular crucibles recovered in Hesse (Photograph used with kind permission from S. Häpe, courtesy of H.-G. Stephan Copyright (2006) H.-G. Stephan)

the copper–arsenic and copper–tin alloys (*ca.* 3500 BC). The usefulness of terra-cotta vessels (pots, beakers, crucibles, etc.), particularly those with a high content of mullite, cannot be over emphasised. Historically, their fabrication was a prerequisite for the smelting of metals, and thus the development of metallurgy per se. The famous Hessian mullite crucibles fabricated since the 12th century at Epterode and Almerode (now; Großalmerode), in the region of Hesse, Germany, were renowned for their refractoriness, and represent the earliest industrial exploitation of mullite in Europe (see Figure 1.2) [7, 8].

Throughout history, terra-cotta vessels of various qualities, have been used for the preparation of many other useful substances, such as, zythum,[6] pigments and glass, as well as for the practices that led to the development of chemistry. Indeed, the very word *chemistry* is considered to have derived from the Arabic *al-kimia*, through the transliteration of the Greek χυμεία (*khymeia*; in English *pouring*) in connection with the *infusion* of plants for pharmaceutical use [9], and the *casting* of metals.[7] The point being stressed here is that in order to prepare a chemical substance, particularly when the process involves liquids or operating at high-temperature, the practitioner

---

[6] *Zythum* was a drink made in ancient times from fermented malt, especially in Egypt.

[7] The popular association of *al-kimia* (in English *alchemy*) with the native name of Egypt, *Khem* ('land of black earth' cf. Plutarch *ca.* 100 AD), by allusion to the black soil of the Nile delta in contrast to the desert sand – hence, 'the art of the black earth' in reference to the pursuits of the Alexandrian alchemists in Hellenistic Egypt– is considered by Mahn [9] to be a subsequent development of the subject, and that the origin of the word is most probably Greek [9].

Figure 1.3 *An Alchemist*, 1661, Adriaen van Ostade (1610–1685). Note the triangular crucibles in the foreground, and the importance of heat for inducing a change in chemical substances (Reproduced with permission. Adriaen van Ostade, An Alchemist © The National Gallery, London.)

must first of all acquire a suitable container in which to conduct the synthesis; an aspect that is of course, still very relevant today.

Chemistry as a scientific discipline grew out of the magical practice of alchemy in Europe during the Middle Ages. One of the main aspects of alchemy involved serious practical attempts at making gold through the transmutation of the less noble metals;[8] and was accompanied by general random experiments with various metals, minerals, salts, acids, gums and chemical extractions, often at elevated temperature. Given the high value of gold, and the heinous punishments meted out by the Church for deviant behaviour, alchemists devised their own mysterious symbolisms for describing their preparative procedures, such that their work was generally shrouded in secrecy. The painting of '*An Alchemist*' by Adriaen van Ostade, casts a satirical portrayal of these futile activities during the 17th century (see Figure 1.3). Coincidentally, Ostade painted '*An Alchemist*' in the same

---

[8] In particular; mercury, copper, tin and lead.

year, 1661, that Robert Boyle wrote '*The Sceptical Chymist*'; which marks a turning point in the transition between alchemy and chemistry.[9] Although these attempts at making gold were of course all in vain, it is through these practical activities that alchemy transformed into scientific pursuits, such as the chemical analysis of minerals and the discovery of the chemical elements and their compounds. These in turn, led to commercial enterprises like the production of European porcelain, and eventually, the modern ceramic and giant electronics industries.

The roots of chemistry, therefore, lie deeply in the intermingled business of synthesis and analysis; in which synthesis can be defined as the art of bringing about purposeful changes in chemical substances. The principle role of the chemist is to investigate the various substances of which matter is composed, and thereby acquire knowledge of their chemical nature, in terms of composition, structure, affinity and reactivity; and develop methods for their synthesis.

Chemistry has often been closely associated with mineralogy. During the 18th and 19th centuries it became very much a gentleman's pursuit to perform chemical analyses on the numerous mineral species that were being discovered during this period of scientific enlightenment. Furthermore, a definitive way to prove the existence of a new mineral species was to synthesise its artificial equivalent. It is also important to remember that all chemical substances are derived ultimately from terrestrial and extraterrestrial resources, of which minerals are by far the major part. Today, although the recycling of spent materials constitutes an important economic resource, minerals still remain the primary source of industrial raw materials; and that includes the feedstock for inorganic chemical reagents.

Minerals constitute a large body of crystalline inorganic compounds, and the mineralogical literature contains valuable information concerning their chemical composition, crystal structure, physical properties and paragenesis.[10] Certain minerals, for example; scheelite, $CaWO_4$; magnetite, Fe_3O_4; and quartz, $\alpha\text{-}SiO_2$; have inspired technological applications directly on account of their luminescent, magnetic and piezoelectric properties, respectively. The rare mineral lazurite, $(Na,Ca)_8[Al_6Si_6O_{24}](Cl,SO_4,S)_{2-\delta}$ has been used as a blue pigment since antiquity and was literally worth its weight in gold. Attempts at synthesising its artificial equivalent presented some considerable challenges in both analytical and preparative chemistry. But eventually, in 1828, with the enticement of a prize worth 6000 francs from the *Société d'encouragement pour l'industrie nationale*, Jean Baptiste Guimet

---

[9] It is interesting to note, that Isaac Newton (1642–1727) spent over 20 years practising alchemy in a garden shed at Trinity College, Cambridge [10].

[10] *Paragenesis* refers to the association of different minerals within the same deposit; and so reflects their origin (i.e., the conditions for their primary crystallisation and subsequent history) either collectively, or in reference to a specific mineral. The word *Paragenesis* is derived from ancient Greek; *Para-* meaning, along side; *–genesis* meaning, creation, origin.

succeeded in making artificial lazurite from kaolin;[11] and his efforts led to the commercial production of the pigment, *ultramarine* [11].

New mineral species continue to be discovered, and many of those that are known already have not yet been prepared artificially, for example, the gemstones; azurite, $Cu_3(CO_3)_2(OH)_2$; and dioptase, $Cu_6Si_6O_{18}·6H_2O$. This is just one of the reasons why the subject is so interesting. Certain workers have even coined the expression, *geomimetics*, to describe the imitation of geological materials, in an attempt to promote the subject from this perspective. In general terms, materials chemistry can be considered as a broad interdisciplinary field that embraces many of the chemical aspects of mineralogy, ceramics and metallurgy, and overlaps with archaeology and solid-state physics. The monograph by Ebelmen, Salvétat and Chevreul, published in 1861 and entitled, '*Chimie, Céramique, Géologie, Métallurgie*', bears testament to this [13].

Today, there are many books published that contain descriptions regarding preparative methods in solid state chemistry; but only a few are dedicated to the subject.[12] Collectively, they represent a very significant source of knowledge and inspiration, but with the notable exception of *Wold and Dwight*, they all focus on methodology and experimental techniques. Examples of real materials, when given, are usually written in the spirit of exemplifying the technique, rather than offering the reader clear and detailed preparative instructions of how to make the material. On the other hand, the preparative methods reported in scientific journals tend to be rather brief, especially those published in recent years. Consequently, many of the descriptions reported in the literature concerning the preparation of inorganic materials can be difficult to follow. So, for students, and postdoctoral workers new to the field, this can present a genuine problem.

This book was written with that problem in mind. Therefore, it can be considered primarily, as a guide to the synthesis of inorganic materials, and offers the reader comprehensive and detailed step-by-step instructions, in order to prepare a selection of materials on a laboratory scale. Since the properties of these materials are of inherent scientific and technological importance, it is necessary to discuss these in reasonable detail. The book describes a series of case examples that involve a wide range of preparative techniques. These have been selected to ensure that they are practically feasible, given normal laboratory facilities, and yet retain a certain challenge to the reader. The subject matter dealing with methodology and experimental

---

[11] Ultramarine can be prepared industrially by heating a blended mixture of kaolin (essentially, kaolinite, $Al_2Si_2O_5(OH)_4$), sodium carbonate, silica, sulfur and coal tar pitch to 750–800 °C for 50–100 h, followed by slow cooling over several days under an oxidising atmosphere [12].

[12] For example; *Inorganic Materials Synthesis and Fabrication*' by Lalena *et al*. [14]; '*Synthesis of Inorganic Materials*,' by Schubert and Hüsing [15]; '*Solid State Chemistry: Synthesis, Structure, and Properties of Selected Oxides and Sulfides*' by Wold and Dwight [16]; and, '*Chemical Synthesis of Advanced Ceramic Materials*' by Segal [17].

techniques are described in many excellent texts and therefore is not covered in great detail here.

The philosophy of the book is to show, through the descriptions of real chemical systems, how the properties of a material can be influenced by modifications to the preparative procedure by which the material is made; and vice versa. For instance, the nature of the chemical reagents, the impurities and the thermal history, can have a profound effect on the product material, especially with regards to chemical composition, homogeneity, crystal structure, morphology, microstructure, and thus, its physical properties. Conversely, the requirement for a specific property within a given material will normally dictate certain details in the course of its preparation. Furthermore, the production of phase-pure material is normally a prerequisite for any meaningful characterisation and exploitation of the material. But in reality, this is often a difficult task, and so, before this goal can be achieved, many skills need to be developed and knowledge of the relevant chemical system acquired.

This book is a cross between a textbook and a monograph. The chapters are named according to chemical composition and preparative method. It was considered rather pointless, however, to place too much emphasis on categorising the materials according to a principle property, since in several cases they exhibit more than one. Although the chapters are essentially self-contained, it is assumed that they will be read in sequential order, for the sake of continuity of the story. In general, the International System of units, is used in this book, but there are occasions when matters of practicality or, common sense, dictate the use of the more appropriate units, such as; degrees Celcius, °C; bar; hour, h; minute, min; ångström, Å; litre, l; Bohr magneton, μ_B; oersted, Oe; and the electromagnetic unit, emu.

The illustrations of the crystal structures in this book were produced with ATOMS, by Shape Software,[13] using data from the Inorganic Crystal Structure Database (ICSD)[14] with reference to the original publications from which these data were compiled. In most cases the atomic and ionic radii are drawn smaller than in reality; thus enabling the reader to look into the interior of, or through, the unit cell. The E_H-pH diagrams were produced using Outokumpu HSC Chemistry® Software.[15] Certain phase equilibria diagrams from The American Ceramic Society[16] have been reproduced in this book and used with permission from the copyright holders of the original publications.

[13] Shape Software, 521 Hidden Valley Road, Kingsport, TN 37663 USA.

[14] Inorganic Crystal Structure Database, Fachinformationszentrum, Hermann-von-Helmholtz-Platz 1, 76344 Eggenstein-Leopoldshafen, Karlsruhe, Germany.

[15] Outokumpu HSC Chemistry® Software, Outokumpu Research Oy, Information Service, PO Box 60, FIN-28101 Pori, Finland.

[16] ACerS-NIST Phase Equilibria Diagrams, The American Ceramic Society, P.O. Box 6136, Westerville, OH 43086-6136, USA.

In order to perform the various syntheses, the reader may wish to use this book simply as a laboratory manual. Alternatively, after reading the book in more depth, the reader should be in a better position to embark on the business of making the inorganic materials that are required for his or her own research, or private interests. Finally, a set of problems is included at the end of each chapter for the reader's perusal. Suggested answers to these, together with formulations for the respective syntheses, can be made available to bona fide course instructors upon request, via; www.wiley.co.uk.

REFERENCES

1. N. L. Bowen, J. W. Greig and E. G. Zies, *Journal of the Washington Academy of Sciences*, **14** (1924) 183–191.
2. R. V. Gaines, H. C. W. Skinner, E. E. Foord, B. Mason, A. Rosenzweig, *Dana's New Mineralogy: The System of Mineralogy of James Dwight Dana and Edward Salisbury Dana*, 8th edn, 1997, Wiley-Blackwell, New York.
3. B. W. D. Yardley, W. S. MacKenzie and C. Guilford, *Atlas of Metamorphic Rocks and their Textures*, Longman, Harlow, Essex, 1990, pp 120.
4. H. Schneider and S. Komarneni (eds.), *Mullite*, Wiley-VCH, Weinheim, (2005) 487pp.
5. D. A. Holdridge and F. Vaughan, *The kaolin minerals (kandites)*, 98–139, in R. C. Mackenzie (ed.), *The differential thermal investigation of clays*, Mineralogical Society, London (1957) 456 pp.
6. F. R. Matson, *A Study of Temperature Used in Firing Ancient Mesopotamian Pottery*, in *Science and Archaeology*, ed. R. H. Brill, MIT Press, Cambridge, Massachusetts (1971) 65–79.
7. M. Martinón-Torres, I. C. Freestone, A. Hunt and T. Rehren, *Journal of the American Ceramic Society* **91** (2008) 2071–2074.
8. M. Martinon-Torres, Th. Rehren and I. C. Freestone, *Nature* **444** (2006) 437–438.
9. C. A. F. Mahn, *Etymologische Untersuchungen auf dem Gebiete der romanischen Sprachen*, F. Duemmler, Berlin, (1854–1864), Specimen XI (1858), Section LXIX, pp 81–85.
10. P. E. Spargo, *South African Journal of Science* **101** (2005) 315–321.
11. E. Bang and S. Harnung, *Dansk Kemi* **72** (1991) 460–465.
12. R. M. Christie, *Colour Chemistry*, Royal Society of Chemistry, Cambridge, 2001, pp 206.
13. J. J. Ebelmen, M. Salvétat, E. Chevreul, *Chimie, Céramique, Géologie, Métallurgie*, Mallet-Bachelier, Paris, (1861) pp 628.
14. J. N. Lalena, D. A. Cleary, E. E. Carpenter and N. F. Dean, *Inorganic Materials Synthesis and Fabrication*, Wiley, New Jersey, 2008, pp 303.
15. U. Schubert and N. Hüsing, *Synthesis of Inorganic Materials*, Wiley-VCH, Weinheim, 2005, pp 409.
16. A. Wold and K. Dwight, *Solid State Chemistry: Synthesis, Structure, and Properties of Selected Oxides and Sulfides*, Chapman & Hall, New York, 1993, pp 245.
17. D. L. Segal, *Chemical Synthesis of Advanced Ceramic Materials*, Cambridge University Press, Cambridge, 1991, pp 182.

2

Practical Equipment

This chapter offers information and practical advice about the equipment needed to perform the syntheses of the materials as described in the book. It does not discuss the principles of the methods employed, at least not in detail, since these are covered elsewhere in many excellent texts; nor does it describe the methods concerning the characterisation of materials, since these are referred to at the relevant places within the main body of the text. This chapter does include, however, certain practical information regarding powder X-ray diffractometry, because this is the principle technique used for the routine phase analysis of crystalline inorganic materials. In addition to the information given here, the reader is strongly advised to follow the manufacture's instructions when using the equipment referred to in this book.

2.1 CONTAINERS

One of the points raised in Chapter 1 is the importance of acquiring a suitable container in which to conduct the synthesis. Fundamentally, the reaction vessel must survive contact, not only with the chemical reagents and precursor compounds, but also with the product material, and any chemical species and intermediate phases produced during the course of the reaction. These requirements generally become more prohibitive as the temperature and pressure of the system is increased. Furthermore, the retrieval of product material that adheres to the walls of the container often presents an additional problem that needs to be considered when devising a preparative method.

Beakers and flasks made of Pyrex™ glass are quite adequate for handling the majority of the aqueous solutions and precipitated material described in

Synthesis, Properties and Mineralogy of Important Inorganic Materials By Terence E. Warner
© 2011 John Wiley & Sons, Ltd

this book. Pyrex™ glass has excellent corrosion and thermal-shock properties that make it ideally suited for general use in preparative chemistry; at least at temperatures below $500\,°C$. For operations above $500\,°C$, a container made from a more resilient material is needed.

Porcelain crucibles find popular use in conventional chemical laboratories, and serve well for many of the preparations performed in analytical chemistry that require moderate heating, for example, with a Bunsen burner. Porcelain is a composite material comprising an entangled mass of acicular crystals of mullite, $Al_6Si_2O_{13}$ dispersed within a silicate glass matrix; which results in a tough material with a high compressive strength and a good resistance to thermal shock. Porcelain crucibles are available; glazed[1] and unglazed. The reader should be aware of substances that can react with the glaze; and indeed with the glass phase within the porcelain itself. For instance, alkali metal oxides can diffuse into the glaze upon heating and thereby degrade the crucible. But in all cases, porcelain crucibles are not recommended for high-temperature or refractory use; especially since more suitable alternatives are commercially available.

From a conceptual point of view, eliminating the presence of the glass phase from a typical porcelain body, leads towards, a ceramic material comprising solely of mullite. Mullite is an important refractory material with a melting point of $1810\,°C$. It has an outstanding resistance to thermal shock, and in this respect is superior to recrystallised alumina, even though alumina has a higher melting point of $2040\,°C$. For these reasons, mullite crucibles are used extensively in the melting of iron alloys, the base metals and their alloys. Nevertheless, one should be aware that mullite contains a significant amount of chemically combined 'SiO_2' that can be reactive towards certain metal oxides at high temperature. The manufacturing of mullite crucibles is generally more demanding than that for porcelain, and so they command a higher price.

This leads to recrystallised alumina, Al_2O_3 as the preferred choice of crucible material for many of the preparative descriptions in this book;[2] see Figure 2.1. Recrystallised alumina crucibles have limitations as well, of course. Although they are densely sintered, they are nevertheless prone to react with certain chemical substances within systems in which free alumina, Al_2O_3 finds itself thermodynamically unstable. For example, Al_2O_3 is unstable in the presence of the copper oxides, CuO, and Cu_2O, with respect to the formation of the ternary phases, $CuAl_2O_4$, and $CuAlO_2$; which can manifest as surface coatings on the inner walls of the alumina crucible. This is

---

[1] A glaze is a thin layer of glass used as a protective coating; commonly used to render pottery impermeable to fluids, and to avoid entrapment of powdered material, dust, etc.

[2] The alumina crucibles used in the practical work as described in this book were supplied by Almath Crucibles Ltd., Epsom Building, The Running Horse, Burrough Green, Nr Newmarket, Suffolk, CB8 9NE, United Kingdom.

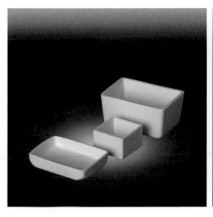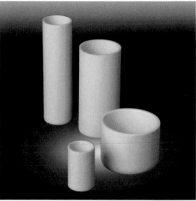

Figure 2.1 Recrystallised alumina crucibles similar to those referred to in the book (Reproduced with permission from Almath Crucibles Ltd Copyright (2010) Almath Crucibles Ltd)

accompanied by an undesirable loss of a corresponding fraction of the copper oxide component from the chemical reactants. The effects of this can be critical in attempts at preparing high-quality material. Furthermore, it is also fairly common for product material to adhere to the inner walls of the alumina crucible, and thereby make it difficult to retrieve the product.

There are alternatives to the above crucibles, but the more specialist crucibles tend to be much more expensive. A few commercial examples are; zirconia,[3] pyrolytic graphite, pyrolytic boron nitride, silicon carbide, gold, and platinum. Pyrolytic graphite moulds are convenient for casting certain metals and silicate glasses (see Figure 2.2). Glassy carbon crucibles, and Pyrex™ and quartz glass crucibles coated with glassy carbon are used for flux-growth methods. There are many other possible materials, for example, copper crucibles are ideal for preparing complex copper metal phosphates under a nonoxidising (inert) atmosphere, in which it is advantageous to maintain a saturation of elemental copper (i.e., a copper activity of unity) in the material throughout the heating process. Crucibles made of other metallic elements might also be useful for a similar reason. But one should be aware of bending metal sheet; hard metals such as tungsten are not recommended, since they splinter readily into razor sharp fragments!

With regards to ampoules, Pyrex™ glass is a convenient material for use as a glass ampoule, below its glass transition temperature of $\sim 500\,^{\circ}$C. These can be evacuated, for example, with the aid of an oil pump to $\leq 10^{-4}$ bar, and then sealed with a blowtorch. Glass ampoules also have the additional benefit of

---

[3] Crucibles trading under the description of 'zirconia', may in fact be, 'fully stabilised zirconia'. This material is a solid solution between yttria and zirconia, with the composition, 8 mole% Y_2O_3 and 92 mole% ZrO_2.

Figure 2.2 Pyrolytic graphite ingot (Photograph used with kind permission from Graphitestore.com, Inc. Copyright (2010) Graphitestore.com, Inc)

being transparent and therefore allowing the contents to remain visible. For use above 500 °C, or with substances corrosive to Pyrex™ glass, one must resort to quartz glass (a glass with $\geq$96% SiO_2). Quartz glass is more expensive, and more difficult to soften and work with, but it is more stable at high temperature and more resistant to chemical attack. It can be used up to 1050 °C for short periods, for example, a couple of hours when not under vacuum.[4] For longer periods, the temperature should not exceed 1000 °C, and this temperature should be reduced if operating under vacuum. All glass ampoules perform poorly with internal pressures exceeding atmospheric pressure; and have a tendency to explode! Their use under these conditions is dangerous and is not recommended. One should be aware of the volatility of the material encapsulated in ampoules, and if necessary, consider the use of an alternative material, such as metal ampoules made of gold, platinum, nickel, copper, etc.; or perhaps the use of more specialist high-pressure equipment.

Autoclaves are used in systems that require the presence of a volatile liquid at elevated temperatures. Chapter 13 describes a hydrothermal synthesis using an autoclave that was built in-house. The autoclave was machined from 18-8 stainless steel (72 mm ID; 94 mm OD; and 200 mm in height) and is equipped with a removable Teflon® container in which the reactants are placed (see Figure 2.3). This autoclave has a safety-valve built into the lid with a release pressure of 15 bar ($\sim$218 psi). One should be aware that the Teflon® container (which is made of polytetrafluoroethylene, PTFE) begins to decompose above 260 °C, with the formation of highly toxic fluorocarbon gases. This particular autoclave is heated by placing it inside a thermostatically controlled electric oven.

---

[4] To avoid any misunderstanding; pure crystalline cristobalite, SiO_2 has a melting point of 1710 °C. Some workers claim that quartz glass can be used as high as 1200 °C for short periods.

Figure 2.3 Stainless steel autoclave (left) with its inner Teflon® container (right) and lid (front-left); built in-house and well used (Photograph used with kind permission from Eivind Skou. Copyright (2010) Eivind Skou)

2.1.1 Instructions for Making a Terra-Cotta Crucible

Terra-cotta crucibles have been used since antiquity and are still useful today. The following account gives a detailed description of how to make a terra-cotta crucible (see Figure 2.4).

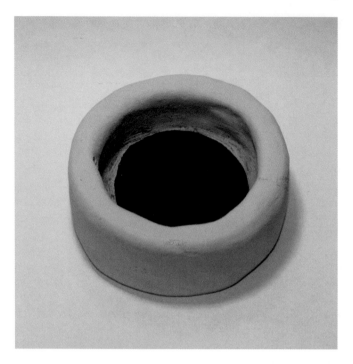

Figure 2.4 A terra-cotta crucible as prepared by the method described in this chapter (internal diameter = 50 mm). The contents of this particular crucible comprise, artificial cuprorivaite $CaCuSi_4O_{10}$ (Egyptian blue) together with the crystallised salt flux, as described in Chapter 3 (Photograph used with kind permission from Simon Svane. Copyright (2010) Simon Svane)

First, obtain a lump of ball clay (1–10 kg) from a local pottery dealer. The clay used by the present author is a local blue-clay that forms a pale yellowish-brown terra-cotta after firing. There are many ways to make a clay pot. The one chosen here is by the coil method as described by Clark [1]. The attraction of this method is that one does not need a potter's wheel. Use unadulterated ball clay, and do *not* add sand or grog (i.e., ignore the advice given by Clark in this respect). The clay should already be sufficiently moist with a high degree of plasticity, so that the addition of water should be avoided throughout the following process. Roll out a lump of the clay (∼200 g) into a long strip ∼12 mm diameter. Begin by forming a flat coil for the base, and then continue upwards in a spiral to form the sides of the crucible, so as to create an internal volume of ∼100 ml (approximately: 60 mm ID, 80 mm OD, and 35 mm in height). Gently smear these coils together to form a coherent and smooth surface. Place a thin plastic sheet loosely over the crucible and leave it to dry slowly in a well-ventilated place away from direct heat. Ideally, this should be left to dry for about one week on a window sill.

Once the clay crucible is dry, it is advisable to place it in an alumina dish (LR94 Almath Ltd) as a precautionary measure, just in case the clay fuses excessively at high temperature. Then place it inside the pottery (or chamber) furnace, and fire (i.e., heat) at 100 °C/h to 1040 °C. Do not heat ball clay >1050 °C, or else the clay will most likely vitrify; and if the temperature is too low (i.e. <950 °C) then mullite is unlikely to form and so the pot will be more fragile. After 10 h at 1040 °C, cool to ambient temperature (20 °C) with a cooling rate of 300 °C/h. This heating cycle will last ~24 h. This method should result in a terra-cotta crucible that can be used, for example, in the synthesis of $CaCuSi_4O_{10}$ as described in the next chapter.

2.2   MILLING

Fine particle size and intimately mixed chemical reactants have a significant and beneficial effect on the quality of the reaction product, especially in terms of phase purity and homogeneity. These aspects also help to maximise the *ceramic density* of ceramic monoliths when prepared by the sintering of powder-compacts as discussed below. Therefore, the grinding and mixing operations of solid-state matter are very important. Before commencing with milling or grinding operations, it is advisable to crush any large or tough lumps of material using, for example, a hardened steel piston-type percussion mortar as shown in Figure 2.5.

Porcelain pestles and mortars are used routinely in the laboratory for grinding particulate matter by hand, and are acceptable for undergraduate laboratory classes. Agate pestles and mortars tend to yield a finer powder, but are more expensive. Ball milling is a preferred alternative, since it offers the combined action of grinding and mixing without the need for manual labour. Ball mills are generally comprised of a hard ceramic vessel that contains a corresponding set of grinding balls and a tightly fastened lid. The operation is normally performed with the addition of an inert liquid medium in order to assist the milling and avoid clumping of the powdered contents. The ball mill is normally left on a rack of rollers overnight, or for more prolonged periods depending on the particular requirements.

A more satisfactory and quicker method than traditional ball milling is the use of a planetary mono mill, for example, the planetary mono mill PULVERISETTE 6 *classic line*, as equipped with a 250-ml zirconium oxide grinding bowl and balls, as shown in Figure 2.6. Cyclohexane (Analar grade) can be used as the liquid medium for the milling of the materials described in this book. The cyclohexane should cover the powders so as to yield, after milling, a consistency similar to that of normal paint. Do not overfill with cyclohexane, and be sure to leave a certain amount of headspace within the bowl beneath the lid. It should be possible, in most cases, to achieve adequate

Figure 2.5   A hardened steel piston-type percussion mortar, internal diameter 35 mm (Photograph used with kind permission from Simon Svane. Copyright (2010) Simon Svane)

milling with this machine within 1 h. After milling, transfer the resultant slurry into a round-bottomed flask. Use a pair of forceps to remove the zirconia grinding balls whilst washing them gently with a small amount of cyclohexane, so as to avoid an unnecessary loss of material. Then attach the flask to a rotary evaporator and a condenser in order to recover the cyclohexane. Retain the zirconia balls for further use.

However, readers working under a restricted budget can still obtain acceptable results. In these circumstances, it is desirable to purchase chemical reagents with a fine particle size. Fine powders can be mixed quite adequately inside a 500-ml polyethylene bottle, with a tightly screwed lid, containing a dozen zirconia grinding balls, together with cyclohexane as the liquid medium. This should be left on a rack of rollers overnight. This method functions purely for the purpose of mixing; the amount of grinding that occurs within a plastic bottle should be considered to be minimal.

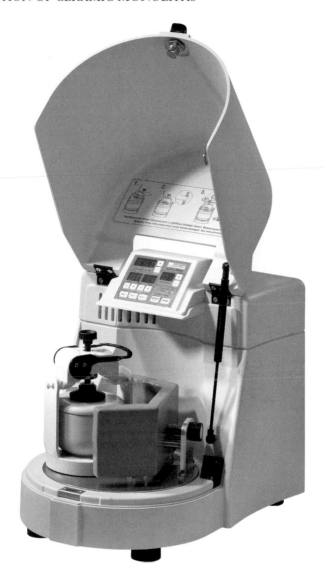

Figure 2.6 Planetary mono Mill PULVERISETTE 6 classic line with a 250 ml zirconium oxide grinding bowl (yellow) (Photograph used with kind permission from Fritsch GmbH. Copyright (2010) Fritsch GmbH)

2.3 FABRICATION OF CERAMIC MONOLITHS

2.3.1 Uniaxial Pressing

Compacted powder monoliths can be made by pressing a finely ground powder inside a Specac hardened stainless steel die; see Figure 2.7. A practical

Figure 2.7   Evacuative hardened stainless steel pellet dies (Photograph used with permission from Specac Ltd. Copyright (2010) Specac Ltd)

diameter is 32 mm. This enables the fabrication of ceramic monoliths, such as, $YBa_2Cu_3O_7$ that are large enough for the purposes of demonstrating the Meissner effect. The addition of organic binders and lubricants are normal practice in the industry. However, satisfactory unfired or green bodies (with an adequate strength for handling prior to sintering) can be obtained for the materials described in this book without recourse to binders and lubricants. Some materials form more coherent bodies upon compaction than others. Metals halides, for instance, compact quite readily, as is the case for $Rb_4Cu_{16}I_7Cl_{13}$.

Other compounds, for example, $YBa_2Cu_3O_7$ and $CuTiZr(PO_4)_3$ form less coherent bodies by this method, and so care must be exercised during handling. Furthermore, one must avoid the formation of lamella; i.e., cracks in the *green body* that lie in a plane perpendicular to the axis of applied pressure.[5] These have a tendency to form above a critical pressure. To this respect certain pressures are recommended for the various materials prepared in the book; although the results can be unpredictable. The die should be inserted into a suitable press, such as, a Specac 15 Ton Manual Hydraulic Press. The reader is advised to follow the manufacture's instructions for operating the press and the die. After pressing, it is wise to transfer the

---

[5] *'Green body'* is an expression used in the ceramics industry to describe the presintered material.

powder-compact directly into the crucible that will be used for the heating process, and avoid unnecessary handling.

2.3.2   Isostatic Pressing

Isostatic pressing produces a higher degree of compaction and coherency of the particles within the green body than can be obtained through uniaxial pressing alone, and avoids the creation of lamella. A powerful hydraulic press is required for this process, for example, the one inch-bore 20-kbar hydraulic press, manufactured by Psika Pressure Systems Ltd, Manchester, England. The process is lengthier, and the hydraulic press is expensive; but it normally results in a superior ceramic body after sintering, compared with uniaxial pressing. The method described here involves the use of silicone rubber moulds, and a description for their fabrication is given below.[6] The mould should be filled tightly with powder so that when the lid is inserted it should just fit. Advice is given at relevant places in the book, as to how much powder for a particular compound to fill the mould with. Each mould should then be inserted into a condom, the air excluded and a knot tied. This should then be placed in the hydraulic press under a pressure of ~15 kbar for ~10 min at ambient temperature, in order to form the pellet.[7]

   After retrieving the condom from the hydraulic press, cut the condom with a pair of scissors and obtain the rubber mould without getting oil on it. With a clean pair of rubber gloves, open the mould and retrieve the pellet; it may be necessary to cut the mould with a scalpel in order to achieve this. It is wise to transfer the pellet directly into the crucible that will be used for the heating process, and avoid unnecessary handling.

2.3.3   Fabrication of Silicone Rubber Moulds

Silicone rubber moulds are fabricated in-house for use in the isostatic pressing of powdered material as described above. The internal dimensions of usable packing space for the mould described here are: 14 mm diameter and 10 mm height. Wacker RTV2 Silicone Rubber, manufactured by Ciba Speciality Chemicals, is ideal for this purpose. It is supplied as a two-part package:

---

[6] There might be occasions when the reader may wish to avoid the use of silicone rubber moulds. So, an alternative practice is to compact the powder by uniaxial pressing at moderate pressure (as described above), then seal the powder-compact inside a condom, and compress under isostatic pressure.

[7] The pressure is transmitted through the liquid (in this case, a mixture of 50 vol.% caster oil and 50 vol.% methanol) in the chamber of the hydraulic press. This deforms the rubber mould and thereby compresses the powder contents. Upon release of the applied pressure, the rubber mould resumes its original shape, whilst the powder remains as a highly compressed powder-compact.

Figure 2.8   An aluminium die (dismantled) built in-house as used in the fabrication of the silicone rubber moulds. NB several such dies can be machined within a longer block of aluminium, thus increasing the production number of the rubber moulds in any one casting (Photograph used with kind permission from Simon Svane. Copyright (2010) Simon Svane)

0.9 kg Elastosil® M4643A and 0.1 kg Elastosil® M4643B. Elastosil® M4643B contains a platinum catalyst, so the operator is advised to wear protective gloves and goggles. Elastosil® M4643B needs to be thoroughly stirred (or shaken) in its bottle before use, since the material tends to settle in the container.

An aluminium die as shown in Figure 2.8, was machined in-house from aluminium alloy, and is used to fabricate the silicone rubber moulds. The aluminium die must be clean, free from all grease, and dry before use. Experience has shown that grease prevents the rubber from curing properly. The nuts and bolts that join the two halves of the aluminium die should be inserted and tightened appropriately. The amount of liquid rubber required depends on how many moulds are to be prepared. The rubber mould produced here has a mass ~10 g. Use a spoon to place 90 g of Elastosil® M4643A into a 250 ml glass beaker, and then add 10 g of Elastosil® M4643B using a disposable plastic pipette; in order to prepare sufficient liquid rubber for about eight moulds including waste. Mix these two liquids thoroughly by stirring with a glass rod. This has a pot life of about an hour. Pour slowly into the aluminium die taking care not to trap any air bubbles. The manufacturer claims that the curing time is about 24 h at room temperature. Experience has shown that two days is best before using the silicone moulds at high pressure. The spoon should be cleaned with tissue paper immediately after use. The unused silicone rubber remaining in the glass beaker is best left alone, since it can be removed very easily once it is cured, and then disposed of.

2.4 FURNACES

The pottery furnace, Aurora Classic Kiln P5923 (Potterycrafts Ltd.) has many attractions. It is marketed as an open-top pottery kiln for the hobbyist and is relatively inexpensive (see Figure 2.9). It is designed to operate with a single phase 220 V domestic electric power supply. The heating elements are Kanthal wire heating coils, enabling a maximum operating temperature of 1260 °C. The furnace functions well for general use below this temperature

Figure 2.9 Aurora Classic Kiln P5923 (Photograph used with permission from Potterycrafts Ltd. Copyright (2010) Potterycrafts Ltd)

and several of the materials described in the book can be prepared in it. The temperature control system as supplied by the manufacture was replaced by a Eurotherm 2408 control unit so as to enable a greater control over the heating and cooling rates, and the dwell times.

There are occasions when a higher temperature, or a greater precision of the temperature, is required, or, when the operator needs to gain access to the contents of the furnace whilst at high temperature, for example, when casting the gold-ruby glass as described in Chapter 17. Hence, there is a need for a high-temperature chamber furnace, such as, model UAF15/10 (Lenton), as shown in Figure 2.10. This furnace operates on a three-phase electric power

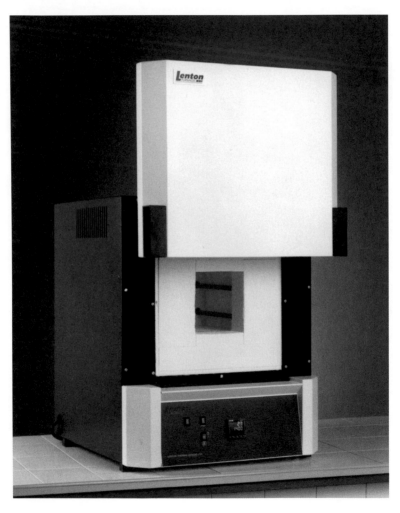

Figure 2.10 High-temperature chamber furnace model UAF15/10 (Courtesy of Lenton. Copyright (2009) Lenton)

supply, and is equipped with silicon carbide heating elements, and is designed to operate up to a maximum temperature of 1500 °C. Heating and cooling rates should not exceed 300 °C/h.

For materials that require annealing under a controlled atmosphere, for example, in flowing hydrogen, as in the case of preparing artificial hackmanite as described in Chapter 16, a tube furnace is required. The tube furnace, model LTF16/50/180 with a recrystallised alumina work-tube (Lenton), is ideal for this purpose, and is shown in Figure 2.11. The manufacture claims a maximum operating temperature of 1600 °C for this particular model. The tube furnace must be operated with a work-tube, for example, a recrystallised alumina (RCA) work-tube 50 mm ID × 900 mm long, as supplied by Lenton. The tube furnace operates on a single-phase electric power supply and is equipped with silicon carbide heating elements that are placed external to the work-tube.

A horizontal tube furnace has a hot zone in the centre, with a negative temperature gradient running towards both ends of the work-tube. Silicone rubber stoppers can be inserted at both ends, together with appropriate water-cooled copper coils attached to the outside ends of the work-tube. Stainless steel tubes can be inserted through purposely made holes in the silicone rubber stoppers, to act as the gas inlet and outlet. Gas tubes, flow meters and taps,

Figure 2.11 Tube furnace model LTF16/50/180 with recrystallised alumina work-tube and attachments (Photograph used with permission from Jan Hutzen Andersen. Copyright (2010) Jan Hutzen Andersen)

should be installed as appropriate, and adequate venting of the exit gas from the building must be ensured.

2.5   POWDER X-RAY DIFFRACTOMETRY

Powder X-ray diffractometry (XRD) is used routinely for the phase analysis of crystalline inorganic materials. The basic theoretical and practical aspects of the method are fairly quick to grasp, and therefore, students should be encouraged to collect their own powder diffraction data at an early stage of their studies. Guidance is given throughout the book concerning the interpretation and use of powder-diffraction data.

Modern powder X-ray diffractometers are equipped with computer-controlled data-capture and data-treatment software systems, which makes the whole procedure quite easy to operate. First, the operator must prepare the sample for analysis. Normally, about 1 g of the powdered material is spread flat within a purposely machined plastic holder, and then inserted into the diffractometer. Alternatively, a smaller quantity ($\sim$50 mg) of material can be ground in a small agate pestle and mortar with a few drops of ethanol. The resulting slurry can then be dispersed with the aid of a disposable plastic teat pipette, to form a film of the material (with a thickness similar to a layer of paint) on either a glass plate or, the back of the plastic sample holder; whichever is considered to be more appropriate.

The reader is advised to study the operating manual for the diffractometer as supplied by the manufacturer. Generally, the diffractometer is installed in a standard mode of operation for routine use, for example, with a copper X-ray tube and a primary monochromator to produce a monochromatic X-ray beam of Cu-$K\alpha_1$ radiation with a wavelength, $\lambda = 1.5405$ Å. The aperture of the various slits in the diffractometer may need to be adjusted. Consideration should also be given to the electrical potential and current of the X-ray generator. The values used here for the Siemens D5000 diffractometer are 35 kV and 35 mA. These values are at the low end of the range, and have been chosen so as to help prolong the life-span of the X-ray tube, for financial reasons. As a general guideline, a typical powder X-ray diffraction pattern can be collected between, 5–90° 2θ, with an incremental step of 0.02° 2θ, and an acquisition time at each step of 2 s. But these parameters should be selected for the specific need of the sample being analysed, and may vary considerably between different models of diffractometer.

After collecting these data, the operator may wish to compare the powder-diffraction pattern with data (d-values and relative peak intensities) held in the 'International Centre for Diffraction Data' (ICDD). These data (available on CD-ROM) are presented for each phase as individually recognised, *Powder Diffraction Files* (PDF). The computer software systems designed for this business can normally interact with these data files. The reader is advised to contact the 'International Centre for Diffraction Data', for further information.[8] Certain *Powder Diffraction Files* (PDF) contain data that have been calculated from other crystallographic sources of information, as complied under the auspices of the *Inorganic Crystal Structure Database* (ICSD).[9] These data have been used in order to construct the drawings for many of the crystal structures as depicted in this book; with appropriate references to the ICSD. The reader may also wish to consult the texts by Weller [2] and West [3] for further information regarding powder X-ray diffractometry.

REFERENCES

1. K. Clark, *The Potter's Manual*, Brown and Company, London, 1993.
2. M. T. Weller, *Inorganic Materials Chemistry*, Oxford University Press, Oxford, 1996.
3. A. R. West, *Basic Solid State Chemistry*, Wiley, Chichester, 2nd edn, 1999.

[8] International Centre for Diffraction Data, 12 Campus Boulevard, Newtown Square, Pennsylvania 19073-3273 United States of America.

[9] Inorganic Crystal Structure Database, Fachinformationszentrum, Karlsruhe, Germany, and the National Institute of Standards and Technology, Technology Administration, U. S. Department of Commerce.

3

Artificial Cuprorivaite CaCuSi$_4$O$_{10}$ (Egyptian Blue) by a Salt-Flux Method

Blue substances are rather unusual in nature and so there has always been an incentive to produce them artificially. 'Egyptian blue' is the name given to a synthetic inorganic pigment comprising crystals of calcium-copper tetrasilicate, CaCuSi$_4$O$_{10}$ (see Figure 3.1). However, the name is also applied to describe a frit[1] in which CaCuSi$_4$O$_{10}$ is the predominant phase but that is invariably contaminated with undesirable phases, such as; glass; quartz, SiO$_2$; wollastonite, CaSiO$_3$; etc.

Egyptian blue frit has been manufactured and used as a pigment in ancient Egypt since at least 2500 BC, and is one of the earliest inorganic materials synthesised by mankind. The pigment is remarkably stable and shows no sign of degradation with time. Practically all the known methods for its synthesis involve the use of a salt flux, in which the reactants, CaO, CuO and SiO$_2$ are partially and transiently dissolved, before exsolving as crystalline CaCuSi$_4$O$_{10}$ whilst still at high temperature. The reactant mixture is held for many hours within the relatively narrow temperature range of 800–900 °C under an oxidising atmosphere, before cooling slowly to room temperature. Given these prerequisites, the synthesis of CaCuSi$_4$O$_{10}$ in antiquity is quite

---

[1] A frit is a sintered material that is formed as a result of diffusion-controlled chemical reactions that take place within a mass of powdered reagents and flux that are in a partially molten state (i.e., normally at a temperature in between the solidus and liquidus of the system).

Synthesis, Properties and Mineralogy of Important Inorganic Materials   By Terence E. Warner
© 2011 John Wiley & Sons, Ltd

Figure 3.1 Specimen of CaCuSi$_4$O$_{10}$ as prepared by the method described in this chapter (Photograph used with permission from Søren Preben Vagn Foghmoes. Copyright (2010) Søren Preben Vagn Foghmoes)

remarkable. It is only comparatively recently, that CaCuSi$_4$O$_{10}$ was found to occur naturally, as the extremely rare mineral, cuprorivaite; and because of its rarity, the material could never have been collected from nature for use as a pigment. Concerning the subject of materials chemistry, it would be difficult to find a more attractive and rewarding introduction to the basic concepts of high-temperature synthesis than through the preparation of this aesthetically important pigment and by a method similar to the way in which it was made originally, more than 4500 years ago.

CaCuSi$_4$O$_{10}$ is a phyllosilicate (sheet silicate) that crystallises with the gillespite (BaFeSi$_4$O$_{10}$) structure, as shown in Figure 3.2, and belongs to the tetragonal crystal system and space group *P4/ncc*. The crystal structure was first determined by Pabst in 1959 using single-crystal X-ray diffractometry [1], and refined by Bensch and Schur in 1995 [2]. The structure has been reviewed recently by Burzo [3]. One important feature of this structure is that four SiO$_4^{4-}$ tetrahedra are linked together so as to form four-membered Si$_4$O$_{10}^{4-}$ rings, which comprise the basic building blocks of the overall structure. The four apical oxygen ions within the Si$_4$O$_{10}^{4-}$ ring all point in the same direction, such that each ring is connected to four other identical rings through their apical oxygen ions; which all point in the opposite direction. This arrangement forms an infinite silicate layer that lies parallel to the (001) plane (see Figure 3.3). Each SiO$_4^{4-}$ tetrahedron within this silicate layer has one unshared corner, such that four of these unshared corners (each one coming from a separate Si$_4$O$_{10}^{4-}$ ring) coordinate to a

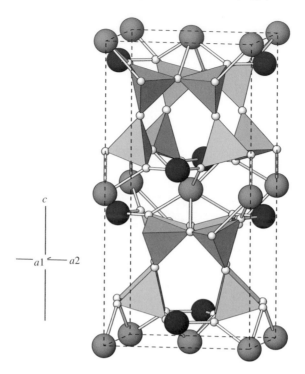

Figure 3.2 Crystal structure of CaCuSi$_4$O$_{10}$ showing the tetragonal unit cell. The Cu^{2+} ions are shown blue, and the Ca^{2+} ions grey. The silicate tetrahedra are shown yellow. This figure was drawn using data from Pabst [1] cf. ICSD 26502

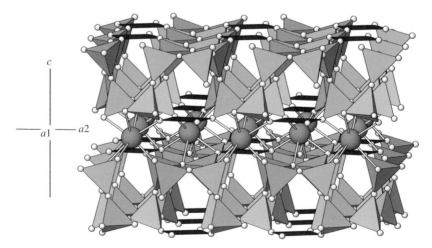

Figure 3.3 Crystal structure of CaCuSi$_4$O$_{10}$ emphasising the square planar coordination of the Cu^{2+} ions (shown blue), and the sheet-like nature of this structure. The Ca^{2+} ions grey, and the silicate tetrahedra are shown yellow. This figure was drawn using data from Pabst [1] cf. ICSD 26502

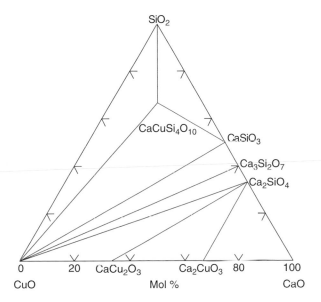

Figure 3.4 Phase relationships in the system CuO–CaO–SiO$_2$ at 1000 °C in air (Reproduced with permission from Z. Krisrallographie, Phase relationships in the system CuO-CaO-SiO$_2$ at 1000 C in air by K.-H. Breuer, W. Eysel and M. Behruzi, 176, 219–232 Copyright (1986) Oldenbourg Wissenschaftsverlag)

copper ion. This gives rise to a copper square-planar[2] coordination lying in the (001) plane. The calcium ions adopt an irregular eight-fold coordination corresponding to a distorted cube, and are located midway between these silicate sheets. It is through the calcium ions that the silicate sheets are held together. Of the eight oxygen ions coordinated around each calcium ion, four are also coordinated to copper, whilst the bridging-oxygen ions within the Si$_4$O$_{10}^{4-}$ rings constitute the other four.

An equilibrium phase diagram of the CaO–CuO–SiO$_2$ ternary system in air at 1000 °C was compiled from experimental data in 1986 by Breuer *et al.* [4], as shown in Figure 3.4. CaCuSi$_4$O$_{10}$ is the only ternary phase that exists under these conditions, and all the phases in this system are reported to be close to ideal stoichiometry with only very limited solid solubility. From this diagram it can be seen that CaCuSi$_4$O$_{10}$ can coexist with CuO, SiO$_2$ and CaSiO$_3$ and O$_2$(g). However, two more ternary compounds were reported to exist, both pertaining to high-pressure environments. Ca$_2$Cu$_3$Si$_6$O$_{17}$ was found to exist as a synthetic phase at 35 kbar by Heinrich-Beda in 1983 [5]. A decade

---

[2] This is an example of a true copper square-planar coordination, in which there are no ligands of any kind lying normal to the [CuO$_4$] plane; and is an unusual coordination symmetry for copper in the solid state.
NB This is *not* an example of a distorted octahedron with grossly elongated apical ligands, as for example, in CuSO$_4$·5H$_2$O.

later, Zöller *et al.* discovered the bluish-green monoclinic mineral, liebauite, $Ca_3Cu_5Si_9O_{26}$ within a very high grade thermally metamorphosed xenolith from the Sattelberg scoria cone near Kruft, in the Quaternary volcanic region of the Eastern Eifel district, Germany [6]. $Ca_2Cu_3Si_6O_{17}$ and $Ca_3Cu_5Si_9O_{26}$ have a similar chemical composition, but whether or not they correspond to the same phase, by way of a solid solution, is not known.

Knowledge of the thermal stability limits (T and P_{O_2}) of $CaCuSi_4O_{10}$ is particularly useful when devising a preparative route for the synthesis of this compound and for the subsequent sintering of ceramic monoliths. The information reported in the literature on this matter is rather sketchy. Bayer and Wiedemann [7] reported that when pure $CaCuSi_4O_{10}$ (i.e., without a flux) is heated in air to 1230 °C, copper(II) is reduced to copper(I); and that upon cooling, copper(I) is reoxidised to copper(II), but the $CaCuSi_4O_{10}$ phase is never fully reformed. Mazzocchin *et al.* [8] reported that $CaCuSi_4O_{10}$ (in the presence of Na_2CO_3) decomposes at >1100 °C, and results in the formation of a green frit. Whilst Jaksch *et al.* [9] reported that $CaCuSi_4O_{10}$ is stable between 850–1050 °C, but decomposes >1050 °C, yielding a mixture of copper oxide (Cu_2O?), SiO_2 and $CaSiO_3$.

The blue colour of $CaCuSi_4O_{10}$ is the principle property of interest in this material, and is a consequence of the Jahn–Teller effect on the Cu^{2+} ion in a square-planar (D_{4h}) environment.[3] The diffuse absorption spectrum of polycrystalline $CaCuSi_4O_{10}$ is shown in Figure 3.5. Absorption of light throughout the near-infrared to green region of the electromagnetic spectrum causes an electron to be promoted from lower energy levels in the *d*-shell to the $d_{x^2-y^2}$ level on the Cu^{2+} ion, corresponding to three absorption bands. From Figure 3.5, the band maxima yield the following values ex-

---

[3] Crystal field theory attempts to account for the electrostatic interaction between the metal ion (in this case, the Cu^{2+} ion) and the surrounding ligands (in this case, O^{2-} ions). For example, if the positively charged metal cation is surrounded by six negatively charged anionic ligands in octahedral symmetry, then the electrostatic interaction between the metal cation and its six anionic ligands cause repulsion between the negatively charged ligands and the negatively charged electrons of the *d*-orbitals associated with the metal cation. These interactions, however, are not all equal. The ligands that lie along the orthogonal axes interact strongly with the $d_{x^2-y^2}$ and d_{z^2} orbitals (known collectively as the e_g orbitals) and so these are raised in energy. Conversely, the d_{xy}, d_{xz} and d_{yz} orbitals (known collectively as the t_{2g} orbitals) interact less strongly since these are directed between the ligands, and so these are lowered in energy. The difference in the energy of the e_g and t_{2g} orbitals is referred to as the octahedral crystal field splitting, and is denoted by the Greek letter, Δ_O. The Jahn–Teller effect concerns distortions in the octahedral symmetry that result in a further degeneracy within the e_g and t_{2g} orbitals. For example, square planar (D_{4h}) symmetry is the extreme case of the elongated tetragonal (T_d) symmetry brought about by the progressive removal, and ultimate absence, of negatively charged ligands along the *z*- and −*z*-axes; as is the case of the oxygen coordination for the Cu^{2+} ion in $CaCuSi_4O_{10}$. Consequently, the d_{z^2} orbital is lowered, and the $d_{x^2-y^2}$ orbital is raised, significantly in energy; whilst the d_{xz} and d_{yz} orbitals are lowered, and the d_{xy} orbital is raised in energy, though to a lesser extent (see Figure 3.6). The reader is referred to Huheey [13] for further details regarding crystal field theory and the Jahn–Teller effect.

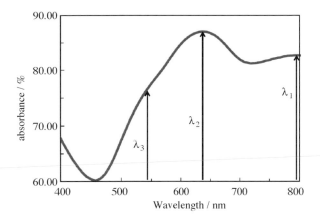

Figure 3.5 Diffuse absorption spectrum of polycrystalline $CaCuSi_4O_{10}$ at 298 K (reference: $BaSO_4$). The arrows point to the band maxima: $\lambda_1 \left(^2B_{1g} \rightarrow {}^2B_{2g}\right) = 800$ nm; $\lambda_2 \left(^2B_{1g} \rightarrow {}^2E_g\right) = 630$ nm; and $\lambda_3 \left(^2B_{1g} \rightarrow {}^2A_{1g}\right) = 540$ nm

pressed here in terms of decreasing wavelength: $\lambda_1 \left(^2B_{1g} \rightarrow {}^2B_{2g}\right) = 800$ nm; $\lambda_2 \left(^2B_{1g} \rightarrow {}^2E_g\right) = 630$ nm; and $\lambda_3 \left(^2B_{1g} \rightarrow {}^2A_{1g}\right) = 540$ nm. These values, and their assignments, are in general agreement with those reported by Ford and Hitchman [10]; Botto *et al.* [11]; and, Kendrick *et al.* [12]. This particular sequence of the energy levels implies that the Jahn–Teller effect on the Cu^{2+} ion in $CaCuSi_4O_{10}$ is so extreme that the d_{z^2} orbital is at a lower energy level than the d_{xy}, d_{xz} and d_{yz} orbitals and is a manifestation of true square planar coordination as opposed to being merely an elongated tetragonal coordination (see Figure 3.6). The blue and violet wavelengths are not absorbed in

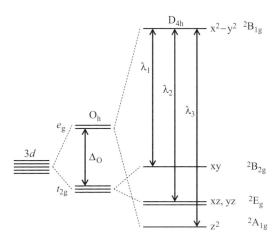

Figure 3.6 The energy diagram for the electronic transitions for the square-planar (D_{4h}) coordinated Cu^{2+} ion in $CaCuSi_4O_{10}$

these processes and so remain essentially visible. However, the human eye is more sensitive to blue than to violet, and so the material appears blue. The intensities of these three absorption bands are relatively weak, because *d-d* transitions are parity forbidden.[4] Nevertheless, a certain amount of absorption does occur through vibronic coupling with *ungerade* (i.e., asymmetric) vibrational modes within the crystal lattice, which results in a pale blue coloration. Vibronic coupling has the additional effect of broadening these absorption bands so as to include substantial parts of the visible spectrum. The reader is referred to the article by Ford and Hitchman for further details concerning the electronic spectrum of $CaCuSi_4O_{10}$ [10].

In 1889, Fouqué [16, 17] observed that crystals of $CaCuSi_4O_{10}$ were distinctly pleochroic (or more specifically, dichroic) when viewed with a transmitted light polarising microscope; whereby the ordinary ray, ω is deep blue, and the extraordinary ray, ε is pale pink.[5] These observations were confirmed by Minguzzi [18]; whereas Mazzi and Pabst [19] describe ε as

---

[4] The parity selection rule forbids electronic transitions between levels with the same parity; for example, within the same *d*-orbital or, *f*-orbital [14]. See also, *Atkins' Physical Chemistry* 8th edn, pages 483–489 [15].

[5] *Pleochroism* is an optical property characteristic of coloured anisotropic crystalline phases. The word 'pleochroism' is taken from the ancient Greek meaning; 'form colour'. Since $CaCuSi_4O_{10}$ belongs to the tetragonal system, its crystals are uniaxial. The unique axis is designated as the *c*-axis, and is referred to as the optic axis. In this case, $CaCuSi_4O_{10}$ is more specifically described as *dichroic* (i.e., a crystal of $CaCuSi_4O_{10}$ displays *two* colours). The actual colour depends on the orientation of the crystal with respect to the electric vector of the illuminating electromagnetic radiation (in this case, transmitted linearly polarised white light). The direction of the *electric vector* should not be confused with the direction of the *vibration* of the linearly polarised light; since these are in fact perpendicular to each other. If a dichroic crystal is viewed by transmitted linearly polarised white light, it will normally be seen to change colour upon rotation, since the absorption spectrum is different for different orientations of the crystal. This is a consequence of double refraction, by which the fast (ω) and slow (ε) rays are absorbed differently as they pass through the crystal. However, when a coloured uniaxial crystal is viewed straight down the *c*-axis (i.e., the optic axis), it gives the appearance of being singly refractive (i.e., $\varepsilon = \omega$), and therefore nonpleochroic; and in the case of $CaCuSi_4O_{10}$ the colour is invariantly deep blue. Hence, this is the colour of the ordinary ray ω (since the vector of the ordinary ray, ω is, by definition, parallel to the *c*-direction in a uniaxial crystal). When a coloured uniaxial crystal is viewed at right angles to the *c*-axis, it appears doubly refractive, and therefore, dichroic; and in the case of $CaCuSi_4O_{10}$ the colours range from pale pink to deep blue. The actual colour, however, is dependent on the particular orientation of the basal $(a - b)$ plane with respect to the direction of the vibration of the linearly polarised light. Hence, the colour changes upon rotating the crystal (by rotating the microscope stage) or, rotating the plane of linearly polarised light (by rotating the polariser). The two extremities of the crystal's colour will arise when the polariser is rotated to two certain orthogonal positions; and thereby allowing for the colour of the extraordinary ray, ε to be deduced from these observations. Intermediate rotary positions of the polariser and intermediate orientations of the crystal show intermediate colorations. Pleochrosim (and dichroism) is described as being weak or strong, depending on the extent of the colour change. The reader is referred to Gribbles and Hall [20] and Nesse [21] for further details of dichroism and pleochroism.

nearly colourless and report the following values for the refractive indices: $\omega = 1.633$ and $\varepsilon = 1.590$.

Concerning the magnetic properties, the Cu^{2+} $(3d^9)$ ions have a ground-state electron configuration $t_{2g}^6 e_g^3$ with a spin state $S = \frac{1}{2}$. The effect of this unpaired spin is that the Cu^{2+} $(3d^9)$ ions are paramagnetic. From the illustrations of the crystal structure in Figures 3.2 and 3.3 it can be seen that the Cu^{2+} ions in $CaCuSi_4O_{10}$ are isolated from each other, such that the nearest-neighbour pathway is via Cu–O–Ca–O–Cu. Because the Cu^{2+} ions are relatively isolated from each other, and the other ions, Ca^{2+}, Si^{4+} and O^{2-} are all diamagnetic, there is very little opportunity for magnetic interaction (or coupling) to take place between the individual magnetic moments on the Cu^{2+} ions. Therefore, the material is expected to be paramagnetic in its macroscopic state.

The measurement of the molar magnetic susceptibility of a single-phase material, such as $CaCuSi_4O_{10}$, is a particularly convenient way to investigate its magnetic properties. These measurements can be obtained using, for example, a SQUID (superconducting quantum interference device) over a typical temperature range of 5–300 K, and if relevant, at different magnetic field strengths. To illustrate this point, Figure 3.7a–c show molar magnetic susceptibility data displayed graphically in three different ways. These data were measured at a constant magnetic field strength of 5 kOe on a polycrystalline sample (85.9 mg) of $CaCuSi_4O_{10}$ as prepared by the method described in the latter part of this chapter.

Figure 3.7a reveals the typical exponential relationship between molar magnetic susceptibility, χ_m (in units of emu mol⁻¹) and temperature, T (in units of K) for a paramagnetic substance. Figure 3.7b shows these data

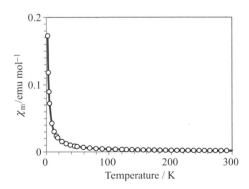

Figure 3.7a A plot of the temperature dependence of the molar magnetic susceptibility (χ_m/emu mol⁻¹) for a polycrystalline sample of $CaCuSi_4O_{10}$ at 5 kOe. These data have been corrected for diamagnetism with the usual Pascal constants. The author is grateful to Professor K. Murray and Dr. B. Moubaraki, Monash University, Australia, for recording these data

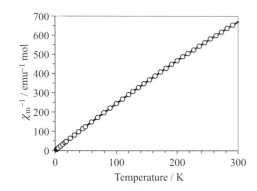

Figure 3.7b A plot of the temperature dependence of the inverse molar magnetic susceptibility (χ_m^{-1}/emu^{-1} mol) for CaCuSi$_4$O$_{10}$

in a form that is more congenial to a quantitative interpretation, i.e., inverse molar magnetic susceptibility, χ_m^{-1} (in units of emu^{-1} mol) as a function of temperature, T (in units of K). Here, it can be seen that these data fit quite closely to the Curie law, and therefore illustrate that CaCuSi$_4$O$_{10}$ is paramagnetic, throughout this temperature range. The numerical value for the Curie constant, C, can be obtained directly from the inverse of the gradient in this type of plot, i.e., $C = (\Delta\chi_m^{-1}/\Delta T)^{-1} \equiv \Delta T/\Delta\chi_m^{-1}$. The effective magnetic moment, μ_{eff} (in units of Bohr magneton, μ_B; in which $\mu_B = 9.27401 \times 10^{-24}$ J T^{-1}), is related to the molar magnetic susceptibility, χ_m, through the following relationship: $\mu_{\text{eff}} = \sqrt{(3k\chi_m T/N_A\mu_B^2)}$, in which k is Boltzmann's constant, and N_A is Avogadro's constant. This can be expressed more simply as; $\mu_{\text{eff}} = 2.83\sqrt{C}$, in which C is the Curie constant; and is a convenient way to obtain the effective magnetic moment, μ_{eff} (in units of μ_B). The effective magnetic moment, μ_{eff} is a very useful parameter for ascribing a spin state to the paramagnetic species, and also in comparing a particular paramagnetic species in different chemical environments. Here, for instance, it demonstrates that the copper ions in CaCuSi$_4$O$_{10}$ are divalent and magnetically isolated. The reader is referred to Orchard [22] and Mabbs and Machin [23] for further details, including the derivation of the above formula.

Figure 3.7c shows a plot of the effective magnetic moment, μ_{eff} (in units of μ_B) as a function of temperature, T (in units of K). This type of plot (which has χ_{mol} embedded in it) is commonly used in the contemporary literature. From this specific plot, it is apparent that the effective magnetic moment, μ_{eff} is slightly dependent on temperature. At very low temperatures ($T \to 0$ K), there is always a deviation from Curie's law due to dipolar coupling through free space. At higher temperatures (>50 K) the system approximates to Curie behaviour; although the gradual increase in μ_{eff} with increase in temperature is attributed to the contribution from temperature-independent

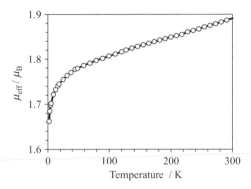

Figure 3.7c A plot of the temperature dependence of the effective magnetic moment (μ_{eff}/μ_B) for CaCuSi$_4$O$_{10}$

paramagnetism (TIP). However, if this linear region (50–300 K) is extrapolated to 0 K (which is equivalent to putting TIP – 0) then $\mu_{eff} = 1.76 \, \mu_B$, which is close to the value for free spin (1.73 μ_B); as expected for a copper(II) ion in a square-planar coordination.

Archaeological evidence reveals that Egyptian blue has been used as a pigment in mural paintings in tombs, temples and palaces in ancient Egypt since at least Dynasty IV of the Old Kingdom (2600–2480 BC) [9]; see Figures 3.8 and 3.9. This makes its early production contemporary with the date for the construction of the great pyramids, Cheops, Chephren and Mycerinus, as well as the Sphinx, at Giza in Egypt. From an estimate of the time required in order to develop the necessary preparative technology, it has been speculated that attempts at preparing CaCuSi$_4$O$_{10}$ may date back even earlier [9]. Archaeological evidence[6] for the actual production of Egyptian blue frit, as opposed to its mere use, comes from the unique discovery of CaCuSi$_4$O$_{10}$ crystals adhering to the inner surfaces of porous ceramic crucibles (made from a very calcareous clay) found in a late bronze age workshop area at Qantir-Pi-Ramesse in the eastern Nile delta, Egypt, ca. 1650–1550 BC (see Figure 3.10) [24]. This discovery is consistent with the various lumps of a 'fashionable' blue pigment, and the numerous blue tiles of glazed faience, as excavated at the same site by Hamza in 1930 [25], and therefore indicates that these were most likely produced on site [24]. To give the reader some idea of the scale of production; Hatton *et al.* [26] have, for instance, estimated the mass of Egyptian blue required to paint a typical temple during the Dynasties XVIII–XX (ca. 1550–1075 BC), such as, Medinet Habu at Thebes, to be ~1.4 tonnes!

---

[6] The author is grateful to Professor Thilo Rehren; University College London, for sharing information regarding this excavation.

Figure 3.8 A small patterned section of the painted 'palace facade' false door of the vizier Ptahhotep (II) at Saqqara, dating to the reign of Unis (ca. 2350 BC) last king of Dynasty V (Photograph used with kind permission from Yvonne Harpur. Copyright (2009) Oxford Expedition to Egypt)

Besides its use as a pigment, Egyptian blue has also been fabricated as a ceramic (albeit on a smaller scale), and so has close associations with the production of faience and glass [27]. It has been speculated by Tite [28] that the development of Egyptian blue frit grew out of the technology surrounding the efflorescence method of faience production in which the quartz body and glazing components are mixed together. Furthermore, Chase [27] considers that a prolonged sintering of compacted bodies of CaCuSi$_4$O$_{10}$ at ~850 °C might enable the production of a ceramic without the glass admixture. The reader is encouraged to explore this aspect; and if the venture can be made successful, it should be quite rewarding, particularly for those who are interested in jewellery and miniature sculpture.

One aspect of early Egyptian-blue technology that is really impressive is that the ancient Egyptians realised the importance of using a flux in their preparations in order to lower the melting point of the reactants and so ensure the presence of a certain amount of liquid phase within the material at high temperature. The use of a flux, when used in moderation, is now understood to be essential for the production of single phase material (or at least a product with a high content of CaCuSi$_4$O$_{10}$). Wiedemann et al. [29] have shown that the chemical nature of the flux influences the tone of the blue coloration in the pigment; a feature that can be important in art work.

Figure 3.9   A view of the Festival hall of Tuthmosis III within the Temple of Amun at Karnak, Egypt, showing Egyptian blue pigment amongst the art decor. New Kingdom, Dynasty XVIII, ca. 1450 BC (Photograph used with permission from Peter Rolfsager. Copyright (2010) Peter Rolfsager)

Figure 3.10   The remains of a batch of CaCuSi$_4$O$_{10}$ (blue crystals) adhering to the inner surface of a fragment (62 mm in width) from a porous and friable ceramic crucible found in a late bronze age workshop area at Qantir-Pi-Ramesse in the eastern Nile delta, Egypt, ca. 1650–1550 BC [24] (Photograph by Grabung Qantir used with kind permission from Thilo Rehren. Copyright (2010) Thilo Rehren)

The analytical work performed by Jaksch *et al.* [9] revealed a significant presence of phosphate within the contaminant glass phase that accompanies the CaCuSi$_4$O$_{10}$ in ancient Egyptian blue frits from the Old Kingdom till the Roman period (Caesar Tiberius, 14–37 AD). Due to the diagnostic association of phosphate with plant material, Jaksch *et al.* [9] interpreted this as evidence that the ancient Egyptians used the ashes from incinerated plants, rather than mineralogical sources of alkali salts (e.g., trona, Na$_3$CO$_3$(HCO$_3$)·2H$_2$O and natron, Na$_2$CO$_3$·10H$_2$O, etc.), as a source for the flux. They also took this notion a stage further and offered analytical evidence to suggest that substantial amounts of flux may have been added on purpose in order to achieve a complete fusion at high temperature, and thus enable a glass phase to be formed upon quenching the melt. Apparently, this glass was subsequently ground and reannealed by the ancient Egyptians in order to exsolve the desired crystalline phase, CaCuSi$_4$O$_{10}$.

Concerning the sources of the other raw materials as used in these preparations, lime (produced by calcining limestone, CaCO$_3$) is presumed to be the source for the CaO component, although this is poorly discussed in the literature. From Dynasty XVIII, the presence of tin oxide (SnO$_2$) within the frit is almost universal (and yet prior to this date it is universally absent) and suggests that a copper–tin bronze was used as the principle source for the copper component; with its subsequent oxidation to CuO (and SnO$_2$) during the heating process [9]. Jaksch *et al.* [9] considered quartz sand to be the likely source of silica as opposed to powdered varieties of chalcedony, etc.

According to Vitruvius Pollio (writing during the first century BC), Vestorius transferred Egyptian-blue technology to the Romans and established its production at Pozzuoli (in Latin: *Puteoli*), near Naples in southern Italy [30]. This choice of location is most likely connected with the fact that Pozzuoli was the principle site for the production of pozzolanic cement[7] during the Roman period. Interestingly, both Egyptian blue and pozzolanic cement involve calcium silicate chemistry, and were used in the building industry. Their production continued to flourish throughout the Roman period.

The present author was fortunate to come into possession of an archaeological specimen of an Egyptian-blue pellet (frit) found during the Nordic excavations of an Imperial villa at the Nemi Lake, Italy [31, 32]. The powder X-ray diffraction pattern (Figure 3.11) indicates that the pigment is essentially CaCuSi$_4$O$_{10}$ with a small fraction of α-SiO$_2$. The villa was constructed

---

[7] This hydraulic cement hardens similarly to modern Portland cement clinker, but contains 5–35% pozzolana; the balance being mainly lime, CaO [33]. Its production was therefore dependent on the local source of pozzolana: a pyroclastic volcanic rock ideally suited for this purpose.

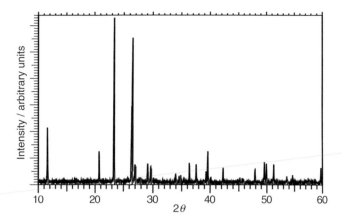

Figure 3.11 Powder XRD pattern (Cu-$K_{\alpha1}$ radiation) obtained from an archaeological specimen of blue pigment found during the Nordic excavations of an Imperial villa at the Nemi Lake, Italy [31, 32]. This sample was kindly provided by the archaeologist, Associate Professor B. Poulsen, University of Aarhus. PDF 81-1238 for CaCuSi$_4$O$_{10}$ is shown in red for comparison. The peaks at 20.66 and 26.35° 2θ correspond to α-quartz

presumably during the Late Republican period (ca. 50 BC) and was deserted about 200 years later.[8] It is very likely that the Egyptian blue pellets were intended for use in wall mosaics, especially in connection with fountains and nymphaea that were popular from the time of Claudius (41–54 AD) [34].

According to conventional wisdom, the use of Egyptian blue faded out along with the decline of the Roman Empire, ca. 500 AD. The blame for the actual loss of the pigment is put by several scholars, including Humphrey Davy [35], on the misleading details given by Vitruvius in his preparative description[9] of *caeruleum* (a Latin expression for 'Egyptian blue'). Vitruvius

---

[8] The author is grateful to Associate Professor Birte Poulsen, University of Aarhus, for kindly providing material and information concerning this excavation.
[9] The following extract is taken from; 'The ten books on architecture' (in Latin: *De architectura libri decem*) by Vitruvius Pollio (written in the first century BC), translated by, M. H. Morgan, Harvard University Press (1914). "Book VII; Chapter XI: Methods of making blue were first discovered in Alexandria, and afterwards Vestorius set up the making of it at Pozzuoli. The method of obtaining it from the substances of which it has been found to consist, is strange enough. Sand and the flowers of natron are brayed together so finely that the product is like meal, and copper is grated by means of coarse files over the mixture, like sawdust, to form a conglomerate. Then it is made into balls by rolling it in the hands and thus bound together for drying. The dry balls are put in an earthen jar, and the jars in an oven. As soon as the copper and the sand grow hot and unite under the intensity of the fire, they mutually receive each other's sweat, relinquishing their peculiar qualities, and having lost their properties through the intensity of the fire, they are reduced to a blue colour.'

lists the ingredients as; sand, natron ($Na_2CO_3 \cdot 10H_2O$) and copper filings, and therefore fails to mention an explicit reagent for the CaO component within $CaCuSi_4O_{10}$. The procedure, as written, would therefore have been quite futile for anyone wishing to follow it. Mazzocchin *et al.* [8] suggest that Vitruvius may have unwittingly overlooked the importance of stating a specific reagent for the lime component in his preparative procedure, because the sand as used at Pozzuoli may have contained a sufficient amount of 'CaO' within it, so as to enable the reaction to take place. Nevertheless, Vitruvius's preparative description is the earliest surviving account for the production of Egyptian blue.

Fresh evidence has come to light through the discovery of Egyptian blue in a Byzantine fresco 'The Ascension of Christ' within the subterranean church of San Clemente, Rome, ca. 850 AD. Lazzarini *et al.* [36] discuss that the poor quality of the pigment would tend to exclude it from being a reclaimed batch of old Roman stock, and that the pigment was most likely produced at a time with the fresco. This discovery extends the period for the use of this pigment by several centuries [36]. But thereafter, and throughout the renaissance, Egyptian blue is apparently absent from the artist's palette, and does not become commercially available again until the late 19th century.

After a dormant period of almost a thousand years, Egyptian blue was resurrected by the French chemist Chaptal in 1809 [37] through his qualitative description of two blue pigments (Nos. 5 & 6) found at Pompeii in Italy.[10] Chaptal considered these to be compounds comprising, *d'oxide de cuivre* (CuO), *chaux* (CaO) and *alumina* (Al_2O_3); and made the association with a blue pigment used in hieroglyphics on ancient Egyptian monuments as observed by Descostils during an expedition to Egypt in 1807 [38]; hence the expression in French, *bleu égyptien*.

During the same year as the battle of Waterloo, Sir Humphry Davy [35], published an article on ancient pigments in which he described a blue pigment used in the baths of Titus at Rome (81 AD) as being a rough but very fine blue powder that does not lose its colour by being heated to red heat (i.e., 500–1000 °C) and is almost inert in acids. Davy determined the SiO_2 content as >60 wt%, and found the material contained CuO together with a considerable quantity of Al_2O_3 but only a small amount of CaO. He also reported that several large lumps of a deep blue frit were found amongst some rubbish in one of the chambers of the baths of Titus, which proved to be the same pigment as that actually used in the baths, but contained significant amounts of soda (presumably, Na_2CO_3). He concluded that this pigment is *caeruleum* (i.e., Egyptian blue).

---

[10] Pompeii was buried during the volcanic eruption of Vesuvius in 79 AD, and has been an important archaeological site since 1748. Vesuvius is still active; the eruption in 1906 brought a new lava flow very close to Pompeii [39].

In May, 1814, a year prior to his publication, Davy visited an excavation at Pompeii, where he came into the procession of a small pot containing a pale blue pigment that had been dug up from the site. He found the contents to be a mixture of calcium carbonate and Alexandrian frit (an alternative expression for Egyptian blue), and concluded that this pigment is probably the same colour as that examined by Chaptal. From Davy's description, it is apparent that he found Egyptian blue *frit* to be a compound comprising; CuO, Al_2O_3, SiO_2 with trace amounts of CaO, which is accompanied occasionally with Na_2CO_3 and $CaCO_3$ as separate discrete phases.

In 1874, Fontenay [40] analysed a specimen of Egyptian blue found within the Roman ruins at Autun in central France,[11] and a fragment of Egyptian blue prepared previously by Darcet.[12] Fontenay also described one of the earliest attempts at preparing Egyptian blue in modern times. His method involved slowly heating a mixture of, 25 parts $CaCO_3$, 15 parts CuO, 70 parts white sand, and 6 parts Na_2CO_3 to 950–1000 °C, then leaving it at this temperature for a few minutes before cooling to room temperature. Fontenay recommended that this heating cycle should be repeated twice more on the product material in order to yield a uniform colour. From the description of the final product, his method appears to have been successful. Fontenay advised against the excessive use of Na_2CO_3, and recommended that this should not exceed 7 wt%.

In 1889, Fouqué [16, 17] determined the chemical composition of Egyptian blue as, $CaO \cdot CuO \cdot 4SiO_2$; which proved to be correct. He described the pigment as a crystalline compound with a specific gravity of 3.04 (cf. 3.09 g cm^{-3} [1]). Fouqué experimented with various methods of synthesis, including the use of K_2SO_4 as a flux, and attempted to replace the CaO component with MgO. Fouqué's reluctance to reveal explicit details, may have been due to a desire to keep these secret on account of his business associations with the French manufacturer, Deschamps Frères; with whom he assisted in the commercial[13] production of Egyptian blue ca. 1893 [41].[14]

Laurie *et al.* [42] published a comprehensive article in 1914 describing the preparation of CaCuSi$_4$O$_{10}$, and their findings are summarised here as

---

[11] 8.35 *chaux* (CaO); 16.44% *oxyde de cuivre* (CuO); 70.25% *silice* (SiO$_2$); 2.36 *fer et alumina* (Al$_2$O$_3$); and 2.83% *soude* (Na$_2$O); cf. stoichiometric CaCuSi$_4$O$_{10}$ equates to: 14.9% CaO; 21.2% CuO; and 63.9% SiO$_2$.

[12] 17.21% *chaux*; 17.92% *oxyde de cuivre*; 60.13% *silice*; 1.02% *fer et alumina*, 3.65% *soude*, and traces of magnesia.

[13] Readers with an interest in antiquarian pigments may wish to know that Egyptian blue is currently manufactured by, Kremer Pigmente GmbH & Co. KG; and Natural Pigments LLC.

[14] Laurie *et al.* [42] writing in 1914, took a less considerate view and criticised Fouqué's preparative description as being simply, *'impossible to follow'*. But these remarks are misconceived, since Fouqué [17] did not report a preparative procedure per se within the pertinent article (as cited by Laurie *et al.*) nor, in his other published work on this subject matter [16]. Instead, Fouqué explored various ways to synthesise the pigment through experimentation, and reported his findings.

follows: The reaction mixture should contain: 3.6 g of calcium carbonate, 4.3 g of copper carbonate (malachite?), 15 g of fine sand and 2 g of a fusion mixture (e.g., trona). Greater quantities of fusion mixture result in glass formation, whilst smaller quantities necessitate a longer reaction time. A reliable temperature range for the synthesis lies between 800–900 °C; preferably, 850 °C with a reaction time of 40 h. At temperatures >900 °C the whole charge becomes molten, resulting in an undesirable[15] product. At 850 °C the quartz is chemically attacked within the *subliquidus–supersolidus* region of the system, resulting in the formation of crystalline CaCuSi$_4$O$_{10}$. For nostalgic reasons, Laurie *et al.* favoured the use of trona as the flux (albeit as an artificial concoction of sodium carbonate, sodium sulfate and sodium chloride), believing, incorrectly, that trona was the flux used by the ancient Egyptians in their preparations. Interestingly, Laurie *et al.* believed that in the presence of an alkali, some of the alkali metal ions (e.g., sodium ions) may displace some of the copper or calcium ions from the CaCuSi$_4$O$_{10}$ compound. Laurie *et al.* suggested a reaction scheme for the formation of CaCuSi$_4$O$_{10}$ that is paraphrased here as follows. A certain amount of molten glass within the charge acts as a diffusion medium for dissolving and transporting the CaO, CuO and SiO$_2$ components, thus enabling them to chemically react together within the melt and thereby exsolve as the crystalline phase, CaCuSi$_4$O$_{10}$. The molten glass then proceeds to dissolve further portions of these components, and then exsolve them as further quantities of crystalline CaCuSi$_4$O$_{10}$, and so on. These features distinguish this mechanism from a true solid-state sintering process, in which diffusion of ionic species occurs exclusively within the solid state.

Several of the conclusions by Laurie *et al.* are adopted in this present work together with various inspirations acquired from other workers. For instance, Mazzocchin *et al.* [8] have explored the use of various crucible materials that included, stainless steel, porcelain, terra-cotta, nickel and zirconium, and concluded that terra-cotta was undoubtedly the best choice; and the least expensive! They also concluded that a good yield of CaCuSi$_4$O$_{10}$ can be obtained using only Na$_2$CO$_3$ as a flux, within the temperature range 800–900 °C, with a reaction time of 12 h. Pradell *et al.* [43] recommend the use of Na$_2$CO$_3$ as a flux (at the equivalent concentration of 3 wt% Na$_2$O with respect to the resultant oxide mixture), and conclude that CaCuSi$_4$O$_{10}$ continues to crystallise down to ~700 °C. Schippa and Torraco [44] favour a prolonged reaction time (~100 h) in order to promote crystal growth. Ford and Hitchman utilised a borax flux (10 wt%) and a slow cooling rate of 100 °C per day, to produce relatively large crystals up to 1 mm in

---

[15] Laurie *et al.* did not report any attempt to reanneal this undesirable product (glass?) at a lower temperature. But had they done so, they might then have exsolved crystalline CaCuSi$_4$O$_{10}$; as inferred from the work by Jaksch *et al.* [9], as discussed above.

diameter [10]; Wiedemann *et al.* [29] also recommend a borax flux (10 wt%) with a long annealing time of 100 h at 880 °C. Tite [28] emphasises the importance of avoiding excess CaO over CuO within the reaction charge so as to avoid undesirable wollastonite, CaSiO$_3$. Bock [41] states that the quartz powder should be very finely ground.

Given the time scale of the events in the history of Egyptian blue, it is comparatively recently, in 1938, that CaCuSi$_4$O$_{10}$ was found to occur naturally as the extremely rare mineral, cuprorivaite. The mineral was discovered by Minguzzi [18], as blue-coloured grains in association with quartz and powdery calcite within a volcanic rock from Mount Vesuvius, Campania, Italy; and was named after the mineralogist, Carol Riva (1872–1902) [45]. Minguzzi performed a chemical analysis of the mineral and described its chemical formula[16] as; 2(Ca,Na)(Cu,Al)(Si,Al)$_4$(O, OH)$_{10}$·H$_2$O. Other localities in Europe are; Wheal Edward, in Cornwall, England; and the Eifel district in Germany [45]. However, it took nearly twenty years before Schippa and Torraco [43] first recognised the analogy between the mineral cuprorivaite and Egyptian blue. This was then confirmed shortly afterwards by Pabst [1] and, Mazzi and Pabst [19]. Until then, slight differences in chemical composition and disagreement over the acceptance of cuprorivaite as a genuine mineral species had impeded their connection. It is now understood that the mineral cuprorivaite is the natural analogue of Egyptian blue.

Very little is known about the paragenesis of cuprorivaite, presumably, on account of its rarity. But one natural occurrence of the barium analogue, effenbergerite, BaCuSi$_4$O$_{10}$ is believed to have formed as a result of a hydrothermal event with a maximum temperature of 450 °C [46]. Perhaps, cuprorivaite is formed similarly in nature, through a hydrothermal process.

The reader should be aware of the close proximity of Mount Vesuvius (where Minguzzi first discovered the mineral cuprorivaite) to Pompeii (where Chaptal found the first archaeological specimens of Egyptian blue frit). All in all, it is a weird coincidence that the extremely rare mineral, cuprorivaite, was first discovered just a few miles up the hill from where its artificial analogue was first discovered in modern times and also within a distance of no more than a day's walk from Pozzuoli, where CaCuSi$_4$O$_{10}$ was allegedly first synthesised in Europe, over 2000 years ago![17]

---

[16] This chemical formula, *as written*, would imply that the mineral is a hemihydrate.

[17] The reader is referred to the following two review articles concerning *Egyptian blue* (written in French) by François Delamare for further reading: F. Delamare, *Le bleu égyptien, essai de bibliographie critique*, 143-162; and F. Delamare, *De la composition du bleu égyptien utilize en peinture murale gallo-romaine*, 177-194; in S. Colinart and M. Menu (eds.) *La couleur dans la peinture et l'émaillage de l'Égypte ancienne* (1998) Edipuglia.

Preparative Procedure

A prerequisite to the preparation of almost all materials is to find a container that is suitable for performing the synthesis. In other words, a crucible or ampoule that will remain chemically inert towards the chemical reagents and the reaction product, as well as, any intermediate phases formed during the process and remain physically stable under the conditions of temperature and pressure deployed throughout the synthesis. This is often a far from trivial matter. This obstacle places limitations on the preparation of many materials and so creates some interesting challenges. Platinum crucibles and ampoules are not always suitable, and even if they were, their cost is prohibitive for routine use such as in an undergraduate laboratory course. A realistic choice of container for the preparation of Egyptian Blue is a terra-cotta crucible; as used since antiquity for precisely this purpose. For those readers not au fait with pottery a description of how to make a terra-cotta crucible is given in Chapter 2.

Prepare 20 g of CaCuSi$_4$O$_{10}$ by the following method: With a precision of ± 0.001 g weigh appropriate amounts[18] of calcium carbonate CaCO$_3$; dicopper dihydroxide carbonate (artificial malachite), Cu$_2$CO$_3$(OH)$_2$; and quartz, SiO$_2$; together with 1.5 g of anhydrous sodium carbonate, Na$_2$CO$_3$; 1 g borax, Na$_2$B$_4$O$_7$·10H$_2$O; and 0.5 g sodium chloride, NaCl that will act as the flux. If possible, all of these chemical reagents should be purchased as fine powders. In particular, the quartz should be of very fine particle size (preferably <325 mesh) so as to enhance the otherwise sluggish interdiffusion reaction between the quartz and the molten flux. Malachite is favoured (over tenorite, CuO) as a chemical reagent for the simple reason that the CuO that is formed as a consequence of the *in situ* decomposition of malachite is finely divided and so presents a reactive surface. If the reader wishes to work with naturally occurring materials, then be aware that the use of quartz with a coarse grain size results in an incomplete reaction. Beach sand is normally too coarse and therefore not appropriate. Desert sand, which is normally quite fine, is more suitable and indeed more relevant when considering that CaCuSi$_4$O$_{10}$ was prepared in ancient Egypt; but even this may require further grinding.

Grind these dry powders together thoroughly using a porcelain pestle and mortar,[19] and place the mixture in a small (~150 ml) terra-cotta crucible. Tap the bottom of the crucible a few times on the laboratory bench so that the powder bed is reasonably compacted. Then place it in a chamber (or a pottery) furnace and heat at 100 °C/h to 875 °C. After 16 h at 875 °C, cool to room temperature (20 °C) with a slow cooling rate of 7 °C/h. This heating cycle will last six days.

---

[18] The reader may wish to consider *Problem 1* before embarking on this calculation.

[19] It is better to use a planetary mono mill (e.g., a *Fritsch Pulverisette 6* with zirconium oxide bowl and balls and cyclohexane as the liquid medium) if the reader has access to such equipment.

Remove the fritted product from the crucible. The product will most likely be adhered to the bottom of the crucible, so it may be necessary to break the terra-cotta crucible with a hammer and remove unwanted pieces of terra-cotta with a hammer and chisel. Using a hand lens ($\times 10$ magnification) note the colour, depth of colour, size and habit of any crystals that may have formed, and note the formation of any obvious glass phase; and if relevant, note its colour.

In order to remove the flux it is necessary to crush the frit very gently (if necessary using a hardened steel percussion impact piston; see Figure 2.5) into loose $\sim$5 mm fragments; *do not* grind the product at this stage. Place these fragments in a 250-ml glass beaker and add 150 ml of distilled water. Stir with a glass rod and leave to stand overnight in order to dissolve the flux.

Decant most of the clear liquor, then pour the remaining slurry onto a wetted sheet of medium speed filter paper (e.g., *Whatman* No. 2) in a Buchner filter funnel and wash it with distilled water. Finally, wash the product with a small amount of acetone (in order to accelerate the drying), and leave it to dry for $\sim$10 min whilst in the Buchner filter funnel under vacuum. Place the product material on a watch-glass and leave it to dry for $\sim$10 min.

With the aid of a hand lens (or a binocular microscope) select a few euhedral crystals and place them in an appropriately labelled specimen jar. These crystals can be analysed using a polarising microscope to observe their dichroism. From the remaining material, remove a 5-g sample and grind this to a fine powder using an agate pestle and mortar; then place it in an appropriately labelled specimen jar. Submit this sample for analysis by powder X-ray diffraction (scan: 10–70° 2θ). Compare the powder pattern with the PDF 85-158 (cf. ICSD 402012) and Figure 3.12. The reader may

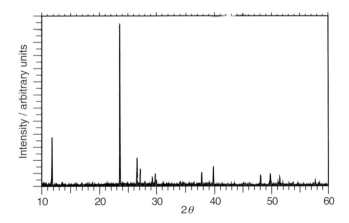

Figure 3.12 Powder XRD pattern (Cu-$K_{\alpha 1}$ radiation) of CaCuSi$_4$O$_{10}$ as prepared by the method described in this chapter. PDF 81-1238 for CaCuSi$_4$O$_{10}$ is shown in red for comparison

wish to use the remaining material to fabricate a ceramic monolith. If so, then press 10 g of finely powdered CaCuSi$_4$O$_{10}$ into a compacted monolith using a 32 mm die and a uniaxial press under a load of 5 tonne. Place the monolith inside an alumina crucible (LR42 Almath Ltd) and heat to 875 °C for 50 h in air with a heating and cooling rate of 100 °C/h.

As a final comment, the literature concerning the synthesis of CaCuSi$_4$O$_{10}$ is dominated by high-temperature methods, almost to the point of exclusion of all other possible routes of synthesis. Therefore, it is interesting to note, that Hein [47] reported the preparation of a calcium-copper silicate by aqueous precipitation, in a German patent in 1964. However, Hein's preparative description is very brief, and the nature of the calcium–copper silicate phase is unspecified in the patent. The reader may wish to explore an aqueous or hydrothermal route, for the synthesis of CaCuSi$_4$O$_{10}$. Success in this area would break the historic dependency on the use of a salt flux, and may provide an insight to the conditions under which cuprorivaite forms in nature.[20]

REFERENCES

1. A. Pabst, *Acta Crystallographica* **12** (1959) 733–739.
2. W. Bensch and M. Schur, *Zeitschrift für Kristallographie* **210** (1995) 530.
3. E. Burzo, *Gillespite Group of Silicates*, in Landolt-Börnstein, New Series, Group III, Volume 27: Magnetic Properties of Non-Metallic Inorganic Compounds Based on Transition Elements, Subvolume I 5α: Phyllosilicates – Part 1, pages 1–15, Springer, Berlin and Heidelberg, 2009.
4. K.-H. Breuer, W. Eysel and M. Behruzi, *Zeitschrift für Kristallographie* **176** (1986) 219–232.
5. A. Heinrich-Beda, *Beiträge zur Kristallchemie des Cu(II) in Silikaten*, PhD Thesis, ETH Nr. 7303, Zürich (1983).
6. M. H. Zöller, E. Tillmanns and G. Hentschel, *Zeitschrift für Kristallographie* **200** (1992) 115–126.
7. G. Bayer and H. G. Wiedemann, *Naturwissenschaften* **62** (1976) 181–182.
8. G. A. Mazzocchin, D. Rudello, C. Bragato and F. Agnoli, *Journal of Cultural Heritage* **5** (2004) 129–133.
9. H. Jaksch, W. Seiple, K. L. Weiner and A. El Goresy, *Naturwissenschaften* **70** (1983) 525–535.
10. R. J. Ford and M. A. Hitchman, *Inorganica Chimica Acta*, **33** (1979) L167–L170.
11. I. L. Botto, E. J. Baran, and G. Minelli, *Anales de la Asociacion Química Argentina* **75** (1987) 429–437.
12. E. Kendrick, C. J. Kirk, and S. E. Dann, *Dyes and Pigments* **73** (2006) 13–18.

---

[20] Within this context it is interesting to note that Holden and Singer [48] describe that in growing crystals of calcium–copper acetate hexahydrate, CaCu(CH$_3$COO)$_4$·6H$_2$O from aqueous solution, there is a need for an excess of calcium ions over copper ions within the aqueous solution. Holden and Singer recommend an initial molar ratio Ca^{2+}(aq)/Cu^{2+}(aq) = 4, in order to avoid the deposition of crystals of copper acetate monohydrate.

13. J. E. Huheey, *Inorganic Chemistry: Principles of Structure and Reactivity*, Harper and Row, New York, 2nd edn, 1978.
14. G. Blasse and B. C. Grabmaier, *Luminescent Materials*, Spinger-Verlag, Berlin, 1994, pp 232.
15. P. Atkins and J. De Paula, *Atkins' Physical Chemistry*, 8th edn, Oxford University Press, Oxford, 2006, pp 1064.
16. F. Fouqué, *Comptes Rendus* **108** (1889) 325–327.
17. F. Fouqué, Bulletin de la Societe Française de Mineralogie **12** (1889) 36–38.
18. C. Minguzzi, *Periodico di Mineralogie* **9** (1938) 333–345.
19. F. Mazzi and A. Pabst, *American Mineralogist* **47** (1962) 409–411.
20. C. D. Gribbles and A. J. Hall, *Optical Microscopy: Principles and Practice*, UCL Press, London, 1999.
21. W. D. Nesse, *Introduction to Mineralogy*, Oxford University Press, New York, 2000, 442 pp.
22. A. F. Orchard, *Magnetochemistry*, Oxford University Press, Oxford, 2003, pp 176.
23. F. E. Mabbs and D. J. Machin, *Magnetism and Transition Metal Complexes*, Chapman and Hall, *London*, 1973, pp 206.
24. Th. Rehren, E. B. Pusch and A. Herold, *Qantir-Piramesses and the organization of the Egyptian glass industry*, Chapter 12, 223–238, in *The Social Context of Technological Change: Egypt and the Near East, 1650–1550 BC*, A. J. Shortland (ed.), Oxbow Books, Oxford, 2001, pp 288.
25. M. E. Hamza, Annales du Service des Antiquités de l'Egypte, **30** (1930) 31–68.
26. G. D. Hatton, A. J. Shortland and M. S. Tite, *Journal of Archaeological Science* **35** (2008) 1591–1604.
27. W. T. Chase, *Science and Archaeology*, ed. R. H. Brill, MIT Press, Cambridge, Massachusetts (1971) 80–90.
28. M. S. Tite, *Archaemetry*, **29** (1987) 21–34.
29. H. G. Wiedemann, G. Bayer and A. Reller, *Egyptian blue and Chinese blue. Production technologies and applications of two historically important blue pigments* 195–203, in, S. Colinart and M. Menu (eds.) *La couleur dans la peinture et l'émaillage de l'Égypte ancienne*, (1998) Edipuglia, Bari, Italy.
30. M. H. Morgan (Translator), *The Ten Books on Architecture* (in Latin: *De architectura libri decem*) by Vitruvius Pollio (written in the first century BC), Book VII; Chapter XI, Harvard University Press, Cambridge, Massachusetts, 1914.
31. B. Poulsen, *Analecta Romana*, **30** (2004) 43–67.
32. P. Guldager Bilde, *Analecta Romana*, **30** (2004) 7–42.
33. P. C. Hewlett (ed.) *Lea's Chemistry of Cement and Concrete*, 4th edn, Butterworth-Heincmann, Oxford, 1998, pp1057.
34. T. Budetta, *Musiva and Sectilia* **2–3** (2005) 43–80.
35. H. Davy, Philosophical Transactions of the Royal Society of London, **105** (1815) 97–124.
36. L. Lazzarini, *Studies in Conservation* **27** (1982) 84–86.
37. J. A. Chaptal, *Annales de Chimie*, **70** (1809) 22–31.
38. F. Delamare, *Le bleu égyptien, essai de bibliographie critique*, 143–162, in, S. Colinart and M. Menu (eds.) *La couleur dans la peinture et l'émaillage de l'Égypte ancienne*, (1998) Edipuglia, Bari, Italy.

39. A. Holmes, *Principles of Physical Geology*, Thomas Nelson and Sons Ltd, 2nd edn, 1965, London pp 1288.
40. H. Fontenay, *Annales de chimie et de physique*, Série 5, **2** (1874) 193–199.
41. L. Bock, *Angewandte Chemie* **29** (1916) 228.
42. A. P. Laurie, W. F. P. McLintock and F. D. Miles, *Proceedings of the Royal Society of London*, Series A, **89** (1914) 418–429.
43. T. Pradell, N. Salvado, G. D. Hatton and M. S. Tite, *Journal of the American Ceramic Society* **84** (2006) 1426–1431.
44. G. Schippa and G. Torraco, *Rassegna Chimica* **96** (1957) 3–9.
45. R. V. Gaines, H. C. W. Skinner, E. E. Foord, B. Mason, A. Rosenzweig, *Dana's New Mineralogy: The System of Mineralogy of James Dwight Dana and Edward Salisbury Dana*, Eight Edition, 1997, Wiley-Blackwell, New York.
46. G. Giester and B, Rieck, *Mineralogical Magazine* **58** (1994) 663–670.
47. R. Hein, *Verfahren zur Herstellung von Schwermetall-Silikaten*, German Patent 1164996 (1964).
48. A. Holden and P. Singer, *Crystals and Crystal Growing*, Heinemann, London, 1960, pp 320.
49. K. Kawamura, and A. Kawahara, *Acta Crystallographica* **33** (1977) 1071–1075.
50. M. T. Weller, *Inorganic Materials Chemistry*, Oxford University Press, Oxford, 1996.
51. A. R. West, *Basic Solid State Chemistry*, Wiley, Chichester, 2nd edn, 1999.

PROBLEMS

1. Write a balanced equation to describe the overall chemical reaction for the formation of $CaCuSi_4O_{10}$ from the chemical reagents as described in this preparative procedure.

2. Pradell *et al.* [43] report that a sufficient quantity of anhydrous sodium carbonate was added to achieve a Na_2O content of 3 wt% with respect to the resultant oxide mixture. Calculate the mass of Na_2CO_3 needed to be added together with a stoichiometric mixture of calcite, malachite and quartz, in order to make 20 g of $CaCuSi_4O_{10}$, in which Na_2CO_3 is the sole component of the flux.

3. State the equilibrium temperature for the following reaction in air (i.e., $P_{O_2} = 0.21$ bar)
   $$2CuO(s) \rightarrow Cu_2O(s) + 0.5O_2(g).$$
   Use thermodynamic data from the literature (or an appropriate software system, for example, Outokumpu HSC Chemistry®) to construct a predominance diagram for the Cu–O system (i.e., P_{O_2} as a function of temperature), in order to help answer this question.

4. In some wood-burning fires, the gas composition creates a reducing atmosphere. What would you expect to happen if the synthesis was attempted under these conditions?

5. Describe the habit of the $CaCuSi_4O_{10}$ crystals in your product material.

6. Upon grinding a sample of your product material to a fine powder, you should have noticed that the blue colour becomes paler. Give a physical explanation for the apparent change in the intensity of the colour.

7. The compound $Na_2CuSi_4O_{10}$ is reported in the literature by Kawamura and Kawahara [49]. Use your knowledge gained from preparing $CaCuSi_4O_{10}$, to briefly describe a method for making 20 g of $Na_2CuSi_4O_{10}$. What colour do you believe the material will be; and why?

8. From a qualitative consideration of the peak widths in your powder pattern, comment on the crystallinity of your material.

9. Use the following data for $CaCuSi_4O_{10}$ (cf. ICSD 402012) in order to help index the powder X-ray diffraction pattern of your product material. Then use your d-values to determine the tetragonal unit cell constants (a and c) corresponding to your specimen of $CaCuSi_4O_{10}$. The following 2θ values relate to Cu-$K_{\alpha 1}$ radiation ($\lambda = 1.5405$ Å). The reader is referred to Weller [50] and West [51] for guidance in relating powder X-ray diffraction data to crystal structure.

hkl	d/Å	$2\theta/^\circ$	Int.
002	7.5840	11.66	24
012	5.2646	16.83	11
110	5.1718	17.13	7
112	4.2728	20.77	0
004	3.7920	23.44	60
020	3.6570	24.32	9
014	3.3664	26.45	74
022	3.2940	27.05	100
121	3.1974	27.88	35
114	3.0581	29.18	33
122	3.0035	29.72	94
123	2.7463	32.58	4

10. Use the molar magnetic susceptibility data in Figure 3.7b together with the formula $\mu_{eff} = 2.83\sqrt{C}$ (where C is the Curie constant) in order to calculate a value for the effective magnetic moment (μ_{eff}) in units of Bohr magneton (μ_B) for the Cu^{2+} ion in $CaCuSi_4O_{10}$. Compare your value with those in Figure 3.7c.

4

Artificial Covellite CuS by a Solid–Vapour Reaction

Copper sulfide, CuS, is quite an enigmatic material for what is in compositional terms a very simple stoichiometric binary compound. It has a unique crystal structure and an unusually complex electronic structure. From an historical perspective, CuS was the first stoichiometric compound to exhibit superconductivity and has a $T_c \sim 1.6\,\mathrm{K}$. Above this temperature, CuS is a p-type metal and at more elevated temperatures it exhibits a significant copper ionic mobility, such that the charge carriers are both positive holes and copper cations. Its optical properties are equally unusual in that it displays very strong reflection pleochroism and very strong anisotropism. Its magnetic properties are not fully understood, but its behaviour at room temperature is consistent with Pauli paramagnetism. CuS is found in nature as the mineral covellite, and is mined in certain deposits for its ore value. The method described here for its synthesis, involves a solid–vapour reaction between elemental copper and sulfur vapour in a sealed evacuated double-bulbed Pyrex™ glass ampoule. The reaction mechanism is strongly influenced by the mixed ionic-electronic conducting properties of both the CuS and the intermediate phase, $Cu_{2-\delta}S$, which results in the mass transport of copper within the solid state, as is evident in the characteristically hollow morphology of the reaction product.

CuS crystallises with the klockmannite (CuSe) structure, as shown in Figure 4.1 and belongs to the hexagonal crystal system and space group $P6_3/mmc$, with cell constants $a = 3.796\,\text{Å}$ and $c = 16.382\,\text{Å}$. The crystal

Synthesis, Properties and Mineralogy of Important Inorganic Materials   By Terence E. Warner
© 2011 John Wiley & Sons, Ltd

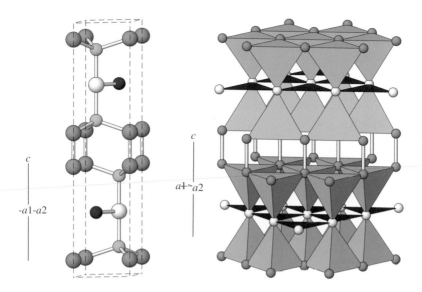

Figure 4.1   Crystal structure of covellite, CuS showing the hexagonal unit cell (left). There are crystallographically, two distinct copper sites designated; Cu(1) (dark blue) and Cu(2) (light blue); and two distinct sulfur sites designated S(1) (white) and S(2) (grey). The linkage of the CuS$_4$ tetrahedral (light blue) and planar CuS$_3$ triangles (dark blue) is shown on the right. This figure was drawn using data from Takeuchi *et al.* [3] cf. ICSD 61793

structure was first determined in 1929 by Roberts *et al.* [1] and refined by Evans and Konnert [2] along with high-pressure measurements up to 33 kbar by Takeuchi *et al.* [3]. The klockmannite structure is unique to the solid solution series, CuSe$_{1-x}$S$_x$ ($0 \leq x \leq 1$); including both its end-members. There are crystallographically, two distinct copper sites, designated Cu(1) and Cu(2), and two distinct sulfur sites, designated S(1) and S(2). Covellite can be described as a layered structure, with the following alternating sequence. One of the layers comprises a net-like plane of [CuS$_3$] triangles within which the Cu(1)–S(1) bonds are remarkably short, 219 pm [2]; and as such, are considered to be metallic [4]. The sulfur atoms at the corner of these triangles form the common apices for pairs of [CuS$_4$] tetrahedra that extend above and below this 'triangular' plane. The bases of these tetrahedra are held to the bases of the next (inverted) layer of tetrahedra by sulfur–sulfur bonds, which lie parallel to the *c*-axis. These S(2)–S(2) bonds have a characteristically short length of 207 pm [2], which is considerably shorter than the S–S bond length of 218 pm in pyrite, FeS$_2$; and are more comparable to 205 pm for the S–S covalent bond length in elemental (orthorhombic) sulfur.

   Generally, within any solid, the various aspects regarding crystal structure, chemical bonding and electronic structure are all strongly intertwined.

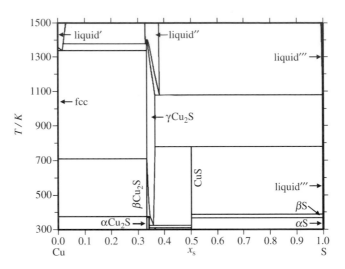

Figure 4.2 Calculated equilibrium phase diagram of the Cu–S binary system (Reprinted with permission from Landolt-Börnstein - Group IV Physical Chemistry, Volume 19B3 Binary system Part 3 'Binary system Cu-S' by, P. Franke and D. Neuschütz Copyright (2005) Springer Berlin Heidelberg)

Covellite is no exception to this; so it is not surprising to find that its electronic structure is also rather peculiar, as will be discussed in detail later in this chapter.

A calculated equilibrium phase diagram for the Cu–S binary system was constructed by Franke and Neushütz [5] as shown in Figure 4.2. This is consistent with the phase diagram by Chakrabarti *et al.* [6] as based on experimental data. The subsolidus phase relationships in the Cu–S binary system are complicated and much uncertainty remains near room temperature. This is further compounded by the existence of several newly discovered minerals, such as; roxbyite, $Cu_{1.78}S$; geerite, $Cu_{1.6}S$; spionkopite,[1] $Cu_{1.4}S$; and yarrowite,[1] Cu_9S_8 [7]. Due to the sluggishness of solid-state reactions near room temperature, it is often very difficult to establish whether a particular phase is truly stable or metastable; and this might well apply to the above-mentioned minerals that so far have been little studied. However, the phase diagram (as it stands) indicates that at room temperature (and up to $\leq 75\,^{\circ}C$), covellite can coexist with either, anilite, Cu_7S_4 or, orthorhombic sulfur, depending, respectively, on whether the system is relatively deficient or excessive in sulfur. But in both cases covellite remains invariably stoichiometric. From the phase diagram, it is also apparent that covellite does *not*

---

[1] Previously, spionkopite, $Cu_{1.4}S$; and yarrowite, Cu_9S_8 were known collectively as blaubleibender covellites (German for: permanent-blue covellites) on account of their behaviour under polarised light with oil immersion [8].

melt[2] at 507 °C (at least not at ambient pressure); but decomposes reversibly into digenite, $Cu_{2-\delta}S$, and sulfur vapour:[3]

$$2CuS(s) \rightleftharpoons Cu_2S(s) + S(vapour)$$

Therefore, this precludes the possibility of growing single crystals of covellite directly from a melt within the Cu–S binary system at ambient pressure. The reader is referred to the review articles by Fleet [7] and Chakrabarti *et al.* [6] for further details of the Cu–S binary system.

At the cold end of the scale, covellite undergoes a second-order phase transition at ∼55 K, which gives rise to a distorted form of the covellite structure with an orthorhombic unit cell at low temperature [9]. Fjellvåg *et al.* [9] attribute this structural distortion to the contraction in the distance between Cu(1)–Cu(2) with respect to the sulfide substructure, and cautiously suggest that this may be due to metal–metal bond formation. This transition coincides with the small anomaly in the heat capacity [10] and a minimum in the Hall coefficients at ∼58 K [11], but it is not particularly noticeable within electrical conductivity and magnetic susceptibility measurements.

CuS occurs in nature as the mineral covellite (in French: covelline), see Figure 4.3. It was named in 1832 after the Italian chemist Niccolo Covelli (1790–1829) who first described the mineral found amongst volcanic sublimates at Vesuvius, Italy [14]. Although Vesuvius was the location of its discovery, covellite rarely occurs as a volcanic sublimate. It is more commonly found as a secondary mineral within the supergene enrichment zone[4] as a product of the alteration (i.e., weathering) of the primary copper-iron sulfide minerals, bornite, Cu_5FeS_4, and chalcopyrite, $CuFeS_2$ [14]. In certain occurrences,[5] this enrichment process can be quite extensive, resulting in covellite being a viable copper ore mineral. Covellite also occurs as a primary mineral within hydrothermal porphyry copper deposits such as those found at Chuquicamata, in northern Chile, which is the world's largest copper ore-body [15]. Covellite of magmatic origin (i.e., crystallised from a melt) is

---

[2] Many authors, including Franke and Neuschütz [5], Chakrabarti *et al.* [6] and Fleet [7], appear to have overlooked this point. The fact of the matter is that the boiling point of sulfur (namely 445 °C) precludes the existence of a liquid sulfur phase >445 °C at ambient pressure (1 bar). At high pressure where sulfur is a liquid, CuS melts, presumably, incongruently.

[3] It is a common falsity to assume that sulfur vapour can be described in a similar fashion to dioxygen gas, i.e., simply as $S_2(g)$; although it is often written so. Sulfur vapour contains a multitude of molecular gaseous species that include predominantly; $S_2(g)$; $S_4(g)$; $S_6(g)$; $S_8(g)$. The reader is referred to the articles (in German) by Braune *et al.* [12, 13] on the dissociation of sulfur vapour.

[4] The reader is referred to Nesse 'Introduction to Mineralogy' [16] for the definition and explanation of common mineralogical terminology and geological phenomena.

[5] The supergene sulfide zone within the Bisha volcanic-associated massive sulfide deposit at western Nakfa terrane, Eritrea, north-eastern Ethiopia, has significant copper enrichment, with predominant chalcocite, Cu_2S, and lesser digenite, $Cu_{1.8}S$, covellite, CuS and bornite, Cu_5FeS_4 [17].

Figure 4.3 Covellite, Leonard Mine, Butte, Silver Bow Co., Montana, U.S.A. Field of view: 74 × 112 mm (Photograph used with permission from Ole Johnsen. Copyright (2010) Ole Johnsen)

unknown to the author; but its absence within this context is consistent at least with the phase relationships within the Cu–S and Cu–Fe–S systems.

Concerning morphology, covellite typically occurs as surface coatings and occasionally as foliated masses, whilst euhedral crystals with well-developed faces are rare. Covellite exhibits a good basal cleavage that can yield thin flexible folia; but these folia do not slip so readily as in molybdenite, MoS_2.[6] Covellite is an opaque phase with a deep blue colour under normal white light. It exhibits very strong reflection pleochroism (deep blue to pale bluish white) under plane-polarised light (see Figure 4.4) and very strong anisotropy (interference colours red-orange to brownish) under crossed polars[7] (see Figure 4.5). Both of these phenomena relate to the hexagonal symmetry of the covellite crystal structure and emphasise its two-dimensionality.

With regards to the chemistry, the formal oxidation states of copper and sulfur in CuS are a matter of some debate. A naïve approach would be to ascribe simply the valences, $Cu^{2+}S^{2-}$. However, both copper and sulfur are very polarisable atoms, such that in covellite there is a strong copper-induced polarisation of sulfur, resulting in a charge transfer from sulfur to copper

---

[6] Crystals of covellite split relatively easily parallel to the {110} plane, and for this reason can result in preferential alignment during analysis by powder X-ray diffractometry. By way of comparison, MoS_2 has a much longer S–S bond length of 347 pm [18], and this emphasises the huge difference in the nature of the S–S bond within these two compounds. In CuS the S–S bond is covalent (or perhaps even metallic), whilst in MoS_2 the S–S bond arises as a consequence of the much weaker van der Waals forces.

[7] The reader is referred to the text 'Ore microscopy and ore petrography' by Craig and Vaughan [8] for further details regarding this topic.

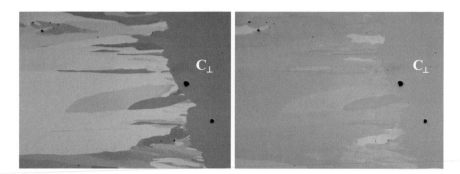

Figure 4.4 Artificial covellite under plane polarised light with the polariser oriented 90° to one another, ×10, air. Width of field = 820 μm. The relatively large crystallite on the right ($C_\perp$) is almost invariantly deep blue indicating that its c-axis is oriented almost perpendicular to the plane of view; in other words, the reader is 'looking-down' the c-axis of covellite in this region. The small black regions are pores in the material

running counter to that in the idealised ionic model, $Cu^{2+}S^{2-}$. Furthermore, there are two very distinct sulfur atoms within the covellite structure; S(1) and S(2), such that the presence of monosulfide (S^{2-}) and disulfide (S_2^{2-}) respectively, might suggest that the compound is more appropriately viewed as $Cu_2S \cdot CuS_2$ [19]. Following on from these ideas, Tossell [20] performed molecular orbital calculations for various copper–sulfur polyhedral clusters. Tossell's results showed that the $4t_2$ molecular orbital (pertaining to the Cu(2) S(2)$_3$S(1) tetrahedral cluster) is at a higher energy than the $4e'$ molecular orbital (pertaining to the Cu(1)S(1)$_3$ triangular cluster), such that there would be a spontaneous flow of charge from $4t_2 \rightarrow 4e'$ (i.e., a 'partial' electron

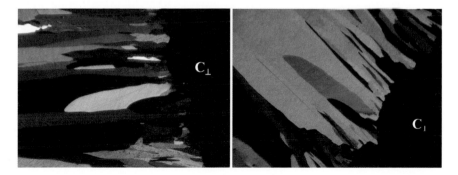

Figure 4.5 Artificial covellite in two orientations at 45° to one another under crossed polars, ×10, air. Width of field = 820 μm. The orientation of the photograph on the left coincides with those in Figure 4.4. The very strong red-orange to brownish anisotropism is clearly visible. The relatively large crystallite on the right ($C_\perp$) is almost invariantly in extinction (i.e., appears almost black in *both* photographs) and therefore is apparently almost isotropic; thus indicating that its c-axis is oriented almost perpendicular to the plane of view

transfer from Cu^+ to Cu^{2+}). Tossell concluded that the electronic configuration, $(Cu^+_2 S^{2-}) \cdot (Cu^{2+} S_2^{2-})$ was unstable, and that the outcome of this electron transfer would result in fractional oxidation states on both of the copper sites within covellite [20]. More recently, studies using X-ray absorption spectroscopy (XAS) [21] and near-edge X-ray absorption fine structure spectroscopy (NEXAFS) [22] indicate[8] that the Cu is essentially in the $3d^{10}$ state, with no indication for any significant presence of the $3d^9$ state within the material. This would imply that all the Cu is nominally monovalent, thus requiring in terms of charge balance, a positive hole to be situated on the sulfur; the question is, which one?

The ionic models, $Cu^+_3 S^- S_2^{2-}$ and $Cu^+_3 S^{2-} S_2^-$ have been described in the literature, with discussion as to which is most realistic [23]. In contradiction to these ionic models, Gainov et al. [24] conclude from nuclear quadrupole resonance measurements on ^{63}Cu nuclei, that the oxidation states regarding both of the copper sites in covellite, Cu(1) and Cu(2), are strictly neither monovalent Cu^+ nor divalent Cu^{2+}, but should be considered as being intermediate and should approximate to $Cu^{1.3+}$. This adds experimental credibility to the earlier theoretical work by Tossell [20] (as discussed above), and shifts the reality away from the ionic model to one based on covalency and electron delocalisation. Quite clearly, the old-fashioned nomenclature, *cupric sulfide*, and its modern equivalent are inaccurate descriptions of CuS. One should also note that in extractive metallurgy, copper sulfides are commonly regarded as alloys, and are often treated within the industry as metallic phases.[9]

It has been known for more than a century that covellite is a very good electrical conductor [25]. Electrical measurements, including a study of the Hall effect, have shown that CuS is a p-type metal [11]. The electrical resistivities at room temperature for c-axis oriented single crystals are, $\rho_{\parallel c} = 3.61 \times 10^{-6} \, \Omega \, m$ and $\rho_{\perp c} = 6.31 \times 10^{-7} \, \Omega \, m$ [4]. These two values are surprisingly similar for a two-dimensional solid, especially when considering that the S–S bonds provide the only linkage between the constituent layers of the structure along the c-axis (see Figure 4.1). The dependencies of the electrical resistivities on temperature over the range 77–298 K are consistent with metallic (rather than semiconductor) behaviour in both orientations. Most interestingly, below ~ 1.6 K CuS becomes a Type-I superconductor[10] as discovered by Meissner in 1929 [26], and confirmed by Buckel et al. in

---

[8] Since the leading absorption peak is at 932 eV this excludes the presence of $3d^9$ in the ground state (i.e., Cu^{2+}), which would be expected at a lower energy within the range; 930.5–931.2 eV, cf. Laan et al. [21]. NB near-edge X-ray absorption fine structure spectroscopy (NEXAFS) is also known as X-ray absorption near-edge structure spectroscopy (XANES).

[9] In the metallurgical industry, Cu_2S is known as, *white-metal*.

[10] A Type-I superconductor shows an abrupt loss of superconductivity when an applied magnetic field exceeds a critical value H_c characteristic of the material; see *Atkins' Physical Chemistry* 8th edn pages 736–738 therein [29]. In CuS, $H_c = 50$ Oe [28].

1950 [27] and more recently by Benedetto *et al.* [28] on a specimen of natural covellite.

The magnetic properties of covellite, as described in the literature, portray a certain degree of contradiction. Nozaki *et al.* [11] reported an almost temperature-independent magnetic susceptibility ($\chi_g \sim 4 \times 10^{-5}$ emu g^{-1}) for synthetic CuS at 10.2 kOe between 4–300 K. Whereas, the magnetic susceptibility measurements by Fjellvåg *et al.* [9] on synthetic CuS over the temperature range 10–195 K reveal Curie–Weiss behaviour with $\theta \sim 10$ K. From these data, Fjellvåg *et al.* [9] suggest that the existence of a magnetic moment ($\mu_{\text{eff}} = 0.27 \pm 0.03 \, \mu_{\text{BM}}$) may indicate $3d^9$ character for some of the Cu atoms or, residual spin density on the covalently bonded S(1) atoms. A plot of molar magnetic susceptibility as a function of temperature at 5 kOe for a sample of CuS as prepared[11] by the method described in this chapter is shown in Figure 4.6a. From this plot it can be seen that CuS is very weakly paramagnetic and that the magnetic susceptibility is almost independent of temperature between 150–300 K; which is consistent with CuS being a Pauli paramagnetic metal, with $\mu_{\text{eff}} = 0.18 \, \mu_{\text{BM}}$ (at 300 K). The low temperature (<150 K) data (as shown in Figure 4.6b) approximates to Curie-like behaviour with $\mu_{\text{eff}} = 0.2 \, \mu_{\text{BM}}$, and is in reasonable agreement with the data by Fjellvåg *et al.* [9] as described above. The low value for μ_{eff} suggests that this magnetic moment relates to a paramagnetic species (perhaps due to copper(II) or a sulfur radical) at a defect level of concentration ~12 at.% per unit formula (CuS). The conclusions drawn from electron paramagnetic resonance spectroscopy on CuS are consistent with these aspects, and indicate that the magnetic moment is more likely due to copper(II) rather than a sulfur radical [30]. The occasional reports of CuS being diamagnetic at room temperature are incorrect.

All of these investigations point to covellite being a metallic, rather than an ionic, compound. It is interesting to note that the converse situation arises in the perception of the chemical bonds within the intermetallic compound caesium–gold [31], which is in fact, ionic, Cs$^+$Au$^-$ and *not* metallic! A simple model for the band structure of CuS was given by Shuey [32], as illustrated in Figure 4.7. The p-type metallic conduction in CuS is attributed to the mobility of positive holes within an incompletely filled valence band comprising copper $3d$ – sulfur $3p$ orbitals [32]. Developing this model further, Vaughan and Tossell [33] suggest that the $4t_2$ and $4e'$ molecular orbitals (as discussed above) interact with each other so as to form the band structure in CuS. Furthermore, the blue coloration of covellite is attributed to electronic transitions associated with this band [33]. Some discrepancy exists in the literature as to whether the top of this valence band is predominantly of copper $3d$ [33] or sulfur $3p$ orbital character [23]. Within this context, it is interesting to consider the hybrid electronic structures for the sulfur atoms that contribute to the band structure in covellite. Vaughan and Craig [19]

---

[11] Using high-purity (99.999% *puratronic*™) elements copper and sulfur as chemical reagents.

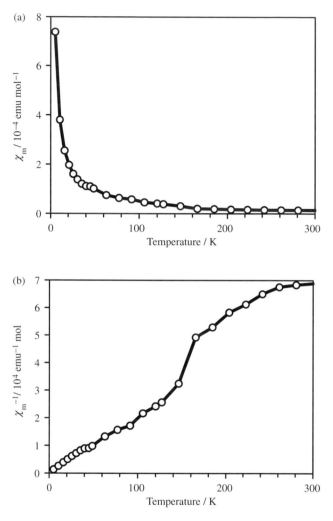

Figure 4.6 (a) A plot of the temperature dependence of the molar magnetic susceptibility ($\chi_m/10^{-4}$ emu mol^{-1}) for a polycrystalline sample of CuS at 5 kOe. These data have been corrected for diamagnetism with the usual Pascal constants. The author is grateful to Professor K. Murray and Dr. B. Moubaraki, Monash University, Australia, for recording these original data. (b) A plot of the temperature dependence of the inverse molar magnetic susceptibility ($\chi^{-1}/10^{4}$ emu^{-1} mol) for CuS corresponding to Figure 4.6(a)

describe the disulfide S(2) atoms as having a sp^3-hybridised tetrahedral quadricovalent argononic structure, and the monosulfide S(1) atoms as having a dsp^3-hybridised quinquecovalent structure with the bonds directed towards the corners of a trigonal bipyramid. Although no longer in vogue

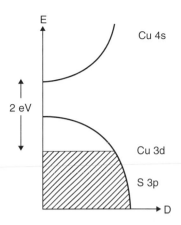

Figure 4.7 Energy, E versus density of states, D for covellite (R. T. Shuey, Semi-conducting ore minerals. 1975, Elsevier, Amsterdam)

the need to invoke *sulfur* 3d orbitals in this electronic structure is an intriguing thought. One should also be aware that whilst sulfur is normally considered to be a nonmetallic element, it transforms into a metallic element at high pressure (950 kbar) [34]; thus raising the spectre for a metallic character in the S(2)–S(2) bonds within covellite, but under less strenuous conditions.

A further complexity concerning the electrical properties of covellite arises on account of the significant mobility of the Cu^+ ions within the crystal structure. In 1950, Buckel *et al.* [27] observed that during the reaction between pieces of copper metal and sulfur vapour at 470 °C, there was an outward diffusion of copper, thus creating an inner hole and an outer CuS mantle, together with an intermediate zone[12] of $Cu_{1.8}S$. Buckel *et al.* [27] attribute this to the outward diffusion of Cu^+ ions and electrons (presumably through both of the phases; digenite[13] and covellite); see Figure 4.8. Furthermore, the copper metal appears to have undergone creep,[14] since the space originally occupied by pieces of metallic copper is replaced by a near vacuum. Presumably, this is caused by the tensile stress induced by the steep chemical potential gradient (with regards to the mobile copper) across the product layer, during the course of the reaction. In other words, the elemental sulfur on the perimeter is literally pulling the elemental copper out of the core. Electrical measurements by Donner *et al.* [35] and Galli *et al.* [36] suggest that

---

[12] Presumably; digenite at 470 °C, before transforming into anilite near ambient temperature.
[13] β-Cu_2S (chalcocite?) is a mixed ionic–electronic conductor with a significant ionic conductivity $\sigma_{ionic} = 0.20\,S\,cm^{-1}$ at 400 °C [37].
[14] Creep is a term used to describe the slow, plastic deformation of metals under stress, particularly at elevated temperatures.

Figure 4.8   Elongated cylindrical holes within a specimen of artificial anilite, Cu_7S_4 prepared in a fashion similar to that for covellite, CuS as described here. These holes correspond to the space occupied previously by the copper wire. Width of field $= 11$ mm

covellite is actually a mixed Cu^+ (ionic)/p-type (metallic) conductor at high temperature, although at room temperature the electrical conduction is dominated by the contribution from the positive holes, such that the transference number for the copper ion, $t_{Cu^+} \ll 1$. Besides these three studies, there is very little information reported in the literature regarding the electrical properties, or for that matter, any other property of covellite above ambient temperature.

Nevertheless, sufficient structural information does exist (albeit at ambient temperature), so as to make the following supposition regarding the mass transport mechanism. Since the bond length of 219 pm for the Cu(1)–S(1) bond is considerably shorter than both the 234 pm for the Cu(2)–S(1) bond and the 230 pm for the Cu(2)–S(2) bond, this would imply in a comparative sense, that Cu(2) is more weakly bonded to sulfur [24]. These structural arguments suggest that Cu(2) should be the more mobile of the two copper ions, provided that a conduction pathway exists to accommodate its movement. One possible route might be through the neighbouring S–S interspatial plane, which is analogous to an intercalation layer; albeit a very narrow one of 207 pm at 298 K. However, this could in principle, facilitate the cooperative migration of copper ions between the normally fully occupied Cu(2) sites, via the vacant trigonal prismatic interstitial sites, within this part of the sulfur substructure. Cu^+ is a highly polarisable ion, and so these transient defect copper ions could intermediately change their valence to divalent Cu^{2+} during transport.[15] This would reduce the activation energy, and their smaller ionic radius would be more suitable for the spatially restricted environment within the S–S interspatial plane.

A similar mechanism implicating the transient existence of Cu^{2+} ions in the activated state (via a reversible charge-transfer process, i.e., $Cu^+ \rightleftharpoons Cu^{2+} + e^-$), has been proposed by Warner and Maier [38] to account for the temperature-independent transference number of the copper ions within the mixed (Cu^+/e^-) conductor, $Cu^I Cu^{II} Mg_3(PO_4)_3$. Given the perplexities of the nondiscrete oxidation states for both the copper and the sulfur within covellite (as discussed above) a similar mechanism might very well apply here too. Whatever the details of the ionic conduction mechanism are in reality, the mass transport of copper through covellite at high temperature ($\sim$400 °C) is certainly real; as the reader will discover through actually preparing the compound and witnessing the formation of the cylindrical hollows within the product material. At present, the mechanism for the mass transport of copper in covellite remains unresolved, and also the crystal structure of CuS at $\sim$400 °C is yet to be confirmed.

Concerning preparative chemistry, artificial covellite can be synthesised through the precipitation of an aqueous solution of a copper(II) salt (e.g., copper(II) sulfate pentahydrate) with a convenient source of sulfide (e.g., sodium sulfide nonahydrate):

$$CuSO_4(aq) + Na_2S(aq) \rightarrow CuS(s) + Na_2SO_4(aq).$$

If the synthesis is conducted at ambient temperature, then the CuS precipitate is initially amorphous. But after aging, or moderate heating ($\geq$40 °C), it transforms into the covellite structure [40]. In the past, the act of employing a copper(II) salt was probably one of the reasons for the common misconception that covellite was described as copper(II) sulfide. The formation of covellite by this method must therefore involve a redox process. Luther et al. [41] have shown recently, by a combined EPR and ^{63}Cu NMR study, that copper(II) reduction by sulfide to copper(I) occurs within an aqueous copper-sulfide complex, $[Cu_4S_6]^{4-}(aq) \rightarrow [Cu_4S_6]^{2-}(aq)$ prior to precipitation. The matter is discussed further by Luther and Richard [42], and by Richard and Luther [43].

Frankel [44], in the year 1917, described a rather bizarre way of preparing artificial covellite by placing chalcocite, Cu_2S in contact with rubber. Frankel observed that after nearly two years under ambient conditions, covellite forms on the surface of the chalcocite. The source of sulfur is attributed to the

---

[15] Maier [39] states in his monograph, *Physical Chemistry of Ionic Materials*; 'It is an obvious and very useful rule of thumb that it is advantageous as far as ionic mobility is concerned, either for the migrating ion itself or for its counter ion to have a high polarisability, i.e., to be soft and deformable. Such dynamic softness can also be associated with intermediate valence change. So it is conceivable that a transition-metal ion (e.g., Cu) could intermediately change its valence during transport and, thus, reduce the activation energy (cf. coordination).'

sulfur content of the rubber used in the experiment. This is perhaps not the preparative method of first choice; but the results have important implications regarding solid state diffusion and the instability of vulcanised rubber.

Saito *et al.* [4] describe a method for the preferential *c*-axis-oriented growth of artificial covellite on prestretched sections of 1–3 mm diameter copper wire. This results in the growth of covellite whereby the *c*-axis of the covellite crystallites are oriented parallel to the longitudinal direction of the precursor copper wire. This approach to produce a textural alignment is adopted here.

Preparative Procedure

Prepare ~10 g of artificial covellite, CuS by the following method: Obtain a sufficient length of ~1 mm diameter copper wire.[16] Whilst wearing goggles, stretch it slightly, by holding one end in a vice whilst pulling the other end with the help of a pair of pliers. Weigh an appropriate amount of elemental copper in the form of ~30 mm lengths of prestretched wire, and the corresponding quantity of sulfur powder; which should include a significant excess above that required stoichiometrically (i.e., ~10% extra). The presence of an excess of sulfur in the system should ensure complete sulfurisation of the copper to CuS, and preclude the formation of sulfur-deficient phases, such as, digenite, $Cu_{2-\delta}S$ and anilite, Cu_7S_4.

Carefully introduce the sulfur powder into the lower compartment of a double-bulbed[17] (~100 and ~200 mm in length) Pyrex™ glass tube (~12 mm ID, ~16 mm OD) see Figure 4.9. Then insert a very small plug of glass wool into the glass tube, so as to just block the entrance to the lower compartment. Then introduce the copper charge into the upper compartment. Attach the glass tube to a vacuum line and *gradually* apply a vacuum. Whilst under vacuum, seal the neck of the tube with a blow-torch so as to create a sealed glass ampoule. It is important that the upper compartment is ~100 mm long, in order to give plenty of space for the copper charge, which will expand drastically in volume upon reacting with the sulfur (cf. *Problem 7*).

Ensure that the copper charge is situated away from the entrance to the lower compartment; otherwise intermediate material, such as, digenite, $Cu_{2-\delta}S$, will tend to grow towards and eventually block the neck between the two compartments, and thus prevent the reaction from running to completion within the time frame of this preparative procedure. Place the

---

[16] Electrical grade copper wire is typically 99.99 + % purity, and is normally preannealed at the factory. This makes it an excellent source for the copper. If relevant, remove the plastic insulation.

[17] A double-bulbed glass tube can be fabricated quite readily from a 300 mm length of Pyrex™ glass tube (~12 mm ID, ~16 mm OD), through the following glass-work procedure. With the use of a blow-torch, close one end of the tube, and then make a necked restriction about 100 mm from this end. Ensure that this restriction is not too tight, but wide enough to enable the sulfur powder to be introduced into the bottom compartment (or bulb). The upper length of 200 mm should be sufficient for vacuum attachment, and for eventually sealing the glass ampoule.

Figure 4.9 The upper photograph shows a sealed Pyrex™ glass ampoule containing the reactants: in this case, copper pellets in the left compartment and orthorhombic sulfur in the right. The glass wool is just visible in the vicinity of glass neck. (Upper photograph: Photograph used with permission from Anne-Mette Nørgaard. Copyright (2010) Anne-Mette Nørgaard. The lower photograph shows the artificial covellite product on the left and the excess sulfur condensed at the far right (Lower photograph: Photograph used with kind permission from Simon Svane. Copyright (2010) Simon Svane))

glass ampoule inside a stainless steel protection sheath and carefully slide this into a horizontally positioned tube furnace such that the copper charge is at the centre of the hot zone. Heat at the rate of 60 °C/h to 445 °C. After 200 h at 445 °C, cool to room temperature (20 °C) with a cooling rate of 20 °C/h. This heating cycle will last ∼279 h (10 days). The reader should be aware that a cooling rate greater than 20 °C/h can result in orthorhombic sulfur condensing in the part of the ampoule containing the covellite, rather than at the cold end, and should therefore be avoided. If carried out properly, this is a convenient way of segregating the excess sulfur from the reaction product.

Whilst wearing goggles, carefully break open the glass ampoule using a glass-cutting tool. Remove the product and carefully observe and note its colour and morphology. The reader may wish to photograph the product material before tampering with it further. Then place about half of the product material (if possible, as an intact lump) in an appropriately labelled specimen jar and close tightly. The reader may wish to analysis this at a later date; for example, by reflected light microscopy, and other analytical techniques as and when available. From the remaining material, submit a small (∼2 g) finely ground sample for analysis by powder X-ray diffractometry (scan: 10–60° 2θ). The XRD pattern should be compared with covellite, CuS, PDF 78-876; anilite, Cu_7S_4, PDF 72-617; and Figure 4.10. If necessary, comparison can also be made with; copper, Cu PDF 4-836; digenite, Cu_9S_5 PDF 47-1748; and chalcocite, Cu_2S, PDF 26-1116.

The reader may wish to experiment further by preparing a slug of artificial digenite, $Cu_{2-\delta}S$, by a method similar to that described here for covellite. The

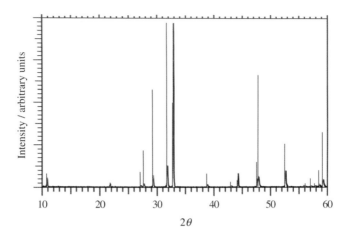

Figure 4.10   Powder XRD pattern (Cu-$K_{\alpha 1}$ radiation) of CuS as prepared by the method described in this chapter. PDF 78-876 for CuS is shown in red for comparison

mechanical properties of digenite are quite extraordinary; it is brittle, in the sense that it can be crushed by a hammer, and yet it is also sectile (i.e., it can be cut by a knife in a similar way as for calcium metal). The reader should study the phase diagram for the Cu–S system shown in Figure 4.2 so as to be aware as to which temperature the charge may need to be quenched so as to avoid undesired phase transformations upon cooling. In principle, many other transition-metal sulfides can be prepared by this method, including, ternary phases, for example, bornite, Cu_5FeS_4 and chalcopyrite $CuFeS_2$. But the reader should be aware that highly exothermic reactions can occur between certain metals and elemental sulfur; some can even be explosive! Another area of interest would be to attempt the crystal growth of covellite using a suitable flux. One possibility might be to allow two eutectic[18] mixtures of molten salts (<507 °C), such as, $CuCl_2$–LiCl–NaCl–KCl and Na_2S–LiCl–NaCl–KCl to gradually interdiffuse with one another inside an evacuated U-shaped glass ampoule, in order to yield euhedral crystals of covellite within the molten salt flux.

REFERENCES

1. H. S. Roberts and C. J. Ksanda, *American Journal of Science* **17** (1929) 489–503.
2. H. T. Evans and J. A. Konnert, *American Mineralogist* **61** (1976) 996–1000.
3. Y. Takeuchi, Y. Kudoh and G. Sato, *Zeitschrift für Kristallographie* **173** (1985) 119–128.

---

[18] The LiCl–NaCl–KCl ternary system has a eutectic at 343 °C. However, the solubility of $CuCl_2$ and Na_2S within this molten eutectic composition is not known.

4. S. H. Saito, H. Kishi, K. Nie, H. Nakamaru, F. Wagatsuma and T. Shinohara, *Physical Review* 55 (1997) 14527–14535.

5. P. Franke and D. Neuschütz, *Binary System Cu-S* in Landolt-Börnstein: Group IV Physical Chemistry, Volume 19B3 Binary system Part 3, Springer, Berlin Heidelberg, 2005.

6. D. J. Chakrabarti and D. E. Laughlin, *Bulletin of Alloy Phase Diagrams* 4 (1983) 254–271.

7. M. E. Fleet, Reviews in Mineralogy and Geochemistry 61 (2006) 365–419.

8. J. R. Craig and D. J. Vaughan, *Ore Microscopy and Ore Petrography* 2nd edn, Wiley, New York, 1994.

9. H. Fjellvåg, F. Grønvold, S. Stølen, A. F. Andresen, R. Müller-Käfer and A. Simon, *Zeitschrift für Kristallographie* 184 (1988) 111–121.

10. E. F. Westrum, S. Stølen and F. Grønvold, *Journal of Chemical Thermodynamics* 19 (1987) 1199–1208.

11. H. Nozaki, K. Shibata and N. Ohhashi, *Journal of Solid State Chemistry* 91 (1991) 306–311.

12. H. Braune, S. Peter and V. Neveling, *Zeitschrift für Naturforschung* 6 (1951) 32–37.

13. H. Braune and E. Steinbacher, *Zeitschrift für Naturforschung* 7 (1952) 486–493.

14. R. V. Gaines, H. C. W. Skinner, E. E. Foord, B. Mason and A. Rosenzweig, *Dana's New Mineralogy: The System of Mineralogy of James Dwight Dana and Edward Salisbury Dana*, 8th edn, 1997, Wiley-Blackwell, New York.

15. G. C. Ossandon, F. C. Guillermo, R. C. Freraut, L. B. Gustafson, D. D. Lindsay and M. Zentilli, *Economic Geology and the Bulletin of the Society of Economic Geologists*, 96 (2001) 249–270.

16. W. D. Nesse, *Introduction to Mineralogy*, Oxford University Press, New York, 2000.

17. T. C. Barrie, W. F. Nielsen and C. H. Aussant, *Economic Geology* 102 (2007) 717–738.

18. T. J. Wieting and J. L. Verble, *Physical Review B*, 3 (1971) 4286–4292.

19. D. J. Vaughan and J. R. Craig, *Mineral Chemistry of Metal Sulfides*, 1978, Cambridge University Press, Cambridge.

20. J. A. Tossell, *Physics and Chemistry of Minerals* 2 (1978) 225–236.

21. G. van der Laan, R. A. D. Pattrick, C. M. B. Hendersen and D. J. Vaughan, *Journal of Physics and Chemistry of Solids* 53 (1992) 1185–1190.

22. S. W. Goh, A. N. Buckley and R. N. Lamb, *Minerals Engineering* 19 (2006) 204–208.

23. W. Liang and M.-H. Whangbo, *Solid State Communications* 85 (1993) 405–408.

24. R. R. Gainov, A. V. Dooglav, I. N. Pen'kov, I. R. Mukhamedshin, N. N. Mozgova, A. V. Evlampiev and I. A. Bryzgalov, *Physical Review B* 79 (2009) 1–11.

25. G. Bodlander and K. S. Idaszewski, *Zeitschrift für Elektrochemie und Angewandte Physikalische Chemie* 11 (1905) 161–182.

26. W. Meissner, *Zeitschrift für Physik* 58 (1929) 570–572.

27. W. Buckel and R. Hilsch, *Zeitschrift für Physik* 128 (1950) 324–346.

28. F. D. Benedetto, M. Borgheresi, A. Canaeschi, G. Chastanet, C. Cipriani, D. Gatteschi, G. Pratesi, M. Romanelli and R. Sessoli, *European Journal of Mineralogy* 18 (2006) 283–287.

29. P. Atkins and J. De Paula, '*Atkins' Physical Chemistry*' 8th edn, 2006, Oxford University Press, Oxford, pp1064.

30. A. M. C. Nørgaard, Unpublished MSc Thesis, 2007, University of Southern Denmark.

31. N. E. Christensen, *Physical Review* B **32** (1985) 207–228.
32. R. T. Shuey, *Semiconducting Ore Minerals*, 1975, Elsevier, Amsterdam.
33. D. J. Vaughan and J. A. Tossell, *Canadian Mineralogist* **18** (1980) 157–163.
34. H. Luo, S. Desgreniers, Y. K. Vohra and A. L. Ruoff, Recent Trends High Pressure Res., *Proc. AIRAPT Int. Conf. High Pressure Sci. Technol.*, 13th (1992), Meeting Date 1991, 374–6.
35. D. Donner and H. Rickert, *Zeitschrift für Physikalische Chemie* **60** (1968) 11–24.
36. R. Galli and F. Garbassi, *Nature* **253** (1975) 720–2.
37. E. Hirahara, *Journal of the Physical Society of Japan* **6** (1951) 428–437.
38. T. E. Warner and J. Maier, *Materials Science and Engineering B* **23** (1994) 88–93.
39. J. Maier, *Physical Chemistry of Ionic Materials: Ions and Electrons in Solids*, Wiley, Chichester, 2004.
40. R. A. D. Pattrick, J. F. W. Mosselmans, J. M. Charnock, K. E. R. England, G. R. Helz, C. D. Garner and D. J. Vaughan, *Geochimica et Cosmochimica Acta* **61** (1997) 2023–2036.
41. G. W. Luther III, M. S. Theberge, T. F. Rozan, D. Richard, C. C. Rowlands and A. Oldroyd, *Environmental Science and Technology* **36** (2002) 394–402.
42. G. W. Luther III and D. Richard, *Journal of Nanoparticle Research* **7** (2005) 389–407.
43. D. Richard and G. W. Luther III, *Reviews in Mineralogy and Geochemistry* **61** (2006) 421–504.
44. J. M. Frankel, *Engineering and Mining Journal* **104** (1917) 252.

PROBLEMS

1. State the boiling point of sulfur.

2. State the partial pressure of pure sulfur vapour in equilibrium with pure liquid sulfur at 445 °C in a closed system of constant volume and in the absence of any external force acting on the system.

3. Pure sulfur vapour at the temperature of 445 °C and a total pressure of 1 bar contains the following molecular species; $S(g)$; $S_2(g)$; $S_3(g)$; $S_4(g)$, $S_5(g)$; $S_6(g)$; $S_7(g)$; $S_8(g)$, such that the partial pressure of $P_{S_2(g)} = 3.6 \times 10^{-2}$ bar. Bearing this in mind:
   (a) Calculate a value for the standard enthalphy of formation (ΔH_f°) at 445 °C for CuS(s).
   (b) How does this value compare with that for ZnS(s) at 445 °C.
   (c) Therefore, what dangerous thing might happen if you attempted to prepare zinc sulfide by the same method as used here to prepare copper sulfide?

4. What is the upper temperature limit for a Pyrex™ glass ampoule as used in this technique? Suggest an alternative material for use as an evacuated ampoule at higher temperature.

5. (a) What would you expect to happen if the reaction mixture as used in your work was heated to 550 °C?
   (b) Would you expect this heat treatment to affect the outcome of the synthesis?

6. At the end of your synthesis, the glass ampoule should in principle contain only covellite and orthorhombic sulfur, together with the corresponding equilibrium vapour pressure for sulfur at room temperature. With regards to this system:
   (a) Comment on the value for the sulfur activity within the orthorhombic sulfur ($a_{S(S)}$).
   (b) Comment on the value for the sulfur activity within the covellite ($a_{S(CuS)}$).
   (c) Comment on the value for the copper activity within the covellite ($a_{Cu(CuS)}$).

7. The sulfurisation of elemental copper (Cu) to covellite (CuS) is accompanied by a large increase in the volume of the phase containing the copper. Calculate the value for this increase expressed as a ratio of the molar volumes: $V_{mol}(CuS)/V_{mol}(Cu)$.

8. (a) Explain why the peak (006) in the powder X-ray diffraction pattern is so predominant.
   (b) Explain how the d_{006} spacing relates to the physical reality of the covellite structure?

9. Describe a method to prepare a sample of artificial klockmannite, CuSe.

10. Describe a method to prepare a sample of artificial villamaninite, CuS_2.

5

Turbostratic Boron Nitride t-BN by a Solid–Gas Reaction Using Ammonia as the Nitriding Reagent

Boron nitride, BN is a colourless stoichiometric binary compound that occurs in several modifications. The amorphous and turbostratic, t-BN modifications are the easiest to prepare.[1] These can be transformed into the hexagonal modification, α-BN, by annealing at high temperature.[2] α-BN has a low density of $2.27\,\text{g cm}^{-3}$ and displays important thermal, mechanical and dielectric properties, and consequently, finds applications as a machinable ceramic, solid lubricant, electrical insulator and refractory material. This chapter describes the preparation of a powder specimen of BN by a solid–gas reaction using commercially available iron boride, FeB as the chemical reagent for the boron component; and ammonia, NH_3 as the chemical reagent for the nitrogen component. FeB powder is nitrided in flowing ammonia gas at $450\,^\circ\text{C}$ to form amorphous BN, followed by in situ annealing at $1000\,^\circ\text{C}$ to convert, ideally, the amorphous BN to crystalline α-BN.[3]

---

[1] *Turbostratic* means literally: 'disturbed layers'. It is a term used to describe a material having a structure intermediate between amorphous and crystalline, consisting of stacked disordered layers.

[2] The hexagonal modification of BN with the graphite-like structure is denoted in this chapter as α-BN. Further notations exist in the literature, for example, h-BN; which can be confused with a high-pressure hexagonal modification (γ-BN) that crystallizes with the wurtzite-type structure.

[3] The use of ammonia as a nitriding reagent, is commonly referred to as, ammonolysis.

Synthesis, Properties and Mineralogy of Important Inorganic Materials   By Terence E. Warner
© 2011 John Wiley & Sons, Ltd

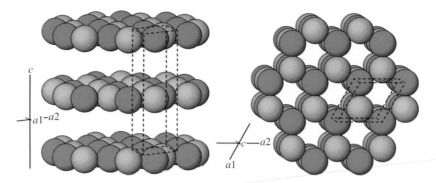

Figure 5.1 Crystal structure of hexagonal α-BN. The boron atoms are shown yellow, and the nitrogen atoms are shown green. The unit cell is shown in dash. This figure was drawn using data from Pease [1] cf. ICSD 24644

α-BN crystallizes with the prototype structure as shown in Figure 5.1. It belongs to the hexagonal crystal system and space group $P6_3/mmc$ with cell constants: $a = 2.504$ Å and $c = 6.661$ Å [1]. The crystal structure was first determined by Hassel [2] in 1927 using powder X-ray diffractometry and refined by Pease [1] in 1952 using single-crystal X-ray diffractometry at 308 K. The α-BN structure is similar to the graphite structure (to which it is isoelectronic), but with a few significant differences. It comprises planar layers of condensed B_3N_3 hexagonal rings, with boron and nitrogen atoms located at alternating positions within each of the rings. These layers lie in a plane parallel to the basal plane of the unit cell, and are stacked directly on top of one another such that the nitrogen atoms in one layer lie directly above the boron atoms in the layer below, and vice versa.[4] This creates an AB stacking sequence in which boron and nitrogen atoms are located at alternating positions along the c-axis. The intraplanar B–N bonds have a bond length of 145 pm and are strongly covalent,[5] whereas the interplanar B–N bonds have a bond length of 333 pm and are of the weak van der Waals type. The nature of the chemical bonding and the extreme crystallographic anisotropy in α-BN have a profound influence on its physical properties, as will be discussed below.

Variations in the stacking sequence are quite common, and these lead to various modifications. For example, rhombohedral boron nitride (r-BN) has a metastable ABC stacking sequence [5], whereas turbostratic boron nitride (t-BN) is a disordered phase with a reduced periodicity in the stacking sequence, and less uniformity in the interplanar distances [6, 7], which leads

---

[4] For the sake of comparison, the planes of carbon atoms in graphite are staggered, so that only half of the carbon atoms in one layer lie directly above the carbon atoms in the layer below; the other half being located directly above the centres of the hexagonal rings.

[5] The intraplanar B–N covalent bonds have a large peak phonon energy at $1370 \, \text{cm}^{-1}$ (corresponding to a B–N stretching mode) thus indicating a high strength for these bonds [3, 4].

ultimately, to amorphous boron nitride. Under very high pressure and temperature these modifications transform into polymorphs with the more
densely packed wurtzite and sphalerite structures.[6] The wurtzite-type phase
(γ-BN) is metastable; whereas, the sphalerite-type phase (β-BN) is stable over a
wide range of temperature and pressure. β-BN is commonly known as cubic
boron nitride, and is used as an abrasive material on account of its extreme
hardness, and is superior to diamond in terms of its enhanced chemical stability
at high temperature [8].[7] It is interesting to note that β-BN is considered to be
the thermodynamically stable polymorph at room temperature and pressure,
thus implying that α-BN is in fact metastable with respect to β-BN under
ambient conditions [8], whereas, the converse situation is true of graphite and
diamond.

The hexagonal modification, α-BN, sublimes at 2330 °C under a nitrogen
partial pressure of 1 bar [10]. But when it is enclosed under a nitrogen partial
pressure of $\geq$ 50 bar, it reveals a melting point of 3100 °C [11]. α-BN is more
resistant to oxidation than graphite, and can survive oxidation in air up to
900 °C [12]. Furthermore, α-BN is chemically stable in the presence of a
number of molten elements, and is used as a crucible material for the fusion
and casting of high-purity metals and alloys; and for containing Si and GaAs
melts as required in the semiconductor industry [13]. α-BN possesses many
other desirable properties; for instance, it is a machinable ceramic that
exhibits high thermal conductivity, excellent resistance to thermal shock,
and a high lubricity, and is used as a refractory material for specialist
components in the metallurgical industry. The crystallographic layers in
α-BN are held together by weakly bonding van der Waals forces, which
makes them easy to cleave (cf. graphite and MoS_2). Therefore, α-BN is ideally
suited as a solid lubricant and is used as a mould-releasing agent in the
metallurgical and glass industries. It is also used in composite materials for the
fabrication of self-lubricating bearings.

α-BN is an intrinsic semiconductor with a very large bandgap ~ 6 eV [14].
It has a correspondingly high electrical resistivity, $\rho = 10^{13}\,\Omega\,cm$ at room
temperature; which in conjunction with a high thermal conductivity ($\kappa_{\|c} =$
$2.9\,W\,m^{-1}\,K^{-1}$ and $\kappa_{\perp c} = 62\,W\,m^{-1}\,K^{-1}$ [12]) and low relative permittivity
($\varepsilon_r \sim 4$), ensures that α-BN has a niche market in the electronics industry as an
electrical insulator.

Boron nitride can be prepared from its constituent elements by reacting
boron and nitrogen together at extremely high temperature. For instance,
Andreev and Lundström [15, 16] have prepared a sample of turbostratic

---

[6] The lower threshold for the direct conversion of pure α-BN to the wurtzite polymorph (γ-BN) is
about 125 kbar at 25 °C [9], and to the sphalerite polymorph (β-BN) is about 113 kbar at
1600 °C. β-BN is normally prepared with the aid of a high melting point flux, for example, alkali
and alkaline earth-metal nitrides that act as a catalyst for its crystallization.

[7] Cubic boron nitride (β-BN) is isoelectronic with diamond; and its sphalerite-type structure can
be derived from the diamond structure.

boron nitride (t-BN) in the laboratory, by heating amorphous elemental boron powder under a flowing atmosphere of nitrogen at 2200 °C in a graphite element resistance furnace for 24 h. Alternatively, a very pure α-BN can be obtained by reacting boron oxide, B_2O_3 with carbon and nitrogen at $\sim$1850 °C [17]. Pyrolytic[8] boron nitride can be prepared as transparent thin films through the pyrolysis of boron trichloride, BCl_3 with ammonia, NH_3 at $\sim$1050 °C [18]. There is a growing technological interest in developing processes for the synthesis of high-purity α-BN through the pyrolysis of molecular and polymeric reagents containing boron (including inorganic polymeric B–N compounds), in order to meet the demands of a wide range of applications. The reader is referred to the review article by Paine and Narula [3] for further details.

Conventional industrial processes for the commercial production of α-BN, as bulk powders, generally involve the nitriding of either boric acid, H_3BO_3 or borax anhydride, $Na_2B_4O_7$ with ammonia or urea, $(NH_2)_2CO$ at high temperature; with the aid of additives, such as, $Ca_3(PO_4)_2$. These are long and complex processes resulting in contamination within the product. In the early 1990s the Kawasaki Steel Corporation in Japan, developed a process for the large-scale production of high-purity α-BN powder with controlled grain size, which involves the nitriding of boric acid or boron oxide with nitrogen and nitrogen-containing organic compounds, such as, urea [12]. The success of their process can be judged by the fact that their '*super-high-purity grade*' α-BN product is pure enough to satisfy the requirements of the cosmetics industry!

Viewing this subject from a slightly different angle, it is important to appreciate that certain metal borides are kinetically more susceptible to nitridation, than elemental boron per se.[9] In 1951, Kiessling and Liu [19] studied the thermal stability of the iron borides, Fe_2B and FeB, under an atmosphere of flowing ammonia as a function of temperature, during a search for ternary phases within the iron–boron–nitrogen system. Although they found none, they concluded nevertheless, that both Fe_2B and FeB were reactive towards ammonia at the relatively mild temperature of 400 °C. They also found that Fe_2B was slightly less kinetically stable than FeB, and attributed this to the difference in reactivity of the isolated interstitial boron atoms in Fe_2B, compared with the B–B atomic chains in FeB, towards ammonia (see Figure 5.2). The products of their reactions comprised boron nitride and various proportions of the iron nitrides: ζ-Fe_2N, ε-Fe_3N and γ'-Fe_4N; and α-Fe; the actual proportions depending upon the

---

[8] *Pyrolytic boron nitride* is a term used to describe heat-moulded boron nitride as prepared by chemical vapour deposition (CVD) [3].
[9] The distribution of the boron atoms within the solid FeB phase facilitates, presumably, a greater rate of reaction, than with the boron atoms in elemental boron.

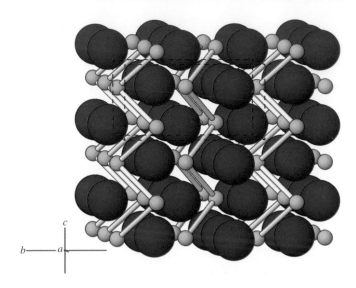

Figure 5.2 Crystal structure of iron boride FeB. The boron atoms are shown yellow, and the iron atoms are shown reddish-brown. The orthorhombic unit cell is shown in dash, with $Z = 4$. The shortest B–B interatomic distances (259 pm) are shown as sticks. This figure was drawn using data from Hendricks and Kosting [23], cf. ICSD 33577

temperature of the reaction. At 400 °C (the lowest temperature reported for an observable reaction) FeB underwent an incomplete nitridation within 24 h yielding, ζ-Fe$_2$N and the unconfirmed presence of BN, alongside unreacted FeB.

The preparative method described in this chapter for the synthesis of BN is based upon the work by Warner and Fray [20]; which in turn was inspired from the original observations by Kiessling and Liu [19], as discussed above. To this respect, it is well known that ammonia has certain kinetic and thermodynamic advantages over nitrogen as a nitriding reagent for metals and alloys. In 1930, Hägg [21] noted that metal nitrides, in general, are formed at a lower temperature when using ammonia (NH$_3$), as opposed to nitrogen (N$_2$), as the nitriding reagent. This is attributed to the lower mean bond enthalpy for the N–H bond in ammonia (388 kJ mol^{-1}) than the N≡N triple bond in the gaseous nitrogen molecule, N$_2$ (946 kJ mol^{-1}) [22]. There is also an appreciable release in Gibbs free energy associated with the dissociation of ammonia (especially at elevated temperature) on account of the large increase in entropy ($\Delta S^{\circ}_{400\,^{\circ}\text{C}} = 227$ J mol^{-1} K^{-1}) accompanying the dissociation reaction:

$$2\text{NH}_3(\text{g}) \rightarrow \text{N}_2(\text{g}) + 3\text{H}_2(\text{g})$$

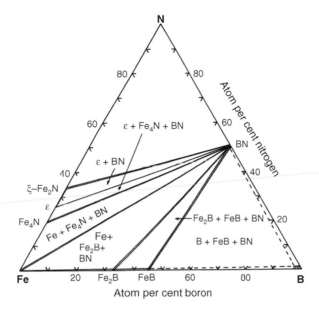

Figure 5.3   An isothermal section of the Fe–B–N ternary system at 400 °C (Reprinted with permission from Phase Diagrams of Ternary Boron Nitride and Silicon Nitride Systems by P. Rogl and J. C. Schuster, ASM International, 218 pp Copyright (1992) ASM International)

Iron boride, FeB is known by metallurgists as ferroboron, and is produced on an industrial scale in the steel industry, and is therefore a convenient source of boron for the preparation of boron nitride. From the phase diagram for the Fe–B–N ternary system at 400 °C (see Figure 5.3) it can be seen that a complete nitriding of FeB with ammonia at 400 °C should result in a two-phase mixture of BN and ζ-Fe$_2$N in accordance with the following reaction:

$$2FeB + 3NH_3(g) \rightarrow 2BN + \zeta\text{-}Fe_2N + 4.5H_2(g)$$

Warner and Fray [20] have performed this reaction, and have shown that it correlates with a predicted mass gain of $\sim$32 wt%. Furthermore, the boron nitride product was found to be amorphous at 400 °C; but then crystallized within 1 h during in situ annealing at 1000 °C. This yielded a product comprising; α-BN together with ε-Fe$_3$N and γ'-Fe$_4$N. The morphology of the product grains comprised an iron nitride core loosely encapsulated in an exfoliated layer of α-BN (see Figure 5.4). The open nature of the product layer was suggested as one of the reasons why only a relatively low temperature (namely 400 °C) was required for the nitridation to run to completion. The ε-Fe$_3$N and γ'-Fe$_4$N were dissolved in a dilute mineral acid solution and removed by filtration, leaving a solid residue of

Figure 5.4 A polished section of the nitrided FeB powder (as prepared by the method described in this chapter) under reflected polarized light. The image reveals a highly reflective core (comprising: ε-Fe$_3$N; and γ'-Fe$_4$N) with a dull exfoliated rim of BN. The dark grey regions are epoxy resin. Width of field $= 150\,\mu$m

α-BN; albeit contaminated with FeB$_{49}$. The FeB$_{49}$ was observed under a reflected light microscope as glistering black crystals, which imparted a grey tone to the product material.[10]

Preparative Procedure

Prepare 1 g of turbostratic BN by the following method: Since the principle chemical reaction is between a gas (ammonia) and a solid (iron boride), it is preferable to acquire a source of iron boride, FeB with a fine particle size, preferably $\leq 45\,\mu$m; although a particle size $< 100\,\mu$m should suffice. The reader should be aware that commercially available ferroboron may contain a fraction of FeB$_{49}$ in addition to FeB.

Determine the mass of the alumina boat (SRX110 Almath Ltd) that will be used for the synthesis to a precision of ± 0.001 g. Then weigh the appropriate amount of iron boride, FeB powder directly into the boat to a precision of ± 0.001 g. This powder should be spread evenly, so as to maximize the surface area of the powder bed in contact with the gas phase.

---

[10] The presence of FeB$_{49}$ is a manifestation of the boron-rich content (18.89 wt%) of the ferroboron alloy (kindly supplied by London & Scandinavian Metallurgical Company Ltd.) as used in the preparation (vis à vis 16.22 wt% in stoichiometric FeB).

Introduce the alumina boat into the central zone of the tube furnace, and attach the silicone rubber stoppers to both ends of the work-tube. Ensure that the exhaust gases from the tube furnace are vented appropriately.[11] Flush the air out of the work-tube using argon (or nitrogen) with a flow rate of 100 ml/minute (at STP) for 1 hour. Then switch the gas flow to ammonia, with a flow rate of 80 ml/min (at STP), and close the argon supply. Heat the contents at a rate of 200 °C/h to 450 °C, whilst maintaining a flow of ammonia with an inlet flow rate of 80 ml/min (at STP).

After 24 h at 450 °C, the temperature should be raised to 1000 °C, with a heating rate of 200 °C/h. After 1 h at 1000 °C, cool to room temperature (20 °C) with a cooling rate of 200 °C/h. This heating cycle will last ~36 h. Once the furnace approaches room temperature, close the ammonia supply and flush the ammonia out of the work-tube using argon (or nitrogen) with a flow rate of 100 ml/minute (at STP) for 1 h. Then close the argon (or nitrogen) supply, and retrieve the product material in the alumina boat. Weigh the product together with the alumina boat. Submit ~100 mg sample for analysis by powder X-ray diffractometry (scan: 10–90° 2θ). Compare the powder pattern with the PDF 73-2095 for α-BN; the PDF 6-627 for γ'-Fe$_4$N; the PDF1-1236 for ε-Fe$_3$N; and the PDF 6-656 for ζ-Fe$_2$N.

Transfer the rest the of product material into a 250-ml glass beaker and add 100 ml of warm 1 mol dm^{-3} hydrochloric acid aqueous solution, in order to dissolve the byproduct (a mixture of iron nitrides). Leave this to stand overnight. Then pour the contents onto a wetted sheet of medium speed filter paper (e.g., *Whatman* No. 2) in a Buchner filter funnel and wash it with distilled water. Finally, wash the product with a small amount of acetone (in order to accelerate the drying), and leave it to dry for ~10 min whilst in the Buchner filter funnel under vacuum. Place the product material (still attached to the filter paper) on a watch-glass and leave it to dry before being transferred into a suitably labelled specimen jar.

Submit ~100 mg sample for analysis by powder X-ray diffractometry (scan: 10–90° 2θ). Compare the powder pattern with the PDF 73-2095 for α-BN, the PDF 45-1171 for r-BN, and the PDF 39-418 for FeB$_{49}$. This can also be compared with Figure 5.5, as well your other powder pattern for the crude (unwashed) product. Keep the remaining material in a suitably labelled specimen jar. The reader may wish to observe the product powder under a reflected light binocular microscope.

The reader may wish to attempt nitriding Fe$_2$B by a method similar to that as described in this chapter for FeB. This should result in a purer BN material without the FeB$_{49}$ phase.

---

[11] The reader is advised to conform to the legal requirements and regulations that may exist in the country in which the work is being conducted. NB ammonia is highly toxic.

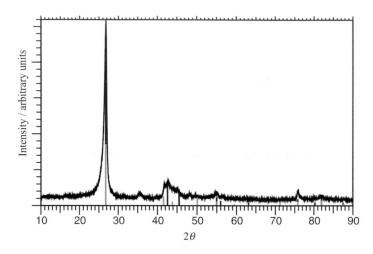

Figure 5.5   Powder X-ray diffraction pattern (using Cu-$K_{\alpha 1}$ radiation) of the product material as prepared by the method described in this chapter. The pattern corresponds to the PDF 45-1171 for rhombohedral BN (red); whilst the additional peaks at 41.7° and 43.9° correspond to the d_{100} and d_{101}, respectively, in the PDF 73-2095 for hexagonal BN (green). This suggests that this material is best described as turbostratic boron nitride

REFERENCES

1. R. S. Pease, *Acta Crystallographica* **5** (1952) 356–361.
2. O. Hassel, *Norsk Geologisk Tidsskrift* **9** (1927) 266–270.
3. R. T. Paine and C. K. Narula, *Chemical Reviews* **90** (1990) 73–91.
4. S. J. Yoon and A. Jha, *Journal of Materials Science* **30** (1995) 607–614.
5. T. *Sato, Proceedings of the Japan Academy B* **16** (1985) 459–463.
6. T. Kobayashi, S. Tashiro, T. Sekine and T. Sato, *Chemistry of Materials* **9** (1997) 233–236.
7. S. Alkoy, C. Toy, T. Gonul and A. Tekin, *Journal of the European Ceramic Society* **17** (1997) 1415–1422.
8. L. Vel, G. Demazeau and J. Etourneau, *Materials Science and Engineering B* **10** (1991) 149–164.
9. F. P. Bundy and R. H. Wentork, *Journal of Chemical Physics* **38** (1963) 1144–1149.
10. N. J. Archer, *Special Publication – Chemical Society* **30** (1977) 167–180.
11. A. V. Kostanovskii and A. V. Kirillin, *International Journal of Thermophysics* **17** (1996) 507–513.
12. T. Funahashi, T. Koitbashi, R. Uchimura, T. Koshida, A. Yoshida and T. Ogasawara, *Kawasaki Steel Technology Report* No. 28, June 1993, 17–25.
13. R. L. Finicle, *Industrial Research & Development* **25** (1983) 113–116.
14. H. P. R. Frederikse, A. H. Kahn and A. L. Dragoo, *Journal of the American Ceramic Society* **68** (1985) 131–135.
15. Y. G. Andreev and T. Lundström, *Journal of Alloys and Compounds* **210** (1994) 311–317.

16. Y. G. Andreev and T. Lundström, *Journal of Alloys and Compounds* **216** (1994) 5–7.

17. W. Büchner, R. Schliebs, G. Winter and K. H. Büchel, *Industrial Inorganic Chemistry*, 1989, VCH, Weinheim.

18. M. Sano and M. Aoki, *Thin Solid Films* **83** (1981) 247–251.

19. R. Kiessling and Y.H. Liu, *Journal of Metals*, August (1951) 639–642.

20. T. E. Warner and D. J. Fray, *Journal of Materials Science* **35** (2000) 5341–5345.

21. G. Hägg, *Zeitschrift für Physikalische Chemie, B* **7** (1930) 339–362.

22. D. F. Shriver, P. Atkins and C. H. Langford, *Inorganic Chemistry*, 2nd edn, 1987, Oxford University Press, Oxford.

23. S. B. Hendricks and P. R. Kosting, *Zeitschrift für Kristallographie, Kristallgeometrie, Kristallphysik, Kristallchemie* **74** (1930) 511–533.

24. P. Rogl and J. C. Schuster, *Phase Diagrams of Ternary Boron Nitride and Silicon Nitride Systems*, ASM International, Materials Park, Ohio, 1992, 218 pp.

PROBLEMS

1. Determine the mass of your product material as retrieved from the tube furnace. Calculate the mass gain as a percentage, and compare your value to the theoretical value for complete nitridation.

6

Rubidium Copper Iodide Chloride $Rb_4Cu_{16}I_7Cl_{13}$ by a Solid-State Reaction

$Rb_4Cu_{16}I_7Cl_{13}$ is a fast ion conductor that exhibits an exceptionally high copper(I) ionic conductivity of $0.34\,S\,cm^{-1}$ at $25\,°C$ and a very low electronic conductivity. $Rb_4Cu_{16}I_7Cl_{13}$ is an important example of a highly disordered complex salt in which the progressive change in the occupancy among the copper sites correlates with the anomalous behaviour in its heat capacity and ionic conductivity. $Rb_4Cu_{16}I_7Cl_{13}$ was first described in the literature[1] by Takahashi *et al.* [1] in October 1979, and is reputed to have the highest ionic conductivity of any known solid at room temperature. Unfortunately, $Rb_4Cu_{16}I_7Cl_{13}$ has a rather low decomposition potential of $0.69\,V$ at room temperature, and this places a limitation on its use as a solid electrolyte within solid-state batteries. The chemical instability of $Rb_4Cu_{16}I_7Cl_{13}$ under atmospheric conditions, together with its relatively low melting point, place additional restraints on the application of this material. The method described here for its synthesis involves the preparation of the precursor

---

[1] According to Geller *et al.* [2] $Rb_4Cu_{16}I_7Cl_{13}$ was reported prior to this date by T. Takahashi, O. Yamamoto, S. Yamada and S. Hayashi, *Solid-state ionics: High copper ion conductivity of the system CuCl–CuI–RbCl*, in the Extended Abstracts No. 6–2 (unpublished) at the 2nd International Meeting of Solid Electrolytes, held at St. Andrews, Scotland, 20–22 September, 1978; and at the 6th Solid Electrolyte Meeting (sponsored by the Japanese Chemical Society) in Tokyo on 19–20 October, 1978 [3]. The compound α-$RbCu_4Cl_3I_2$ was first described by Geller *et al.* [2] in May 1979, and mentioned therein to be part of the same solid solution phase as the compound $Rb_4Cu_{16}I_7Cl_{13}$ reported by Takahashi *et al.* in Scotland.

Synthesis, Properties and Mineralogy of Important Inorganic Materials   By Terence E. Warner
© 2011 John Wiley & Sons, Ltd

compound CuCl via a heterogeneous redox reaction. $Rb_4Cu_{16}I_7Cl_{13}$ is then synthesised through the solid-state reaction of a compacted stoichiometric mixture of anhydrous binary metal halides at 200 °C under an inert atmosphere.

Fast ion conductors are a class of materials in which the electric current is conveyed almost exclusively by the transport of ions within the solid state, thereby enabling the passage of a significant ionic current density at a practically relevant temperature. The equally valid expression, *solid-electrolytes* is also used to describe this class of materials. Examples of both fast-cation and fast-anion conductors exist. To avoid confusion, *mixed ionic–electronic conductors* are a separate category of related materials in which the electric current is conveyed by a mixture of ions and electrons (or holes); and digenite, $Cu_{2-\delta}S$ is a good example, as discussed previously in Chapter 4.

In the case of crystalline materials, fast ion conductors usually comprise an ordered, rigid framework structure with an ionic substructure which is only partially occupied and thereby has a high concentration of ionic vacancies pertaining to the mobile ionic species. This allows for a substantial degree of disorder with regards to the distribution of the mobile ions among these crystallographic sites. Furthermore, these sites are energetically similar and in close proximity with one another, so as to enable the ions to hop from site to site with relative ease through the open channels created by the framework structure. In certain materials, the ionic mobility and the charge carrier concentration can be very high, such that this part of the ionic substructure can be perceived as having become prematurely 'molten' at a temperature many degrees below the actual melting point of the compound. But having stated that, this analogy must not be taken too strongly, since the ions are believed to hop between discrete crystallographic sites, rather than being free to move around with true liquid-like motion as in molten-salt electrolytes [4, 5].

One way to illustrate this is through the example of silver iodide.[2] β-AgI is a poor Ag^+ ionic conductor that exists at room temperature with the wurtzite structure. But at 147 °C it undergoes a solid-state phase transition to the slightly denser cubic phase, α-AgI, which is a fast Ag^+ ionic conductor up to its melting point at 557 °C.

$$\beta\text{-AgI} \xrightarrow{\Delta S^{\circ}_{147\,^{\circ}C}=14.5\ J\ mol^{-1}\ K^{-1}} \alpha\text{-AgI} \xrightarrow{\Delta S^{\circ}_{fusion(557\,^{\circ}C)}=11.3\ J\ mol^{-1}\ K^{-1}} \text{melt}$$

It is rather fascinating that the entropy change for the solid-state phase transition (β-AgI $\rightarrow$ α-AgI) is similar to that for the fusion of α-AgI. This indicates that the increase in disorder associated with the solid-state phase

---

[2] This example is paraphrased here from the descriptions given by West [5], Kudo [6] and Maier [7]; using the original data from O'Keeffe and Hyde [8].

transition is comparable with that for the fusion of α-AgI. It is as if the Ag^+ ionic substructure has become quasimolten, whilst the I^- ionic substructure remains rigid, thus allowing the material to remain essentially solid. It is also reassuring that the summation of these two entropies is similar to the standard entropy of fusion for NaCl ($24\,J\,mol^{-1}\,K^{-1}$). These highly conductive materials with low activation energies (≤ 0.1 eV) have been referred to by certain workers as *superionic* conductors;[3] α-AgI and $Rb_4Cu_{16}I_7Cl_{13}$ are two prime examples. They may be considered to be *superionic conducting* in so far as they have an exceptionally large concentration of mobile ions, rather than the ions themselves being exceptionally mobile. It is also interesting to note that they tend to comprise extremely polarisable ions. The reader is referred to the following texts by: Bruce [9], Gellings and Bouwmeester [10], Maier [7], Reichart [11], Geller [12] and Kudo and Fueki [13], for further information regarding the subject of solid-state electrochemistry.

An interest in the ionic conducting properties of metal halides can be traced back to Michael Faraday's discovery of the first solid electrolyte, PbF_2 in 1838.[4] In 1920, Tubandt and Eggert [14] demonstrated electrolytically that α-AgI (as discussed above) was an Ag^+ ion conductor at $150\,°C$. In 1967, superionic conduction was extended to near-ambient[5] temperature through the discovery of the related phase, $RbAg_4I_5$ [15]. But attempts at preparing the copper analogue, $RbCu_4I_5$, only went to prove that this compound does not exist [15]. However, the partial substitution of chloride ions for iodide ions within the hypothetical compound, '$RbCu_4I_5$' lead to the simultaneous discoveries of $Rb_4Cu_{16}I_7Cl_{13}$ by Takahashi *et al.* [1] and α-$RbCu_4Cl_3I_2$ by Geller *et al.* [2].

Soon after the discovery of $Rb_4Cu_{16}I_7Cl_{13}$ and α-$RbCu_4Cl_3I_2$, these compositions were shown to be part of the same solid-solution series, $Rb_4Cu_{16}I_{7+x}Cl_{13-x}$, but with some disagreement concerning the solubility limits. With regards to the expression, $Rb_4Cu_{16}I_{7+x}Cl_{13-x}$ Takahashi *et al.* [16] reported the limits as $-0.2 \leq x \leq 0.5$; Tokumoto *et al.* [17] as $0 \leq x \leq 0.5$; and Nag and Geller [18] as $-0.6 \leq x \leq 1$, although Nag and Geller hastened to mention that the iodide-rich composition with $x = 1$ (i.e., α-$RbCu_4I_2Cl_3$), was stable only within the very narrow temperature range $200 \pm 15\,°C$; which makes it something of an oddity.[6] However, a variation in the solubility limits with regards to temperature is quite normal for a solid solution. A typical discontinuous solid solution, such as $Rb_4Cu_{16}I_{7+x}Cl_{13-x}$, normally becomes broader in composition at higher temperature on account

---

[3] For further information on superionic conductors, the reader is referred to *Section 6.2.1: Ion Mobility* in the monograph by Maier [7].

[4] PbF_2 is a F^- ion conductor.

[5] $RbAg_4I_5$ is metastable below $27\,°C$.

[6] At that time, the existence of α-$RbCu_4I_2Cl_3$ as a single-phase material was somewhat contentious; Takahashi *et al.* [1] reported this composition to comprise mixed-phase material; i.e., CuCl and a new uncharacterised cubic phase.

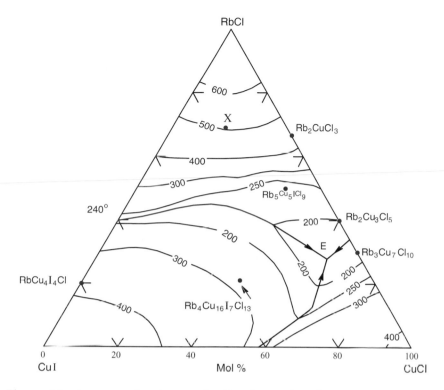

Figure 6.1 An approximation to the liquidus surface for the ternary system CuI-CuCl-RbCl (Reprinted with permission from Solid State Ionics, Solid-state ionics-the CuCl—CuI—RbCl system by T. Takahashi, R. Kanno, Y. Takedo et al., 3–4, 283–287 Copyright (1981) Elsevier Ltd)

of the contribution from the entropy of mixing towards the Gibbs free energy of formation for the solid-solution phase.[7]

A phase diagram for the CuCl–CuI–RbCl ternary system was constructed by Takahashi et al. [19] in 1981, from a series of differential thermal analyses on various samples contained within evacuated Vycor-glass ampoules, as shown in Figure 6.1. From the diagram it is clear to see how the temperature of the *liquidus* decreases progressively upon entering the ternary field (which is a typical feature for a multicomponent system), with the occurrence of a ternary eutectic at 144 °C near the CuCl–RbCl binary edge. The composition $Rb_4Cu_{16}I_7Cl_{13}$ has an incongruent melting point (coinciding of course, with the *solidus*) at 234 ±5 °C, and a corresponding *liquidus* at 293 ±5 °C [1]. However, very little *subsolidus* information is reported for this system besides

---

[7] Since $\Delta S^\circ_{\text{mixing}}$ is normally positive, then above a certain temperature the negative term $(-T\Delta S^\circ)$ will predominate in the Gibbs free energy function $(\Delta G^\circ = \Delta H^\circ - T\Delta S^\circ)$ irrespective of the value for ΔH° and, thus, ΔG° will be negative.

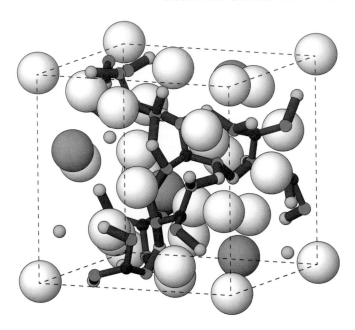

Figure 6.2 Crystal structure of $Rb_4Cu_{16}I_{7.2}Cl_{12.8}$ at 300 K showing the unit cell with $a = 10.0134$ Å and $Z = 1$. The rubidium ions are shown grey and the halide ions are shown white. The Cu(1) sites are shown green, the Cu(2) sites are shown yellow, and the Cu(3) sites are shown red. The faster conducting Cu(1)–Cu(2)–Cu(1) pathway is shown with thicker sticks. This figure was drawn using data from Kanno *et al.* [20], cf. ICSD 73811

the existence of the two quaternary phases; $RbCu_4I_4Cl$, which is stable between 231–244 °C, and $Rb_5Cu_5ICl_9$, which is stable ≤ 225 °C [19].

Ironically, the first composition within the $Rb_4Cu_{16}I_{7+x}Cl_{13-x}$ solid solution to be structurally determined was the one that is least stable, i.e., α-$RbCu_4I_2Cl_3$ by Geller *et al.* [2] in 1979 using powder X-ray diffraction (cf. ICSD 200657). Nevertheless, it was shown to crystallise with the KCu_4I_5 structure,[8] and belongs to the cubic crystal system and space group $P4_132$ (or its enantiomorph, $P4_332$). Several years later in 1993, Kanno *et al.* [20] performed a neutron-diffraction study on the composition $Rb_4Cu_{16}I_{7.2}Cl_{12.8}$ over the low-temperature range 50–300 K, with an interest in determining the site-occupation factors for the copper ions as a function of temperature. Their data confirm the structure proposed earlier by Geller *et al.*, [2] and were used here to draw the crystal structure for this phase at 300 K as shown in Figure 6.2.

---

[8] KCu_4I_5 is stable only between 242 °C and its incongruent melting points at 332 °C at ambient pressure, and has a Cu^+ ionic conductivity of 0.61 S cm^{-1} at 267 °C [21].

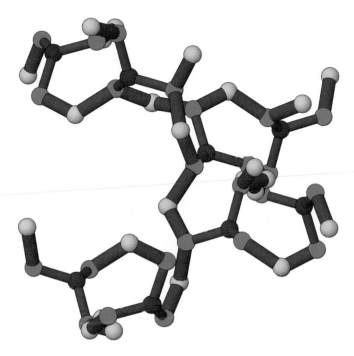

Figure 6.3 The possible interconnections between the copper sites in $Rb_4Cu_{16}I_7Cl_{13}$ as viewed directly down the a-axis. The Cu(1) sites are shown green, the Cu(2) sites are shown yellow, and the Cu(3) sites are shown red. The faster conducting Cu(1)–Cu(2)–Cu(1) pathway is shown with thicker sticks. This figure was drawn using data from Kanno *et al.* [20], cf. ICSD 73811

Concerning the crystal structure of the phase $Rb_4Cu_{16}I_{7+x}Cl_{13-x}$ there are three sets of crystallographically nonequivalent sites for the copper ions, one 8-fold set nominated Cu(3), and two 24-fold sets nominated Cu(1) and Cu(2). This gives rise to a total of 56 sites for only 16 copper ions per unit cell; therefore a surprisingly large number ($\sim71\%$) of the copper sites within the structure are actually vacant. The site occupation factors vary gradually with temperature, such that a net displacement from Cu(3) into Cu(2) occurs with an increase in temperature.[9] This coincides with the nonlinear (curved) region in the Arrhenius conductivity plot (see Figure 6.5), whereupon the apparent activation energy decreases pro rata with the enhanced population of the Cu(2) site, culminating with the remarkably low value of $\sim0.1\,eV$ at (and above) ambient temperature. Possible conduction pathways for the copper ions are shown in Figures 6.3 and 6.4. Alternating Cu(1) and Cu(2)

---

[9] There is a 42% drop in the site-occupation factor for Cu(3) upon warming from 110 to 300 K [20]. This tallies with a broad anomaly of the excess heat capacity over this temperature range (peaking at 190 K); revealing a noncooperative mechanism for the redistribution of the Cu^+ ions among the three copper sites in $Rb_4Cu_{16}I_7Cl_{13}$ [22], [20].

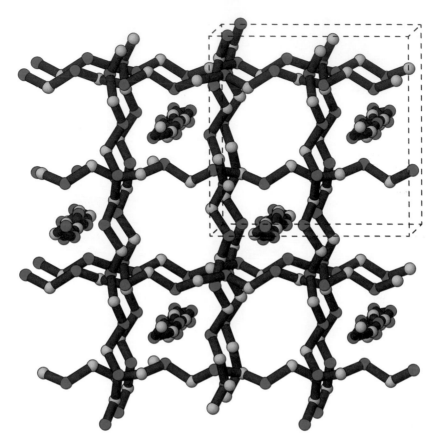

Figure 6.4 The Cu(1)–Cu(2)–Cu(1) conduction pathways in $Rb_4Cu_{16}I_7Cl_{13}$ in relation to the unit cell. The Cu(1) sites are shown green, and the Cu(2) sites are shown yellow. This figure was drawn using data from Kanno *et al.* [20], cf. ICSD 73811

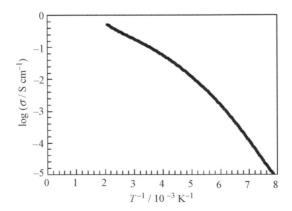

Figure 6.5 Arrhenius conductivity plot for $Rb_4Cu_{16}I_7Cl_{13}$ redrawn using data from Kanno *et al.* [20]

sites form tunnels that run parallel to the three unit cell axes, whilst the Cu(3) site forms interconnections between these tunnels. Kanno et al. [20] offer crystallographic evidence to suggest that the conduction pathway is along Cu(1)–Cu(2)–Cu(1); whilst other pathways, such as those involving the Cu(3) site, do not contribute significantly to the total conduction in the structure.

Chen and Zhao [23] measured the ionic conductivity of a densely sintered specimen of $Rb_4Cu_{16}I_7Cl_{13}$ at 23 °C as a function of applied pressure over the range 0.5 to 11.6 kbar. Their results showed a linear decrease in ionic conductivity with increasing pressure over this range. This restriction on the ionic conduction was inferred as a contraction or, distortion, of the conduction pathways with increasing pressure; which is reasonably intuitive. The temperature dependence of the ionic conduction was also determined at various pressures over the temperature range −80 to 80 °C. The activation energies showed an increase with increasing pressure that is consistent with the above inference.

One of the best experimental methods for obtaining direct evidence for the migration of a specific mobile ionic species within a solid electrolyte is by electrolysis. In other words to observe and quantitatively measure the macroscopic transport of a specific ion through a solid electrolyte by the Tubandt (or Hittorf) method [24].[10] Warner et al. [25] transported copper ions by electrolysis through a sintered monolith of $Rb_4Cu_{16}I_7Cl_{13}$ sandwiched between two pieces of copper foil (that served as the electrodes) at 200 °C under an atmosphere of argon. A constant current density of 37.7 mA cm^{-2} was passed through the cell for 3040 s corresponding to a theoretical mass transport of 100 mg of copper. The extensive deposit of dendritic copper on the cathode copper foil was visible evidence that $Rb_4Cu_{16}I_7Cl_{13}$ is truly a copper ionic conductor. In hindsight; if the mass of the copper deposit had been determined, then this might have served to verify the transference number, as well as the oxidation state, of the mobile copper ion.

With regards to thermodynamic stability, $Rb_4Cu_{16}I_7Cl_{13}$ has a rather low decomposition potential of 0.69 ±0.01 V at 22 °C [1]. The root of this instability lies in the relatively high nobility[11] of elemental copper, and also the ease with which the iodide ion can be oxidised. Likewise, copper's high nobility imposes a rather low limit on the potentials (typically ∼0.5 V) that can in practice be generated by an electrochemical cell based upon copper chemistry. Nevertheless, these aspects have not hindered the suggestion of numerous solid-state rechargeable batteries that incorporate $Rb_4Cu_{16}I_7Cl_{13}$ as the solid electrolyte. One of the most promising examples is the cell; $Cu_{3.8}Mo_6S_{7.8}|Rb_4Cu_{16}I_7Cl_{13}|Cu_{0.2}Mo_6S_{7.8}$ as proposed by Kanno et al. [26], which utilises the Chevrel phase, $Cu_{3.8-x}Mo_6S_{7.8}$ as the copper intercalation

---

[10] The review on multivalent cationic conduction in crystalline solids by Köhler et al. [24] gives a valuable insight into the methodology in this subject.

[11] The expression, *high nobility*, is synonymous here with the expression, *low electropositivity*.

compound for both the anode and the cathode. Similar cells have been reported in the literature utilising, for example, TiS_2 [27], NbS_2 [28], and the Wadsley-Roth phase, $V_{1+x}W_{1-x}O_5$ ($0 < x < 0.8$) [29] as copper intercalation materials for the cathode; but with inferior performances.

A more realistic application is in the field of microbatteries. Within this context, $Rb_4Cu_{16}I_7Cl_{13}$ has also been patented as a solid electrolyte for use in double-layer (or electrochemical) capacitors [3, 30]. These devices have a high capacity and low leakage current, and have been proposed as standby power sources for RAM devices [31]. Another potential application for $Rb_4Cu_{16}I_7Cl_{13}$ is in solid-state voltage memory devices; akin to those based on Ag^+ solid electrolytes [32].

$Rb_4Cu_{16}I_7Cl_{13}$ was prepared originally by Takahashi et al. [1]. Importance was placed on the pretreatment stage for the drying of the chemical reagents (anhydrous RbCl, CuI and CuCl,) under vacuum, in order to eliminate water from the system. Appropriate quantities of RbCl, CuI and CuCl were ground together and preheated at $130\,^{\circ}C$ under a flowing nitrogen atmosphere for 17 hours. This mixture was then encapsulated inside an evacuated Pyrex™ ampoule, then melted and allowed to cool slowly to room temperature. The product was ground in a planetary ball mill, and pressed into a pellet under a pressure of 3 kbar, then annealed at $130\,^{\circ}C$ under a flowing nitrogen atmosphere for 17 h. The procedure was repeated until no further changes were detected in the powder X-ray diffraction pattern. The description in the patent by Yamamoto and Takahashi [33] that appeared in the following year, omits the encapsulation and fusion; presumably, these aspects are not essential.

More recently, Miyata et al. [34] prepared $Rb_4Cu_{16}I_7Cl_{13}$ by mechanically impacting a mixture of RbCl, CuI and CuCl inside a high-energy ball mill at room temperature for at least 20 h. Although this preparative technique has certain advantages over conventional thermal treatment in the synthesis of metastable materials, this method seems difficult to justify in the case of $Rb_4Cu_{16}I_7Cl_{13}$.

Preparative Procedure

The method described here for the synthesis of $Rb_4Cu_{16}I_7Cl_{13}$ involves a solid-state reaction between RbCl, CuCl and CuI in a similar fashion to that described originally by Takahashi et al. [1]. With regard to ionic materials, solid-state reactions involve the interdiffusion of cations and anions between the various reactant phases. The rate of reaction is normally dependent on the mobility of the ions within the phases concerned. Therefore, this method is very well suited for the preparation of a polycrystalline monolith of $Rb_4Cu_{16}I_7Cl_{13}$, since the copper ions are highly mobile within the product phase, and they also show appreciable mobility in the reactants CuCl and CuI at elevated temperature (see Figure 6.6).

Figure 6.6 Fractured surface of the annealed disc of $Rb_4Cu_{16}I_7Cl_{13}$ as prepared by the method described in this chapter. Width of field $= 10$ mm

Warner *et al.* [25] have reported a modification to the synthesis that excludes the vacuum pretreatment, and dispenses with the need for encapsulation in evacuated glass ampoules. It relies instead on suitably anhydrous commercially available chemical reagents,[12] and with the reaction taking place under a flowing atmosphere of argon inside a tube furnace. Nevertheless, their method results in the formation of single-phase material and is adopted here. However, the preparation of the precursor compound, copper(I) chloride, involves a classical piece of heterogeneous redox chemistry, and for this reason it is included in the description.

First, prepare an adequate amount ($\sim$10 g) of CuCl powder using copper(II) chloride dihydrate ($CuCl_2 \cdot 2H_2O$) and copper wool (99.9%, 0.03–0.06 mm diameter wire, nonlacquered; The Metallic Wool Co Ltd, UK) as chemical reagents. Place $\sim$20 g of $CuCl_2 \cdot 2H_2O$ in a 500-ml glass beaker and add 100 ml of 6 mol dm^{-3} hydrochloric acid solution. Place the beaker on a hot plate inside a fume cupboard. To this blue-green solution add an excess amount of copper wool ($\sim$10 g) and then boil the mixture for $\sim$5 min. Make a note of any changes in colour that you observe within the solution during this process. The elemental copper should reduce the complex copper(II) chloride aqueous ions to copper(I) through the following redox reaction:

$$[CuCl_4]^{2-} (aq) + Cu(s) \rightarrow 2[CuCl_2]^- (aq).$$

Allow the reaction mixture to cool somewhat, then decant the brown-coloured liquor directly into a 1-l glass beaker containing 500 ml of distilled water at room temperature. Take care so as to avoid transferring any fragments of copper wool into the second beaker. The dilution process

---

[12] This places the onus on the manufacturer as the purveyor of chemical reagents with a certified degree of purity; particularly the water content.

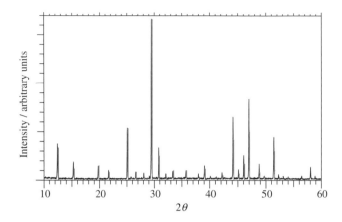

Figure 6.7 Powder XRD pattern (Cu-$K_{\alpha 1}$ radiation) of $Rb_4Cu_{16}I_7Cl_{13}$ as prepared by the method described in this chapter. PDF 37-1206 for $Rb_4Cu_{16}I_7Cl_{13}$ is shown in red for comparison

will result in the precipitation of colourless CuCl through the equilibrium reaction:

$$[CuCl_2]^-\,(aq) \rightleftharpoons CuCl(s) + Cl^-\,(aq)$$

This precipitate should be separated promptly using a Buchner filter funnel with a medium speed filter paper (e.g., *Whatman No. 2 Qualitative*). Wash the CuCl powder-cake with distilled water in order to remove the acid solution, and finally wash with liberal amounts of acetone so as to remove the excess water and thereby accelerate the drying process. Continue the suction for several minutes so as to produce a reasonably dry powder-cake. Remove the filter paper together with the CuCl powder-cake still intact, and place it on a large watch glass. Dry the product by heating gently with a hot air blower. If prepared correctly, the product copper(I) chloride should be a dry *'snow-white'* powder. Place your dried sample in a specimen jar and fasten with a tightly secured lid; then store in an evacuated desiccator.

Prepare 15 g of $Rb_4Cu_{16}I_7Cl_{13}$ by the following method: Weigh appropriate amounts (to a precision of ± 0.001 g) of rubidium chloride, copper(I) iodide and copper(I) chloride (as prepared above). Grind them together thoroughly with a porcelain pestle and mortar. Compress the total mixture into a single disc using a 32-mm diameter Specac stainless steel die and a uniaxial press with a load of ~5 tonne. Place the compacted disc in a round shallow alumina crucible (LR42 Almath Ltd) and place this in the central zone of a tube furnace. Place a ball of copper wool inside the furnace worktube at a position in between the gas inlet and the alumina crucible. This will act as an oxygen getter (or scavenger) and should enable a low partial pressure of oxygen to be maintained in the flowing argon atmosphere.[13] Attach the silicone rubber stoppers accordingly. Flush the air out of the work-tube with a medium flow rate of argon ($\sim$1 l/min at STP) for $\sim$30 min.[14] Then reduce the argon flow rate to $\sim$0.1 l/min at STP and heat at 60 °C/h to 200 °C. After 36 h at 200 °C, cool to room temperature at 60 °C/h. This heating cycle will last 2 days.

Determine the mass of your annealed disc of $Rb_4Cu_{16}I_7Cl_{13}$ and measure its diameter and thickness as precisely as you can. Use these data to determine the density of the annealed material. Then use a knife to cut a small piece ($\sim$100 mg) of the material in order to be analysed by powder XRD (scan: 10–60° 2θ). Compare your powder pattern with the PDF 37-1206 for $Rb_4Cu_{16}I_7Cl_{13}$ (see Figure 6.7). Place the remaining disc in a specimen jar and fasten with a tightly secured lid; then store in an evacuated desiccator.

REFERENCES

1. T. Takahashi, O. Yamamoto, S. Yamada and S. Hayashi, *Journal of the Electrochemical Society* **126** (1979) 1654–1658.
2. S. Geller, J. R. Akridge and S. A. Wilber, *Physical Review* **19** (1979) 5396–5402.
3. S. Sekido and Y. Ninomiya, *Solid state double layer capacitor*, US Patent 4363079 (1982).
4. H. L. Tuller, *Highly Conductive Ceramics* Chapter 7 in *Ceramic Materials for Electronics*, R. C. Buchanan (ed.), Marcel Dekker Inc., New York (1991) 379–433.
5. A. R. West, *Crystalline Solid Electrolytes I: General Considerations and the Major Materials*, 7–42, in P. G. Bruce (ed.) *Solid State Electrochemistry*, 1995, Cambridge University Press, Cambridge.
6. T. Kudo, *Survey of Types of Solid Electrolytes*, Chapter 6, 195–221, in P. J. Gellings and H. J. M. Bouwmeester (eds.), *The CRC Handbook of Solid State Electrochemistry*, 1997, CRC Press, Boca Raton, Florida.
7. J. Maier, *Physical Chemistry of Ionic Materials: Ions and Electrons in Solids*, 2004, Wiley, Chichester, 537pp.
8. M. O'Keeffe and B. G. Hyde, *Philosophical Magazine* **33** (1976) 219–224.
9. P. G. Bruce (ed.) *Solid State Electrochemistry*, Cambridge, Cambridge University Press, (1995), pp344.
10. P. J. Gellings and H. J. M. Bouwmeester (eds.), *The CRC Handbook of Solid State Electrochemistry*, CRC Press, Boca Raton, Florida, (1997) 630 pp.
11. H. Rickert, *Solid State Electrochemistry: An Introduction*, (1986) Springer-Verlag, Berlin, Heidelberg, New York.
12. S. Geller (ed.) *Solid Electrolytes: Topics in Applied Physics*, **21** (1977) Springer-Verlag Heidelberg, pp229.

---

[13] If the work-tube has just been used with an internal atmosphere of oxygen or air, then the work-tube should be pretreated by heating in flowing argon at $\sim$1000 °C for 1 h, with a heating and cooling rate of 300 °C/h.

[14] This flow rate relates to a work-tube with dimensions: 60 mm ID, 70 mm OD and 0.9 m length.

13. T. Kudo and K. Fueki, *Solid State Ionics* (1991) VCH, Weinheim, pp241.
14. C. Tubandt and S. Eggert, *Zeitschrift für Anorganische und Allgemeine Chemie* **110** (1920) 196–236.
15. J. N. Bradley and P. D. Green, *Transactions of the Faraday Society* **63** (1967) 424–430.
16. T. Takahashi, R. Kanno, Y. Takeda and O. Yamamoto, *Solid State Ionics* **3–4** (1981) 283–287.
17. M. Tokumoto, N. Ohnishi, Y. Okada and T. Ishiguro, *Solid State Ionics* **3–4** (1981) 289–293
18. K. Nag and S. Geller, *Journal of the Electrochemical Soci*ety **128** (1981) 2670–2675
19. T. Takahashi, R. Kanno, Y. Takeda and O. Yamamoto, *Solid State Ionics* **3–4** (1981) 283–287.
20. R. Kanno, K. Ohno, Y. Kawamoto, Y. Takeda, O. Yamamoto, T. Kamiyama, H. Asano, F. Izumi and S. Kondo, *Journal of Solid State Chem*istry **102** (1993) 79–92.
21. S. Hull, D. A. Keen, D. S. Sivia and P. Berastegui, *Journal of Solid State Chemistry* **165** (2002) 363–371.
22. T. Atake, H. Kawaji, R. Kanno, K. Ohno and O. Yamamoto, *Solid State Ionics* **53–56** (1992) 1260–1263.
23. L.-Q. Chen and Z.-Y. Zhao, *Solid State Ionics* **9–10** (1983) 1223–1226.
24. J. Köhler, N. Imanaka and G. Adachi, *Chemistry of Materials* **10** (1998) 3790–3812.
25. T. E. Warner, P. P. Edwards, W. C. Timms and D. J. Fray, *Journal of Solid State Chem*istry **98** (1992) 415–422.
26. R. Kanno, Y. Takeda, Y. Oda and O. Yamamoto, *Materials Research Bulletin* **22** (1987) 1283–1290.
27. R. Kanno, Y. Takeda, M, Imura and O. Yamamoto, *Journal of the Electrochemical Society* **12** (1982) 681–685.
28. R. Kanno, Y. Takeda, Y. Oda, H. Ikeda and O. Yamamoto, *Solid State Ionics* **18–19** (1986) 1068–1072.
29. K. Kuwabara, S. Arai and K. Sugiyama, *Solid State Ionics* **39** (1990) 283–288.
30. M. Matthews, *Method for producing multi-cell solid state electrochemical capacitors and articles formed thereby*, Canadian Patent 2105124 (1994).
31. C. Julien, '*Solid State Batteries*' in P. J. Gellings and H. J. M. Bouwmeester (eds.), *The CRC Handbook of Solid State Electrochemistry*, CRC Press, Boca Raton, Florida, (1997) Chapter 11, 371–406.
32. O. Yamamoto, *Applications*, in P. G. Bruce (ed.) *Solid State Electrochemistry*, Cambridge University Press, Cambridge, (1995) 292–332.
33. O. Yamamoto and T. Takahashi, *Copper ion-conductive solid electrolyte*, Japanese Patent 55116629 (1980).
34. S. Miyata, N. Machida, H. Peng and T. Shigematsu, *Funtai oyobi Funmatsu Yakin* **48** (2001) 836–842.

PROBLEMS

1. Explain what is meant by the expression: 'crystallographically nonequivalent sites'.
2. State the value for the transport number for the copper(I) ion in $Rb_4Cu_{16}I_7Cl_{13}$ at 25 °C?
3. Concerning the following solid-state battery; $Cu_{3.8}Mo_6S_{7.8}|Rb_4Cu_{16}I_7Cl_{13}|Cu_{0.2}Mo_6S_{7.8}$

(a) Which material forms the 'positive' terminal of the battery in open-circuit, i.e., which end of the battery should be marked: $+$?

(b) If the above battery is fabricated so as to comprise initially an equal molar amount of $Cu_{3.8}Mo_6S_{7.8}$ and $Cu_{0.2}Mo_6S_{7.8}$ as the two electrodes, respectively, what will the stoichiometry of the two electrode materials be once the battery is fully discharged?

4. State the value of the pressure in kbar that corresponds to a load of 5 tonne as applied to an area corresponding to a circle with a diameter of 32 mm.

5. (a) From the mass and spatial dimensions of your annealed polycrystalline monolith, calculate its density ($\rho_{monolith}$) in units of $g\,cm^{-3}$.

(b) From the crystallographic data given in the text for $Rb_4Cu_{16}I_{7.2}Cl_{12.8}$ calculate the crystallographic density of $Rb_4Cu_{16}I_{7.2}Cl_{12.8}$ ($\rho_{crystal}$) in units of $g\,cm^{-3}$.

(c) From your answers to *Problems* 5(a) and 5(b) determine the fractional density ($\rho_{monolith}/\rho_{crystal}$) as a percentage, assuming that $Rb_4Cu_{16}I_{7.2}Cl_{12.8}$ and $Rb_4Cu_{16}I_7Cl_{13}$ have similar values for the cell constant a.

6. Find appropriate crystallographic data in order to calculate the percentage volume change at 25 °C and 1 bar associated with the following reaction:

$$4RbCl + 9CuCl + 7CuI \rightarrow Rb_4Cu_{16}I_7Cl_{13}.$$

7. Consult the phase diagram for the RbI-CuI binary system as reported by Bradley and Green [32]. Use this information to suggest what would happen if you attempted to prepare the compound $Rb_4Cu_{16}I_{7+x}Cl_{13-x}$ corresponding to $x-13$, at ambient pressure.

8. Copper wool is used as an oxygen getter (or scavenger) so as to enable a low partial pressure of oxygen to be maintained in the flowing argon atmosphere. Use your diagram relating to *Problem* 4 in Chapter 3, regarding the binary Cu–O system, in order to obtain the value for the equilibrium partial pressure of oxygen P_{O_2} in equilibrium with elemental copper at 200 °C.

9. What do you think would happen to your product $Rb_4Cu_{16}I_7Cl_{13}$ if it was left exposed to the air at ambient temperature? Write chemical equations where relevant.

7

Copper Titanium Zirconium Phosphate CuTiZr(PO$_4$)$_3$ by a Solid-State Reaction Using Ammonium Dihydrogenphosphate as the Phosphating Reagent

CuTiZr(PO$_4$)$_3$ is the midintermediate member of the solid-solution series CuTi$_{2-x}$Zr$_x$(PO$_4$)$_3$ $(0 \leq x \leq 2)$ that crystallises with the NASICON-type structure and has an orange–brown coloration.[1] It is an example of a three-dimensional open-framework structure that exhibits fast Cu$^+$ ion conduction. The end members, CuTi$_2$(PO$_4$)$_3$ and CuZr$_2$(PO$_4$)$_3$, have been described in the literature with regard to applications as solid electrolytes and heterogeneous catalysts, but very little is reported for the intermediate compositions. Ceramic monoliths of CuTi$_{2-x}$Zr$_x$(PO$_4$)$_3$ have sufficient refractoriness for use up to at least 1300 °C; but below ∼500 °C the material is unstable in air with respect to the binary assemblage: Cu$_{0.5}$Ti$_{2-x}$Zr$_x$(PO$_4$)$_3$

---

[1] During an alternative method for preparing CuTiZr(PO$_4$)$_3$ by sintering a compacted mixture of CuTi$_2$(PO$_4$)$_3$ and CuZr$_2$(PO$_4$)$_3$, the powder X-ray diffraction pattern of partially reacted material indicates the existence of a continuous solid solution between CuTi$_2$(PO$_4$)$_3$ and CuZr$_2$(PO$_4$)$_3$ [3].

Synthesis, Properties and Mineralogy of Important Inorganic Materials   By Terence E. Warner
© 2011 John Wiley & Sons, Ltd

and CuO. This chapter describes the preparation of ceramic[2] CuTiZr(PO$_4$)$_3$ by a high-temperature method. The constituent metal oxides are first reacted together with NH$_4$H$_2$PO$_4$ as the phosphating reagent at 300 °C, and then calcined in air at 1250 °C.[3] The product is ground and sintered in air at 1250 °C, and then quenched during the cooling cycle at 700 °C so as to avoid oxidation.

The events that led to the discovery of CuTi$_{2-x}$Zr$_x$(PO$_4$)$_3$ began in 1976, when Hong, Goodenough and Kafalas [1, 2] discovered the solid electrolyte Na$_{1+x}$Zr$_2$(SiO$_4$)$_x$(PO$_4$)$_{3-x}$ ($0 \leq x \leq 3$) during a search for three-dimensional open-framework structures that display extraordinary fast alkali ion transport. Na$_{1+x}$Zr$_2$(SiO$_4$)$_x$(PO$_4$)$_{3-x}$ displays fast Na$^+$ ion transport and is referred to by the acronym NASICON (Na$^+$ superionic conductor).[4] This phase allows for phosphorus and silicon to substitute for one another, resulting in a fine control of the sodium stoichiometry. An optimal ionic conductivity occurs with $x \sim 2$ due to a compromise between maximising the concentration of the Na$^+$ ions, and yet retaining a sufficiently high concentration of vacancies for these ions to move through; together with a large enough 'bottleneck' opening between the respective sites.[5]

The NASICON prototype structure belongs to the trigonal/rhombohedral crystal system and space group $R\bar{3}c$ (see Figure 7.1). The structure consists of a three-dimensional substructure comprising [ZrO$_6$] octahedra that share all their corners with six SiO$_4$^{4} (or PO$_4$$^{3-}$) tetrahedra. The Na$^+$ ions are accommodated within two large, but distinct, crystallographic sites within this substructure, designated M(1) and M(2); as represented by the formula M(1)$_1$M(2)$_3$[Zr$_2$(SiO$_4$)$_3$]; see Figures 7.3 and 7.4. The octahedrally coordinated M(1) site is completely occupied in both end-members of the solid solution, whilst the occupancy of the irregularly eight-coordinated M(2) site varies across the system from 0 to 100%. What is particularly interesting is that these two sites are interconnected with one another in a three-dimensional manner, such that each M(1) site is connected to six M(2) sites, and each M(2) site is connected to two M(1) sites, as shown in Figure 7.5. This connection leads to the infinite sequence: M(1)−M(2)−M(1)−M(2)−...etc. in three dimensions, which forms the conduction pathway for the mobile Na$^+$ ions.

---

[2] The word *ceramic* is used here to distinguish this material from single-crystal and powdered specimens of the same phase.

[3] By comparison, the normal temperature range for firing bone-china (i.e., typical English porcelain with a content of ∼50 wt% calcium phosphate) is 1250−1350 °C.

[4] The term *superionic conductor* is controversial with regards to this material, but is still used for historical reasons. The existence of Na$_4$Zr$_2$(SiO$_4$)$_3$ as a crystalline phase with a rhombohedral structure was reported by D'Ans and Löffler in 1930 [4].

[5] Ceramic Na$_3$Zr$_2$(SiO$_4$)$_2$(PO$_4$) has $\sigma_{Na^+} = 0.20$ S cm^{-1} at 300 °C with an activation energy $E_a = 0.29$ eV [2]; cf. the ceramic near end-member Na$_{3.8}$Zr$_2$(SiO$_4$)$_{2.8}$(PO$_4$)$_{0.2}$ has $\sigma_{Na^+} = 0.018$ S cm^{-1} at 300 °C with an activation energy $E_a = 0.24$ eV [2].

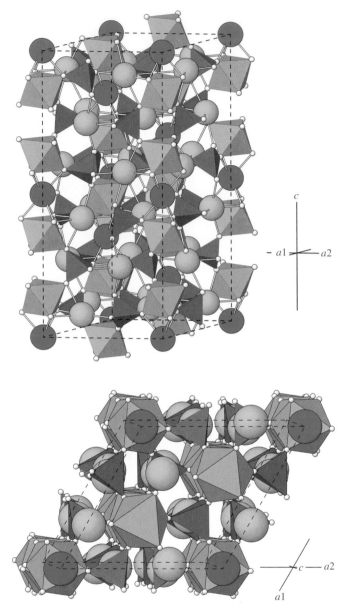

Figure 7.1   Crystal structure of NASICON, $Na_4Zr_2(SiO_4)_3$. The trigonal symmetry is evident in the lower illustration. The M(1) sites are shown green, and the M(2) sites are shown yellow. This figure was drawn using data from Kohler *et al*. [35] cf. ICSD 38055

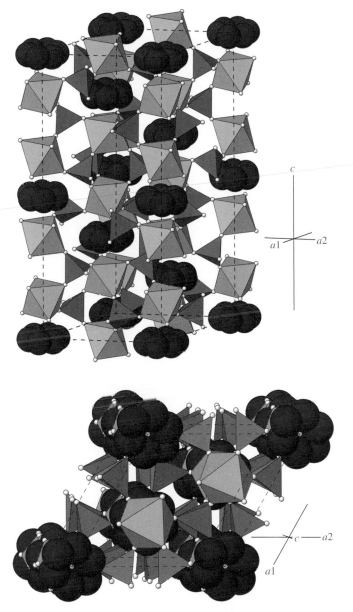

Figure 7.2 Crystal structure of $CuZr_2(PO_4)_3$ emphasising the six off-centred equivalent positions in the M(1) site (shown red) that are partially occupied by the Cu^+ ions with an ionic radius of 96 pm (drawn to scale). The distorted nature of these copper sites appear like 'daisy-heads' running parallel to the $a(1)$-$a(2)$ basal plane. This figure was drawn using data from Bussereau *et al.* [36] cf. ICSD 71881

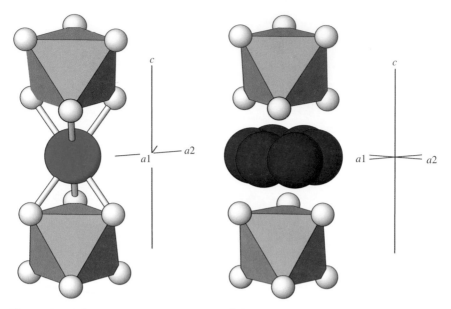

Figure 7.3   The antiprismatic geometry of the M(1) site (green); oxygen atoms are shown white. The distorted positions of the Cu$^+$ ion occupancy in the M(1) site are shown red

Owing to its fast ionic conducting properties there has been a technological interest in NASICON for use as a solid electrolyte in sodium batteries and electrochemical sensors. However, inherent problems concerning the instability of NASICON in contact with molten sodium have restricted the exploitation of this material, at least in so far as sodium batteries are concerned. Nevertheless, NASICON and its analogues present a rich and diverse range of model compounds for exploring the chemistry of ionic conduction in three-dimensional open-framework structures. But it is worth noting that even though the possibilities for substitutional chemistry in this material are vast, geological occurrences of NASICON-type phases are unknown. The reader is referred to the review article by Kreuer *et al.* [5] for further details concerning sodium-ion transport in NASICON materials.

Since 1976, many other NASICON-related phases have been discovered. Taking the pure phosphate end-member, NaZr$_2$(PO$_4$)$_3$ as an example, it should appear quite reasonable on topological grounds to consider substituting Cu$^+$ for Na$^+$, since these ions share very similar ionic radii.[6] Indeed, the synthesis of the copper analogue CuZr$_2$(PO$_4$)$_3$ was first reported in the published literature[7] by Yao and Fray [6] in 1983, and was shown to be a fast

[6] Pauling crystal radii: Na$^+$ = 95 pm and Cu$^+$ = 96 pm.
[7] CuZr$_2$(PO$_4$)$_3$ was first reported by J. McInerney and D. R. Morris, at the Canadian Ceramic Society Meeting, at Calgary, February 1982.

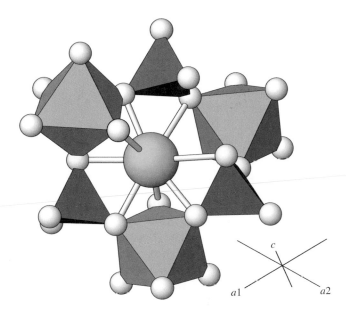

Figure 7.4   The irregular 8-fold coordination of the M(2) site (yellow); oxygen atoms are shown white

Cu$^+$ ion conductor (namely CUSICON). The intermediate solid solution, (Na$_{1-x}$Cu$_x$)Zr$_2$(PO$_4$)$_3$ $(0 \leqslant x \leqslant 1)$ was then prepared by Polles *et al.* in 1988 [7].

The crystal structure for CuZr$_2$(PO$_4$)$_3$ was confirmed by Bussereau *et al.* [8] in 1992 using X-ray and neutron powder diffraction and is shown in Figure 7.2. This figure shows how the Cu$^+$ ions are distributed between six off-centred equivalent positions in the M(1) site, which gives rise to the daisy-like features in the illustration. Stoichiometric CuZr$_2$(PO$_4$)$_3$ is colourless and displays a transference number for the Cu$^+$ ions of unity, which suggests that the Cu–O bonds in CuZr$_2$(PO$_4$)$_3$ are essentially ionic in character [3].[8] The M(2) site is vacant; at least at ambient temperature.[9] In the M(1) site there are two short Cu–O bonds of 205 pm and four longer Cu–O bonds of 270 pm. This is a result of the unusual copper(I) coordination that appears to be a compromise between the usual linear (180°) coordination and the antiprismatic geometry of the M(1) site created by the [Zr$_2$(PO$_4$)$_3$]$^-$ substructure [8] (see Figure 7.3), and is consistent with the ionic character of the Cu–O bonds in this material.[10]

---

[8] Hong [1] has interpreted the whiteness of AgZr$_2$(PO$_4$)$_3$ as an indication of ionic Ag–O bonding.
[9] The M(2) site is also vacant in NaZr$_2$(PO$_4$)$_3$ at ambient temperature.
[10] In contrast to the covalent bonds in the red-coloured semiconductor copper(I) oxide, cuprite, Cu$_2$O.

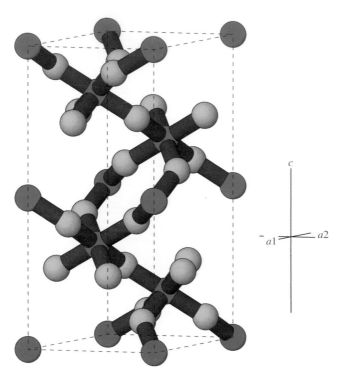

Figure 7.5 The conduction pathways between the M(1) sites (green) and the M(2) sites (yellow) in NASICON, in relation to the unit cell

As a consequence of the large size of the M(1) site, it is possible to have two Cu$^+$ ions within the same M(1) site. From a study using, extended X-ray absorption fine structure (EXAFS), Fargin *et al.* [9, 10] showed that $\sim$20% of the M(1) sites are occupied by Cu$^+$–Cu$^+$ pairs, $\sim$60% are singularly occupied by Cu$^+$ ions and the remaining $\sim$20% are empty; thus illustrating the disordered nature of CuZr$_2$(PO$_4$)$_3$. The Cu$^+$ ions within these Cu$^+$–Cu$^+$ pairs are separated by a remarkably short distance of 240 pm; which is shorter than the Cu–Cu interatomic distance of 256 pm in copper metal. The presence of a Cu$^+$–Cu$^+$ pair is considered to be the source of the intense green luminescence (540 nm emission) of CuZr$_2$(PO$_4$)$_3$ under ultraviolet light at ambient temperature, corresponding to transitions between the 3d^{10} ground-state level and the lowest levels of the 3d^9 4s^1 electronic configuration [11], [7]. By comparison, Cu$^+$–Cu$^+$ pairs do not exist in CuTi$_2$(PO$_4$)$_3$ on account of the smaller M(1) site in this material [10]; which may explain why this phase is nonluminescent.

The blue-green copper(II) analogue, Cu$_{0.5}$Zr$_2$(PO$_4$)$_3$ has been prepared by the oxidation of CuZr$_2$(PO$_4$)$_3$ in oxygen at 500 °C [12]; and by two different sol-gel routes, with subsequent annealing at ~800 °C [13, 14]. The occurrence of monoclinic distortions within these materials (in relation to the rhombohedral unit cell of the NASICON prototype structure) is dependent upon the preparative route and thermal history. It is interesting to note, that neutron diffraction data for the high-temperature (rhombohedral) form of $Cu_{0.5}Ti_2(PO_4)_3$ indicate that the M(1) site is vacant, with the Cu^{2+} ions being located solely at the M(2) site [15]. This is in sharp contrast to $CuTi_2(PO_4)_3$, in which the M(1) site is occupied by the Cu^+ ions, and the M(2) site is vacant.

A controversial issue for many years has been the question concerning the possible existence of a solid solution between $CuZr_2(PO_4)_3$ and $Cu_{0.5}Zr_2(PO_4)_3$. Although stoichiometric $CuZr_2(PO_4)_3$ is colourless, this phase has often been prepared with various shades of pale-green, which indicates the presence of copper(II) in the solid state. However, an electron paramagnetic resonance (EPR) study by Christensen et al. [16] regarding certain compositions within the $CuZr_2(PO_4)_3-Cu_{0.5}Zr_2(PO_4)_3$ system has shown that $Cu_{1-\delta}Zr_2(PO_4)_3$ contains Cu^{2+} ions, but these are only present at a dilute ($\delta \ll 0.5$) and dispersed level of concentration, such that the extent of any solid solution between $CuZr_2(PO_4)_3$ and $Cu_{0.5}Zr_2(PO_4)_3$ is extremely limited. The absence of a solid solution may be due to the peculiar coordination environments of the Cu^+ and Cu^{2+} ions in these materials. Although the substructural frameworks of $CuZr_2(PO_4)_3$ and $Cu_{0.5}Zr_2(PO_4)_3$ are closely related, the coordination of Cu^+ and Cu^{2+} ions are very different, which leads to local distortions. This might explain why Cu^+ and Cu^{2+} ions cannot be accommodated simultaneously within a single CUSICON-type phase, except when present at a defect level of concentration; which is an interesting point, since Cu^+ and Na^+ ions coexist within $(Na_{1-x}Cu_x)Zr_2(PO_4)_3$ [7], and Cu^+ and Cu^{2+} ions coexist within the alluaudite-type phase, $Cu_2Mg_3(PO_4)_3$ [17].

The divalent Cu^{2+} ions in $Cu_{0.5}Zr_2(PO_4)_3$ are comparatively immobile as a consequence of their higher charge, such that this material is essentially an insulator.[11] Therefore, in technological applications where $CuZr_2(PO_4)_3$ is to be used as a Cu^+ solid electrolyte, for example, in an electrochemical sensor, effort should be taken to avoid the alteration of $CuZr_2(PO_4)_3$ to $Cu_{0.5}Zr_2(PO_4)_3$. This alteration can in principle, arise from either oxidation $(Cu^+ \rightarrow Cu^{2+} + e^-)$ or, from disproportionation[12] $(2Cu^+ \rightarrow Cu^{2+} + Cu^0)$ [16].

---

[11] Electrochemical impedance spectroscopy was performed on a ceramic monolith of $Cu_{0.5}Ti_2$ $(PO_4)_3$ over the temperature range 494–783 °C in air by the present author at the Max-Planck-Institut für Festkörperforschung, Stuttgart, Germany, 1991. It is unlikely that $Cu_{0.5}Zr_2(PO_4)_3$ behaves differently in this respect.

[12] Through contact with metallic substances that display a higher affinity for copper than what the oxidised phase $Cu_{0.5}Zr_2(PO_4)_3$ has for copper itself, at a given temperature.

In the Cu$-$O binary system, Cu$_2$O is oxidised reversibly to CuO in air below 1029 °C according to the following reaction:

$$2Cu^IO(s) + O_2(g) \xrightarrow[]{< 1029\,°C(air)} 4Cu^{II}O(s)$$

By comparison, CuZr$_2$(PO$_4$)$_3$ exists in air down to the remarkably low temperature of 475 °C before it is oxidised reversibly to Cu$_{0.5}$Zr$_2$(PO$_4$)$_3$ and CuO:

$$4Cu^IZr_2(PO_4)_3(s) + O_2(g) \xrightarrow[]{< 475\,°C(air)} 4Cu^{II}_{0.5}Zr_2(PO_4)_3(s) + 2Cu^{II}O(s)$$

The presence of 'P$_2$O$_5$' (sensu stricto, the zirconium phosphate substructure) has a stabilising effect on copper(I) in CuZr$_2$(PO$_4$)$_3$. Whilst 'SiO$_2$' appears to have quite the opposite effect, in that attempts at preparing the copper(I) silicate, 'Cu$_4$Zr$_2$(SiO$_4$)$_3$' have so far failed.[13] This is consistent with the fact that there are no crystalline copper(I) silicate phases reported in the literature;[14] and that includes the silica-rich compositions within the hypothetical silicate$-$phosphate solid solution, Cu$_{1+x}$Zr$_2$(SiO$_4$)$_x$(PO$_4$)$_{3-x}$. The factors that govern the relative stability of metal silicates are a complex matter; as is the chemical nature of the silicon$-$oxygen bond, which is considered to be partly ionic and partly covalent. The reader is referred to the monograph by Liebau [18] for further details of the structural chemistry of silicates.

However, various metal cations can be substituted for zirconium(IV) within the copper(I) phosphate, CuZr$_2$(PO$_4$)$_3$, as exemplified by the following phases: CuTi$_2$(PO$_4$)$_3$ [21]; CuHf$_2$(PO$_4$)$_3$ [22]; CuSn$_2$(PO$_4$)$_3$ [23]; CuScNb (PO$_4$)$_3$ [24]; Cu$_2$CrTi(PO$_4$)$_3$ [25]; Cu$_2$CrZr$_2$(PO$_4$)$_3$ [26]; and Cu$_2$ScZr (PO$_4$)$_3$ [3].[15] Several copper(II) analogues have also been reported, including: Cu$_{0.5}$Ti$_2$(PO$_4$)$_3$ [27]; Cu$_{0.5}$Hf$_2$(PO$_4$)$_3$ [22]; and Cu$_{0.5}$Sn$_2$(PO$_4$)$_3$ [23]. In addition, it is very likely that many of these phases are capable of forming solid solutions with one another. For an example, CuTiZr(PO$_4$)$_3$ was discovered independently, using different reaction routes by, Berry *et al.* [28] and Warner *et al.* [3] in 1992, and indicates the existence of the

---

[13] The present author has attempted to prepare the material, 'Cu$_4$Zr$_2$(SiO$_4$)$_3$' by heating a stoichiometric mixture of Cu$_2$O, ZrO$_2$ and SiO$_2$ at high temperature under an atmosphere of flowing argon. The product material comprised: 2Cu$_2$O + 2ZrSiO$_4$ + SiO$_2$.

[14] In contrast to copper(I), several anhydrous copper(II) silicate phases do exist. But they are confined to phases that also incorporate an alkali or alkaline-earth metal oxide, Y$_2$O$_3$ or PbO within their structure. It is interesting to note that besides PbO (which is an amphoteric oxide), all of these are oxides of strongly electropositive metals [19]. Furthermore, no anhydrous copper silicates are known from primary silicate rocks [20].

[15] CUSICON-type phases containing more than two copper ions per unit formula have not been reported in the literature; for example, Cu$_3$Sc$_2$(PO$_4$)$_3$ has not yet been prepared.

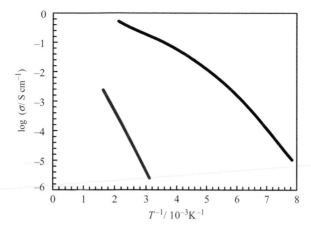

Figure 7.6 Arrhenius conductivity plot for CuTiZr(PO$_4$)$_3$ (shown red); after Warner *et al.* [3]; and Rb$_4$Cu$_{16}$I$_7$Cl$_{13}$ (shown black) after Kanno *et al.* [37]

solid solution CuTi$_{2-x}$Zr$_x$(PO$_4$)$_3$. The reader may wish to explore other possible combinations.

CuTi$_{2-x}$Zr$_x$(PO$_4$)$_3$ has definite technological advantages over the copper(I) halides (cf. Chapter 6) in that it is a refractory material, and so can operate at high temperature (up to ~1300 °C) and importantly, in air >500 °C (see Figure 7.6). CuZr$_2$(PO$_4$)$_3$ has been exploited as a solid electrolyte within electrochemical sensors designed for measuring high concentrations of copper within copper-rich molten metallurgical phases, such as, copper mattes and alloys [29]. Whilst CuTi$_2$(PO$_4$)$_3$ has been studied for its catalytic behaviour for the oxidation of propene to acrolein [30, 31]. However, the performance of CuTiZr(PO$_4$)$_3$ with regards to these applications has not yet been verified.

The metastability of CuTi$_{2-x}$Zr$_x$(PO$_4$)$_3$ under ambient conditions presents the following scenario. When CuTiZr(PO$_4$)$_3$ is heated in air from ambient temperature, it first undergoes an oxidation reaction to form a binary mixture of Cu$_{0.5}$TiZr(PO$_4$)$_3$ and CuO (see Figure 7.7). This is followed by a reduction reaction above 500 °C, whereupon this material loses oxygen and reverts to single-phase CuTiZr(PO$_4$)$_3$. Upon slow cooling in air, CuTiZr(PO$_4$)$_3$ undergoes oxidation <450 °C to form a binary mixture of Cu$_{0.5}$TiZr(PO$_4$)$_3$ and CuO. Alternatively, if CuTiZr(PO$_4$)$_3$ is quenched in air from above 450 °C, then it can exist as a metastable phase under ambient conditions.

Concerning the preparation of CUSICON-type materials, CuZr$_2$(PO$_4$)$_3$ was prepared originally by Yao and Fray in 1983 [6]. Stoichiometric amounts of Cu$_2$O, ZrO$_2$ and NH$_4$H$_2$PO$_4$ were milled for 16 h using an alumina ball mill with ethanol as a dispersant. The mixture was dried and then heated to 300 °C for 4 h in order to decompose the NH$_4$H$_2$PO$_4$, before calcining at

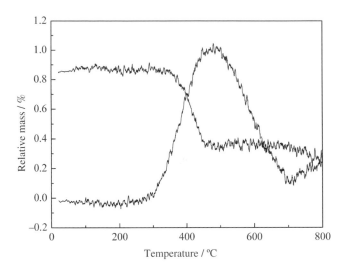

Figure 7.7   Thermogravimetric hysteresis loop for CuTiZr(PO$_4$)$_3$ in air with heating and cooling rates of 30 °C/h. CuTiZr(PO$_4$)$_3$ is oxidised initially at 300 °C; then reduced at 500 °C; and then reoxidised at 450 °C. This cycle corresponds to the reversible reaction: $4Cu^{I}TiZr(PO_4)_3(s) + O_2(g) \rightleftharpoons 4Cu^{II}_{0.5}TiZr(PO_4)_3(s) + 2Cu^{II}O$ (s). The relative mass change suggests that this solid-state redox reaction is incomplete (Courtesy: Professor Eivind M. Skou.)

850 °C in air for 16 h. The partially sintered mixture was ground for 5 days using an alumina ball mill with ethanol as a dispersant. The powder was dried, and then cold pressed isostatically under a pressure of 7.6 kbar into monoliths. These were then heated in air to 1200 °C for ~5 h before cooling to ambient temperature.[16] Powder X-ray diffractometry revealed that the product material, CuZr$_2$(PO$_4$)$_3$ had a crystal structure analogous to NaZr$_2$(PO$_4$)$_3$.

   Shortly after this, the titanium analogue, CuTi$_2$(PO$_4$)$_3$ was prepared by Mbandza *et al.* in 1985 [21]. Stoichiometric amounts of Cu$_2$O (or CuO), TiO$_2$ (anatase) and NH$_4$H$_2$PO$_4$ were mixed and ground in an agate mortar. The mixture was then heated to 300 °C for 3 h in order to decompose the NH$_4$H$_2$PO$_4$. This material was then pressed into pellets and calcined at 850 °C in either nitrogen or air. The product material was pale brown. Gravimetric analysis confirmed that the stoichiometry is CuTi$_2$(PO$_4$)$_3$. Powder X-ray diffractometry showed that the product material, CuTi$_2$(PO$_4$)$_3$ had a crystal structure analogous to NaTi$_2$(PO$_4$)$_3$, but was actually closer to AgTi$_2$(PO$_4$)$_3$. Calcining at the higher temperature of 1000 °C resulted in a darker shade of brown; this is undoubtedly a consequence of increased grain size.

---

[16] Warner *et al.* [32] have shown that a temperature of at least 1150 °C is required to form CuZr$_2$(PO$_4$)$_3$ by this synthesis route.

CuTi$_2$(PO$_4$)$_3$ has also been prepared by a hydrothermal process by McCarron *et al.* in 1987 [33]. A small (5 g) sample of CuTi$_2$(PO$_4$)$_3$ was prepared by encapsulating stoichiometric quantities of Cu$_2$O, TiO$_2$ and H$_3$PO$_4$ in a gold ampoule and heating at 500 °C for 12 h in an autoclave at ∼3 kbar, and slowly cooling at 10 °C/h to ambient temperature. The product material comprised red euhedral crystals of millimetre dimensions. These were structurally determined by single-crystal X-ray diffraction and shown to be isostructural with NaTi$_2$(PO$_4$)$_3$ except for a distortion in the copper positions in an analogous way to that now known in CuZr$_2$(PO$_4$)$_3$.

In 1992, Berry *et al.* [28], prepared CuTiZr(PO$_4$)$_3$ by reacting together an appropriate mixture of Cu$_2$O, TiO$_2$ (anatase), ZrO$_2$ and NH$_4$H$_2$PO$_4$ at 300 °C for 3 h in air. This material was then reground and heated at 1200 °C for 36 h in air. But the heating and cooling rate were not disclosed. The product material, CuTiZr(PO$_4$)$_3$ was described as pale brown and the powder X-ray diffraction pattern is published in ICDD-PDF 48-659, with cell constants; $a = b = 8.733(3)$ Å and, $c = 22.042(30)$ Å.

In the same year, Warner *et al.* [3] prepared CuTiZr(PO$_4$)$_3$ by sintering an intimately finely ground and compacted (isostatically pressed under a pressure of ∼3 kbar) equimolar mixture of CuTi$_2$(PO$_4$)$_3$ and CuZr$_2$(PO$_4$)$_3$, at 1300 °C for 16 h under a flowing atmosphere of argon in an alumina boat. Both heating and cooling rates were 180 °C/h. Analysis by powder X-ray diffractometry indicated that the material comprised a wide range of solid-solution compositions CuTi$_2$ $_x$Zr$_x$(PO$_4$)$_3$ ($0 \le x \le 2$). The homogeneous composition CuTiZr(PO$_4$)$_3$ was obtained only after regrinding and sintering a newly compacted pellet as above (see Figure 7.8). The slow rate of this reaction reflects the slow counterdiffusion of the highly charged Ti^{4+} and Zr^{4+} ions between these two isostructural phases, whilst their crystal structures offer no intrinsic vacancies to facilitate this diffusion process.

Preparative Procedure

Prepare 20 g of CuTiZr(PO$_4$)$_3$ by the following method: Weigh appropriate amounts[17] of copper(II) oxide, CuO (Aldrich 99 + %)[18], titanium(IV) oxide, rutile, TiO$_2$ (Aldrich 99.9 + % <5 μm), zirconium(IV) oxide, ZrO$_2$ (Aldrich 99.9% <1 μm) and ammonium dihydrogenphosphate, NH$_4$H$_2$PO$_4$ (Aldrich 99.99 + %), to a precision of ±0.001 g. Grind these powders together thoroughly in their dry state using a porcelain pestle and mortar; whilst taking care to avoid any spillage.

Place the powdered mixture in a large alumina crucible (CC62 Almath Ltd) and place a corresponding alumina lid on top of the crucible. Then insert the

---

[17] The reader may wish to consider *Problems 1 and 2* before embarking on these calculations.
[18] CuO is preferred over Cu$_2$O since it is commercially available as a stoichiometric compound with a more precise copper content. If Cu$_2$O is used, it will merely oxidise to CuO during the early stage of the reaction.

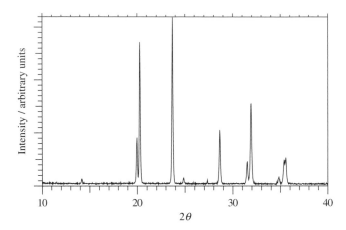

Figure 7.8   Powder XRD pattern (Cu-$K_{\alpha 1}$ radiation) of CuTiZr(PO$_4$)$_3$ as prepared by sintering a 1:1 molar mixture of CuTi$_2$(PO$_4$)$_3$ and CuZr$_2$(PO$_4$)$_3$ in argon at 1300 °C followed by a cooling rate of 180 °C/h. This material is considered by the author to be single-phase CuTiZr(PO$_4$)$_3$; $a = b = 8.686(1)$ Å and, $c = 21.762(5)$ Å

crucible in a chamber furnace and heat in air at 60 °C/h to 300 °C. The pyrolysis of NH$_4$H$_2$PO$_4$ results in the formation of solid P$_2$O$_5$ that sublimes >300 °C. Fortunately, P$_2$O$_5$ is sufficiently reactive at 300 °C towards these metal oxides so as to from a nonequilibrium mixture of metal phosphates.[19] After 3 h at 300 °C, continue heating at 200 °C/h to 1250 °C. After 48 h at 1250 °C, cool at 200 °C/h to ambient temperature (20 °C). This complete heating cycle will last ∼67 hours (i.e., almost 3 days).

Note the colour and morphology of your intermediate product material. Take care to physically extract all of this material from the crucible. It may be necessary to scrape this from the walls of the crucible with a metal spatula. This material should then be ground thoroughly to a very fine powder using an agate pestle and mortar.[20] Submit a small (1 g) specimen of this powder for analysis by powder X-ray diffraction (scan: 10–40° 2θ); since it should be informative to see which phases are produced during this stage of the process.

---

[19] The material produced at this early stage of the process comprises; Cu$_2$P$_2$O$_7$, TiP$_2$O$_7$, ZrP$_2$O$_7$, TiO$_2$ and ZrO$_2$ in various amounts. The alumina crucibles are sufficiently densely sintered so as to avoid any significant reaction with P$_2$O$_5$ under these conditions, and so prevent the formation of AlPO$_4$.

[20] If available, the reader is advised to use a planetary mono mill (e.g., a *Fritsch Pulverisette* 6 with zirconium oxide bowl and balls and cyclohexane as the liquid medium). The reader should be aware that CuTi$_2$(PO$_4$)$_3$ can readily be prepared at 850 °C; whilst CuZr$_2$(PO$_4$)$_3$ and CuTiZr (PO$_4$)$_3$ require a much higher temperature, and normally require repeated grinding and sintering before single-phase material is produced.

The remaining powder will now be used to prepare two ceramic monoliths of CuTiZr(PO$_4$)$_3$. Compress 10 g of the above powder into a disc using a 32-mm diameter Specac stainless steel die and a uniaxial press with a load $\sim$10 tonne. Very gently, place the *green*[21] monolith directly into an alumina boat (SRX110 Almath Ltd). This procedure should be repeated on the remaining powder (i.e., $\sim$9 g), with this *green* monolith placed adjacent to the other one in the same alumina boat.

Insert the alumina boat into the central region of the tube furnace and heat in air to 1250 °C with a heating rate of 200 °C/h. After 55 hours at 1250 °C, cool to 700 °C with a cooling rate of 200 °C/h. Leave the furnace to dwell at 700 °C[22] for 12 h before undergoing further cooling to ambient temperature (20 °C) with a cooling rate of 200 °C/h. It is important to remember that the alumina boat must be removed from the tube furnace whilst it is at 700 °C, using a hooked steel wire probe, whilst wearing a visor and thermally insulated gloves. Then as quickly as possible, use a pair of steel forceps to plunge the red-hot ceramic monoliths into water and thereby quench them. Retrieve the ceramic monoliths from the water and dry them with a warm air blower. Note the colour and morphology of your product. Crush and grind one of the monoliths using an agate pestle and mortar and submit a small sample (1 g) for analysis by powder XRD (scan: 10–40° 2θ). Compare your powder pattern with those shown in Figures 7.8 and 7.9. By comparing your powder pattern with the PDF 81-526 for CuZr$_2$(PO$_4$)$_3$ and the PDF 84-1355 for CuTi$_2$(PO$_4$)$_3$ it should be clear to see that the cell constants for CuTiZr(PO$_4$)$_3$ lie in between those for CuTi$_2$(PO$_4$)$_3$ and CuZr$_2$(PO$_4$)$_3$. Then place the rest of the product material in an appropriately labelled specimen jar and close tightly, and keep it for further use (see below).

Depending on the outcome of this analysis, the reader may wish to regrind and repeat the sintering process if deemed necessary. However, if the material is essentially phase pure, then the second ceramic monolith of CuTiZr(PO$_4$)$_3$ (as prepared in concert) should therefore be suitable for characterisation, for example, by electrochemical impedance spectroscopy.[23]

Finally, the reader may wish to prepare Cu$_{0.5}$TiZr(PO$_4$)$_3$ by chemically oxidizing CuTiZr(PO$_4$)$_3$ to Cu$_{0.5}$TiZr(PO$_4$)$_3$. This can be performed by placing the remaining ($\sim$8 g) powdered specimen of CuTiZr(PO$_4$)$_3$ in the alumina boat (SRX110 Almath Ltd) as used previously, and heating in a tube

---

[21] The expression *green* is a common technological expression in reference to the nonfired condition of a compacted powder held together by the cohesive forces resulting from compaction alone or with the assistance of a temporary binder.

[22] It should take $\sim$64 h (i.e., $\sim$2⅔ days) to reach this specific point in the heating cycle. As soon as possible thereafter, the operator should arrange to retrieve (i.e., quench) the sample from the furnace before the tube furnace embarks on cooling further to ambient temperature.

[23] The reader is referred to the textbook on impedance spectroscopy edited by Barsoukov and Macdonald [34].

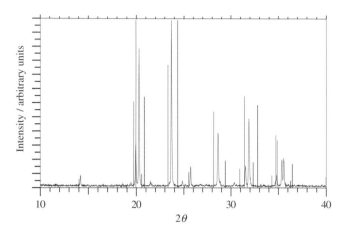

Figure 7.9   Powder XRD pattern (Cu-$K_{\alpha 1}$ radiation) of CuTiZr(PO$_4$)$_3$ as prepared by the method described in this chapter; sintered in air at 1250 °C and then quenched from 700 °C. PDF 81−526 for CuZr$_2$(PO$_4$)$_3$ and PDF 84−1355 for CuTi$_2$(PO$_4$)$_3$ are shown green and red, respectively, for comparison. The peak at 25.72° 2θ is attributed to an impurity phase; possibly the d_{101} in anatase, cf. PDF 75-1537

furnace under flowing oxygen (500 ml/min) at 450 °C for 48 hours; with heating and cooling rates of 200 °C/h. The product should then be ground, and the exsolved CuO removed with warm (50°C) 4 mol dm^{-3} HNO$_3$ solution. Then filtered using a Buchner filter funnel; washed first with distilled water, then ethanol, and finally left to dry. The product can be analysed by powder X-ray diffraction; and comparisons made with the PDF 47-75 for CuZr$_4$P$_6$O$_{24}$ and the PDF 86-555 for Cu$_{0.5}$Ti$_2$(PO$_4$)$_3$, as well as the powder pattern for the precursor compound, CuTiZr(PO$_4$)$_3$, as shown in Figures 7.8 and 7.9.

REFERENCES

1. H. Y.-P. Hong, *Materials Research Bulletin* **11** (1976) 173–182.
2. J. B. Goodenough, H. Y.-P. Hong and J. A. Kafalas, *Materials Research Bulletin* **11** (1976) 203–220.
3. T. E. Warner, W. Milius and J. Maier, *Berichte der Bunsengesellschaft für Physikalische Chemie* **96** (1992) 1607–1613.
4. J. D'Ans and J. Löffler, *Zeitschrift für Anorganische und Allgemeine Chemie* **191** (1930) 1–35.
5. K.-D. Kreuer, H. Kohler and J. Maier, *High Conductivity Ionic Conductors: Recent Trends and Application*, 242–279, ed. T. Takahashi, World Scientific Publishing Co., Singapore (1989).
6. P. C. Yao and D. J. Fray, *Solid State Ionics* **8** (1983) 35–42.

7. G. Le Polles, C. Parent, R. Olazcuagal, G. Le Flem and P. Hagenmuller, *Comptes Rendus de l'Academie des Sciences* Série II **306** (1988) 765–769.
8. I. Bussereau, M. S. Belkhiria, P. Gravereau, A. Boireau, J. L. Soubeyroux, R. Olazcuagal and G. Le Flem, *Acta Crystallographica C* **48** (1992) 1741–1744.
9. E. Fargin, I. Bussereau, G. Le Flem, R. Olazcuagal, C. Cartier and H. Dexpert, *European Journal of Solid State Inorganic Chemistry* **29** (1992) 975–980.
10. E. Fargin, I. Busserean, R. Olazcuagal and G. Le Flem, *Journal of Solid State Chemistry* **112** (1994) 176–181.
11. B. Maihold and H. Wulff, *Wissenschaftliche Zeitschrift der Ernst-Moritz-Arndt-Universität Greifswald, Mathematisch-naturwissenschaftliche Reihe* **36** (1987) 27–29.
12. A. El Jazouli, M. Alami, R. Brochu, J. M. Dance, G. Le Flem and P. Hagenmuller, *Journal of Solid State Chemistry* **71** (1987) 444–450.
13. I. Bussereau, R. Olazuagal, G. Le Flem and P. Hagenmuller, European, *Journal of Solid State Inorganic Chemistry* **26** (1989) 383–399.
14. E. Christensen, J. H. von Barner, J. Engell and N. J. Bjerrum, *Journal of Materials Science* **25** (1990) 4060–4065.
15. R. Olazcuaga, G. Le Flem, A. Boireau and J. L. Soubeyroux, *Advanced Materials Research* **1** (1994) 177–188.
16. R. H.-W. Christensen and T. E. Warner, *Journal of Materials Science* **41** (2006) 1197–1205.
17. T. E. Warner and J. Maier, *Materials Science and Engineering B* **23**, (1994) 88–93.
18. F. Liebau, *Structural Chemistry of Silicates: Structure, Bonding, and Classification*, Springer, Berlin, 1985, 347 pp.
19. C. H. L. Goodman, *Mineralogical Magazine* **56** (1992) 373–383.
20. K. H. Breuer, W. Eysel and M. Behruzi, *Zeitschrift für Kristallographie* **176** (1986) 219–232.
21. A. Mbandza, E. Bordes and P. Courtine, *Materials Research Bulletin* **20** (1985) 251–257.
22. R. Ahmamouch, S. Arsalane, M. Kacimi and M. Ziyad, *Materials Research Bulletin* **32** (1997) 755–761.
23. A. Serghini, R. Brochu, R. Olazcuaga and P. Gravereau, *Materials Letters* **22** (1995) 149–153.
24. G. Ohene-Saforo, *A study of copper containing silicates with the NASICON type structure*, Unpublished MSc Thesis, 2005, University of Southern Denmark.
25. A. El Jazouli, A. Serghini, R. Brochu, J.-M. Dance and G. Le Flem, *Comptes Rendus de l'Academie des Sciences* Série II **300** (1985) 493–496.
26. A. Boireau, J. L. Soubeyroux, P. Gravereau, R. Olazcuaga and G. Le Flem, *Journal of Alloys and Compounds* **188** (1992) 113–116.
27. A. El Jazouli, J. L. Soubeyroux, J. M. Dance, and G. Le Flem, *Journal of Solid State Chemistry* **65** (1986) 351–355.
28. F. J. Berry, G. Oates, L. E. Smart, M. Vithal, R. Cook, H. G. Ricketts, R. Williams and J. F. Marco, *Polyhedron* **11** (1992) 2543–2547.
29. A. J. Davidson and D. J. Fray, *Solid State Ionics* **136–137** (2000) 613–620.
30. F. Oudet, A. Vejux, T. Kompany, E. Bordes and P. Courtine, *Materials Research Bulletin* **24** (1989) 561–570.
31. Y. Kousuke and A. Yoshihiro, *Materials Research Bulletin* **35** (2000) 211–216.

32. T. E. Warner, P. P. Edwards and D. J. Fray, *Materials Science and Engineering B* **8** (1991) 219–224.

33. E. M. McCarron, J. C. Calabrese and M. A. Subramanian, *Materials Research Bulletin* **22** (1987) 1421–1426.

34. E. Barsoukov and J. R. Macdonald (eds.), *Impedance Spectroscopy: Theory, Experiment and Applications*, 2nd edn, 2005, Wiley-Blackwell, New York, 616 pp.

35. H. Kohler, H. Schulz and O. K. Mel'nikov, *Materials Research Bulletin* **18** (1983) 589–5925.

36. I. Bussereau, M. S. Belkhiria, P. Gravereau, A. Boireau, J. L. Soubeyroux, R. Olazcuaga and G. le Flem, *Acta Crystallographica* C **48** (1992) 1741–1744.

37. R. Kanno, K. Ohno, Y. Kawamoto, Y. Takeda, O. Yamamoto, T. Kamiyama, H. Asano, F. Izumi and S. Kondo, *Journal of Solid State Chemistry* **102** (1993) 79–92.

PROBLEMS

1. Write a balanced chemical equation to describe the pyrolysis of ammonium dihydrogenphosphate.

2. Write a balanced equation to describe the overall chemical reaction for the formation of CuTiZr(PO$_4$)$_3$ from the chemical reagents as described in this preparative procedure.

3. Discuss what you would expect to produce if a large excess of ammonium dihydrogenphosphate was used in an attempt to prepare CuTiZr(PO$_4$)$_3$ by the method described here.

4. Use the PDF 81-526 for CuZr$_2$(PO$_4$)$_3$ and the PDF 84-1355 for CuTi$_2$(PO$_4$)$_3$ as guides to help you index the powder pattern corresponding to your specimen of CuTiZr(PO$_4$)$_3$. Then use these data to determine the cell constants for CuTiZr(PO$_4$)$_3$, and thus calculate the volume of the unit cell.

5. Measure the density of your ceramic monolith of CuTiZr(PO$_4$)$_3$. Then use this value, in conjunction with your answer to *Problem 4*, in order to determine the fractional density ($\rho_{monolith}/\rho_{crystal}$) as a percentage.

6. Suggest a reason why CuTiZr(PO$_4$)$_3$ has an orange–brown colouration.

7. Consult the literature in order to find a NASICON-type phase in which all the M(1) and M(2) sites are vacant.

8. Explain why (Na$_{0.5}$Cu$_{0.5}$)Zr$_2$(PO$_4$)$_3$ is expected to have a significantly lower ionic conductivity than either of the end-members, NaZr$_2$(PO$_4$)$_3$ and CuZr$_2$(PO$_4$)$_3$ at a given temperature.

9. Describe what you would expect to happen if CuTiZr(PO$_4$)$_3$ was heated with elemental sulfur in an evacuated sealed glass ampoule at 510 °C. Write a balance chemical equation to describe the pertinent chemical reaction.

10. Describe a method for making the silver analogue, AgTiZr(PO$_4$)$_3$.

8

Cobalt Ferrite CoFe$_2$O$_4$ by a Coprecipitation Method

Cobalt ferrite, CoFe$_2$O$_4$, is an intermediate member of the solid solution series, Co$_{3-x}$Fe$_x$O$_4$ ($0 \leq x \leq 3$) that crystallises with the spinel structure. At high temperature, the iron and cobalt cations are disordered among the tetrahedral and octahedral sites, whilst there is a tendency towards the ordered, inverse spinel, Fe^{III}[Co^{II}Fe^{III}]O$_4$, at ambient temperature. An interest in Co$_{3-x}$Fe$_x$O$_4$ as a ferrimagnetic[1] ceramic material arose in Japan during the early 1930s, on account of its high remanence,[2] coercivity,[3] and electrical resistivity. But its commercialisation as an

---

[1] A ferromagnetic phase exhibits very strong magnetism, and therefore has a very large positive magnetic susceptibility. This effect is caused by the cooperative alignment of a large number of spins with parallel orientations, commonly referred to as domains. A *ferrimagnetic* phase is similar to a ferromagnetic phase, except that the saturation magnetisation, M_s at 0 K, is less than that corresponding to complete parallelism of all the spins (as arises in a ferromagnetic phase). In other words, some of the spins may be aligned in an antiparallel arrangement (so as to cancel each other out). Or alternatively, the spins may be canted.

[2] *Remanence*, M_r, is the magnetisation persisting in a ferromagnetic or, ferrimagnetic material when the exciting magnetising force is removed. It is also termed residual magnetisation or, remanent magnetisation.

[3] *Coercivity*, H_c, is the magnetising force necessary to demagnetise a previously magnetised ferromagnetic or, ferrimagnetic, material at a given temperature. Therefore, it is a measure of how well a material can retain its magnetisation, whilst under the influence of an opposing magnetic field. *Coercivity*, relates to a certain point on the hysteresis loop that occurs in the plot of induction, B as a function of an applied magnetic field, H. (The closely related term, *coercive field*, relates to a certain point on the hysteresis loop that occurs in the plot of magnetisation, M as a

---

Synthesis, Properties and Mineralogy of Important Inorganic Materials   By Terence E. Warner
© 2011 John Wiley & Sons, Ltd

electrically insulating permanent magnet, was superseded during the 1950s by the mass production of the cheaper hexagonal barium ferrite, $BaFe_{12}O_{19}$, which has been used extensively in electric motors, television tubes and loudspeakers.[4] Recently, technological interest in $CoFe_2O_4$ has re-emerged, in the form of nanometre-scale particles for potential applications in magnetic data storage systems, electromechanical transducers, and biomedicine. This chapter describes the preparation of a ceramic monolith of the near-inverse spinel $CoFe_2O_4$ through the coprecipitation of $CoFe_2O_4$ particles from an aqueous solution of cobalt(II) chloride and iron(III) chloride, followed by compaction, sintering and annealing processes. Finally, the ceramic monolith is magnetised to yield a permanent magnet.

$CoFe_2O_4$ crystallises with the spinel ($MgAl_2O_4$) structure and belongs to the cubic crystal system and space group $Fd\bar{3}m$ (see Figures 8.1 and 8.2). The crystal structure was determined by Natta and Passerini in 1929 using powder X-ray diffractometry [3], and refined by Yunus et al. in 2008 using powder neutron diffractometry [4]. Before contemplating on the crystal structure of $CoFe_2O_4$, it is necessary to be familiar with the general aspects of the $MgAl_2O_4$ prototype structure. The oxygen substructure in $MgAl_2O_4$ comprises a cubic close packed (ccp) structure of oxide ions.[5] Within this substructure lie a set of octahedral interstices, together with a set of smaller, but more numerous, tetrahedral interstices. $MgAl_2O_4$ is referred to as a *normal* spinel, since by definition it has an ordered structure in which $^1/_8$ of the tetrahedral interstices are occupied solely by Mg^{2+} ions, and $^1/_2$ of the octahedral interstices are occupied solely by Al^{3+} ions. Therefore, the structure can be expressed as, $(Mg^{II}\square\square\square\square\square\square)^{tet}[Al^{III}Al^{III}\square\square]^{oct}O_4$, in which the squares represent unoccupied interstices with respect to the oxygen lattice; emphasising the vacant nature of the spinel structure.

An important feature of the spinel structure is that it can accommodate many different metal cations, many of which can display variable oxidation states, such that they can be distributed in various ways among the tetrahedral and octahedral sites. This creates a multitude of substitutional possibilities, that include ordered (i.e., normal and inverse), disordered, and partially

---

function of an applied magnetic field, H). Therefore, H_c, is the numerical value of H corresponding to the point where the 'demagnetising part of the hysteresis loop' intercepts the H-axis (i.e., where the induction, B (or magnetisation, M) $= 0$).The so-called hard magnets have a high coercivity, whilst soft magnets have a low coercivity. Pure magnetite, Fe_3O_4 is a soft-magnet, whereas, $CoFe_2O_4$, is a hard magnet.

[4] The most common figure of merit for a permanent magnet is the maximum energy product $(BM)_{max}$. Sintered $CoFe_2O_4$ has a maximum energy product $\sim$22 kJ m^{-3} [1] that is comparable to commercial $BaFe_{12}O_{19}$ $\sim$25 kJ m^{-3} [2].

[5] The reader is referred to West [5] for a detailed description of the ccp structure, which belongs to the face-centred cubic (fcc) Bravais lattice. A helpful way to visualise the ccp structure is with a collection of polystyrene balls although originally, Kepler's approach was to use cannon balls!

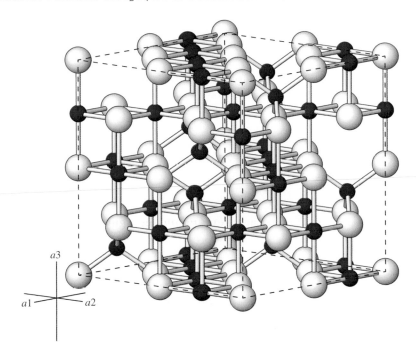

$a3$

$a1$ $a2$

Figure 8.1   Idealised crystal structure of spinel, MgAl$_2$O$_4$ showing the cubic unit cell with $Z = 8$. The tetrahedral coordinated Mg^{2+} ions are shown red and the octahedral coordinated Al^{3+} ions are shown blue. From this illustration, the oxygen substructure can be seen to consist of eight face-centred cubic (*fcc*) unit cells

disordered structures.[6] In order to appreciate the subtle features of the ternary compound CoFe$_2$O$_4$, it is wise to first consider the binary end-members; Fe$_3$O$_4$ and Co$_3$O$_4$.

Fe$_3$O$_4$ occurs naturally as the rock-forming mineral, magnetite, and is a common constituent of mafic and ultramafic rocks;[7] although it is unstable in air <1330 °C with respect to hematite, Fe$_2$O$_3$. Magnetite adopts the *inverse* spinel structure, in which the Fe^{2+} ions have a strong preference for octahedral sites as represented by the empirical formula, Fe^{III}[Fe^{II}Fe^{III}]O$_4$. The magnetic moments on the high-spin Fe^{3+} ($3d^5$) ions in the tetrahedral

---

[6] A normal spinel corresponds to (M^{II})^{tet}[M^{III}M^{III}]^{oct}O$_4$. An inverse spinel corresponds to (M^{III})^{tet}[M^{II}M^{III}]^{oct}O$_4$. The majority of spinels display nonconvergent disorder among the cations within the tetrahedral and octahedral sites and can be described by the inversion parameter i, in relation to the formula, (M$^{II}_{1-i}$M$^{III}_{i}$)^{tet}[M$^{II}_{i}$M$^{III}_{2-i}$]^{oct}O$_4$. At the extremities; $i = 0$ corresponding to a normal spinel; and $i = 1$ corresponding to a completely inverse spinel. With $i = {}^2/_3$ corresponding to a fully disordered spinel.

[7] *Mafic and ultramafic rocks* are essentially quartz-free igneous rocks rich in ferromagnesian minerals; the latter particularly so.

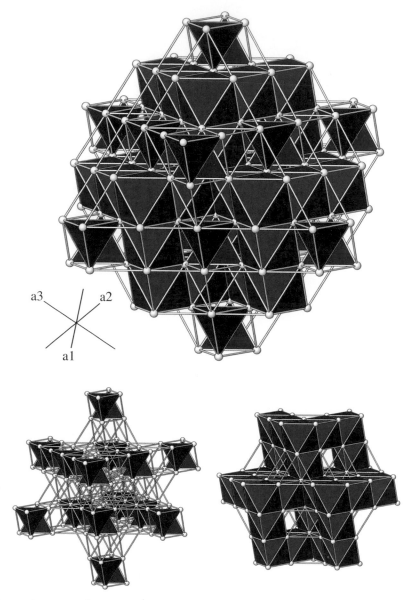

Figure 8.2 An alternative view of the spinel, MgAl$_2$O$_4$ crystal structure. The tetrahedral coordinated Mg^{2+} substructure is red; and the octahedral coordinated Al^{3+} substructure is shown blue. The sticks connecting the oxide ions (shown as small white spheres) emphasise their ABC stacking sequence

Figure 8.3 The magnetic attraction between the CoFe₂O₄ ceramic monolith and a single crystal of naturally occurring magnetite, Fe₃O₄. The specimen of magnetite is from the author's private collection. Width of field = 30 mm (Photograph used with kind permission from Simon Svane. Copyright (2010) Simon Svane)

sites are coupled antiferromagnetically with the magnetic moments on both the high-spin Fe^{3+} ($3d^5$) and the high-spin Fe^{2+} ($3d^6$) ions in the octahedral sites, which results in an overall ferrimagnetic behaviour. With reference to bulk material,[8] Fe₃O₄ has a Curie temperature, $T_c = 858$ K [6], a reasonably high saturation magnetisation, M_s (at 0 K) = 98 A m² kg^{-1} [6], with a corresponding magnetic moment, μ (at 0 K) = 4.1 μ_B [6]; see Figure 8.4. But, Fe₃O₄ has a very low remanence and coercivity; see Figure 8.5.

Fe₃O₄ exhibits a significant electronic conductivity at ambient temperature ($\sigma = 2.5 \times 10^4 \Omega^{-1}$ m^{-1}) such that it is almost metallic [7]. This is a consequence of charge transfer between the Fe^{2+} and Fe^{3+} ions in the octahedral sites via the bridging oxide ions, which results in a hopping mechanism ($Fe^{2+} + Fe^{3+} \rightarrow Fe^{3+} + Fe^{2+}$), such that the 'extra electron' associated with Fe^{2+} is delocalised throughout the octahedral sites. Apparently, the tetrahedral Fe^{3+} ions do not play a direct role in the electronic conduction.

---

[8] The expression, *bulk material*, is commonly used in Materials Science to signify a material that is in its macroscopic state, i.e., as a single crystal; ceramic; or as a powder with a particle size > 0.1 μm.

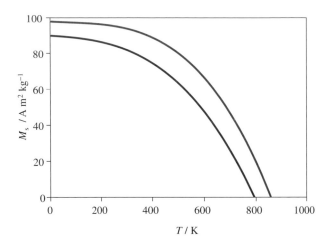

Figure 8.4   Temperature dependence of saturation magnetisation for CoFe$_2$O$_4$ (red line) and Fe$_3$O$_4$ (blue line). After Smit and Wijn [6]

Upon cooling, however, Fe$_3$O$_4$ undergoes the Verwey transition at 119 K, whereby its conductivity drops by two orders of magnitude. This corresponds to an ordering of the Fe^{2+} and Fe^{3+} ions within the octahedral sites, since there is no longer sufficient thermal energy to enable electron hopping between them. The reader is referred to Orchard [7] and Goodenough [8] for further details concerning the magnetic and electrical properties of magnetite.

Pure magnetite has a very low remanence and therefore can not act as a permanent magnet. But, there exists a peculiar type of rock called *lodestone*[9] that possesses a magnetic polarity and therefore behaves as a permanent magnet. One of the earliest references to lodestone, is in the story about a herdsman called Magnes; as recorded by Pliny (*Gaius Plinius Secundus, 23–79 AD*).[10] Apparently, whilst Magnes was taking his herds to pasture upon Mount Ida (on Crete or Asia Minor?) he found that the nails of his shoes and the iron ferrule of his staff adhered to the ground [9]. Nowadays, this magnetic rock is recognised as lodestone, which consists essentially of magnetite, Fe$_3$O$_4$. Yet most importantly, lodestone also contains minute inclusions of the defective spinel, maghemite, γ-Fe$_{2.67}$O$_4$, and is often accompanied by an inhomogeneous dispersion of impurity metal ions, such as, titanium, aluminium and manganese [10]. The rock's abnormally high remanence and coercivity is attributed to this unique microstructure.[11] It has been speculated that lodestones may have become magnetised through the

---

[9] An alternative spelling of *lodestone* is; *loadstone*.

[10] Incidentally, Pliny died at Pompeii during the volcanic eruption of Vesuvius in 79 AD, as referred to in Chapter 3.

[11] Microstructure is a fundamentally important aspect of permanent magnets in general.

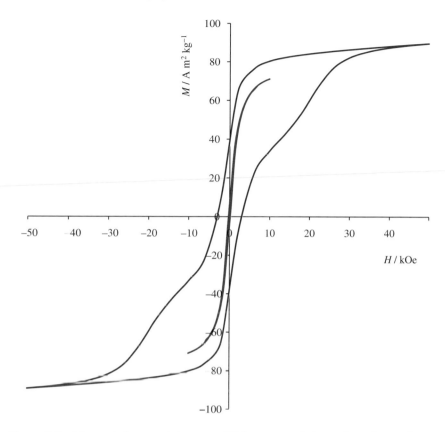

Figure 8.5   Magnetic hysteresis loops at 5 K for a ground sintered specimen of near-inverse CoFe$_2$O$_4$ as prepared by the method described in this chapter (red line). Saturation magnetisation, $M_s = 89.5$ A m^2 kg^{-1}; remanence, $M_r = 38.5$ A m^2 kg^{-1}; and coercive field, $H_c = 2.94$ kOe. Natural magnetite is shown in blue for comparison (The author is grateful to Profesor K. Murray and Dr B. Moubaraki, Monash University, Australia for recording these data)

strong magnetic fields surrounding lightning bolts; since the earth's magnetic field (~ 0.5 G) is too weak for this purpose [10]. But either way, lodestone has been used since the 12th century to magnetise the steel needles as used in magnetic compasses.

In sharp contrast to Fe$_3$O$_4$, the cobalt analogue Co$_3$O$_4$ adopts the *normal* spinel structure, in which the Co^{2+} ions have a strong preference for tetrahedral sites; Co^{II}[Co^{III}Co^{III}]O$_4$ [11]. Mixed valency does not occur on the octahedral sites within Co$_3$O$_4$, and so this phase is an insulator [7]. Furthermore, the low-spin 3d^6 configuration of the octahedral coordinated Co^{3+} results in a zero magnetic moment, such that the cobalt(III) substructure is essentially diamagnetic.[12] Conversely, the high-spin $3d^7$ configuration of the tetrahedral coordinated Co^{2+} results in a magnetic moment on these ions only. But the distance between two tetrahedral sites is too great for their magnetic moments to be coupled at ambient temperature, therefore, Co_3O_4 is paramagnetic.[13] At low temperature, however, these magnetic moments order gradually, such that below the Néel temperature, $T_N = 40$ K, Co_3O_4 is an antiferromagnetic[14] phase [11]. Natural occurrences of Co_3O_4 are unknown.

From the above descriptions, it should be apparent that the distribution of the iron and cobalt cations in $CoFe_2O_4$ could, in principle, be quite complex. ^{57}Fe-Mössbauer spectroscopy [12], X-ray absorption near-edge structure spectroscopy (XANES) and extended X-ray absorption fine structure spectroscopy (EXAFS) [13] reveal that $CoFe_2O_4$ has a disordered structure in between that of normal and inverse spinel, and that the cation distribution is dependent upon the annealing temperature and cooling rate. $CoFe_2O_4$ that has been annealed in a platinum crucible at 1050 °C for 80 h then quenched in water, has an inversion parameter, $i = 0.636$, and is close to being fully disordered [13].[15] Whilst, slow cooling (4 °C/h) results in a preference of Co^{2+} for octahedral coordination, with $i = 0.93$, and is a near-inverse spinel, $(Co^{II}_{0.07}Fe^{III}_{0.93})^{tet}[Co^{II}_{0.93}Fe^{III}_{1.07}]^{oct}O_4$ [14], [15]. This is consistent with the theory in which a zero ligand field stabilisation energy (LFSE = 0) for high-spin Fe^{3+} ($3d^5$), should favour the Co^{2+} ($3d^7$) ions occupying the octahedral sites in the spinel structure [16]. This progressive disorder–order phase transition involves the net movement of metal cations into their energetically more favoured sites, as the temperature of the material is lowered. It occurs through the rather slow process of solid-state diffusion; but there are plenty of unoccupied tetrahedral and octahedral interstices within the oxygen substructure that can provide pathways for the cation rearrangement process. Generally, as the temperature decreases, the probability of a cation hopping successfully between two adjacent sites diminishes exponentially, and therefore, a low cooling rate is essential for cation ordering to be achieved.

As in the above case for Fe_3O_4, the magnetic moments on the high-spin Fe^{3+} ($3d^5$) ions in the tetrahedral sites are coupled antiferromagnetically with the magnetic moments on both the high-spin Fe^{3+} ($3d^5$) and the high-spin Co^{2+} ($3d^7$) ions in the octahedral sites, which results in an overall ferrimagnetic behaviour for near-inverse spinel, $CoFe_2O_4$, at ambient temperature.

---

[12] A diamagnetic phase is repelled by the magnetising force, and therefore exhibits a negative magnetic susceptibility.

[13] A paramagnetic phase is attracted by the magnetising force, and therefore exhibits a positive magnetic susceptibility. But, the effect is much weaker than in a ferromagnetic or ferrimagnetic phase.

[14] In an antiferromagnetic phase, the spins are aligned in an antiparallel arrangement, such that the net magnetic moment sums up to zero, and therefore the phase *appears* to be nonmagnetic.

[15] The author is grateful to Professor Henderson for providing certain details.

Charge transfer between Co^{2+} and Fe^{3+} is apparently restricted, such that CoFe$_2$O$_4$ is an insulator, with a resistivity, $\rho \sim 10^5 \, \Omega$ m [7]. With reference to the bulk material, CoFe$_2$O$_4$ has a Curie temperature, $T_c = 793$ K [6]; a saturation magnetisation, M_s (at 0 K) $= 90$ A m^2 kg^{-1} [6], with a corresponding magnetic moment, μ (at 0 K) $= 3.7 \, \mu_B$ [6]; see Figures 8.4 and 8.5.

Since both the grain size and the inversion parameter, i, of CoFe$_2$O$_4$ are dependent on the temperature and cooling rate employed during the synthesis, it is difficult to control these factors independently of one another. Therefore, it is hard to determine the absolute effect that variations in grain size may have on the magnetic properties of CoFe$_2$O$_4$, whilst at the same time ensuring constancy in the inversion parameter, and vice versa. This problem applies to both precipitated powders and sintered ceramic material alike. Nevertheless, the material has been well studied, and it is commonly accepted that remanence and coercivity are strongly dependent on the size of the crystallites.[16] According to Wang and Ren [17], millimetre-sized single crystals of CoFe$_2$O$_4$ exhibit a negligible coercive field at ambient temperature. Whereas, Liu et al. [18], show that CoFe$_2$O$_4$ with a grain size of 120 nm has a coercive field, H_c (300 K) $= 453$ Oe. Whilst, Chinnasamy et al. [19] report a coercivity, H_c (300 K) $= 4.65$ kOe for a powder specimen of (Co$^{II}_{0.25}$-Fe$^{III}_{0.75}$)^{tet}[Co$^{II}_{0.75}$Fe$^{III}_{1.25}$]^{oct}O$_4$ with a grain size of 40 nm. In fact, these crystallites are so small that they are considered to be single magnetic domains. But when the crystallite size shrinks to < 5 nm, CoFe$_2$O$_4$ becomes superparamagnetic[17] [20], and is then useless as a permanent magnet.

In general, CoFe$_2$O$_4$ has a lower Curie temperature and a lower saturation magnetisation than Fe$_3$O$_4$, but a significantly higher remanence and coercivity. It is the high value of the remanence (M_r (298 K) ~ 35 A m^2 kg^{-1} [18]), that enables CoFe$_2$O$_4$ to act as a permanent magnet. The microstructure is an important aspect for optimising the performance of CoFe$_2$O$_4$, as indeed, for all ferrimagnetic and ferromagnetic materials. The reader is referred to the monographs by McCurrie [21] and Valenzuela [22] for further information on the subject of ceramic magnets.

One of the earliest descriptions for the preparation of CoFe$_2$O$_4$ was given by Holgersson in 1927 [23]. This involved the reaction between CoO and Fe$_2$O$_3$ in the presence of molten KCl to yield a strongly magnetic black powder. Analysis by X-ray diffractometry indicated that CoFe$_2$O$_4$ crystallised with the spinel structure with the cell constant, $a = 8.395 \pm 0.005$ Å. In 1929, Natta and Passerini [3] prepared the spinel, CoFe$_2$O$_4$ by calcining a stoichiometric mixture of Co(NO$_3$)$_2$ and Fe(NO$_3$)$_3$ at $\sim$950 °C.

---

[16] Saturation magnetisation, M_s, is generally independent of particle size, since it is essentially a thermodynamic property; cf. heat capacity, etc.

[17] The phenomenon of *superparamagnetism* arises in magnetic materials that comprise very small crystallites (typically < 10 nm), such that the orientations of the crystallites become randomised by the ambient thermal energy. In other words, these materials behave as if they are paramagnetic, at temperatures below the Curie (or Néel) temperature.

In 1933, Kato and Takei [24] discovered that a solid solution of magnetite and cobalt ferrite, Fe$_3$O$_4$·3CoFe$_2$O$_4$ (i.e., Co$_{0.75}$Fe$_{2.25}$O$_4$) was strongly magnetised at 300 °C. The material was prepared through the precipitation of a mixed aqueous solution of Co(NO$_3$)$_2$ and Fe(NO$_3$)$_3$ with alkali, followed by heating the precipitate at 600 °C. In 1935, Kato and Takei [24] patented this material for use as a metal oxide magnet [25], and it was marketed subsequently in the United States under the name, 'Vectolite'. A few technical details of Vectolite are given by Finke [26]; for example, that a magnetising force of 3 kOe is required for complete magnetisation of this material.

Preparative Procedure

The method described in this chapter for the preparation of near-inverse spinel CoFe$_2$O$_4$ involves the coprecipitation of CoFe$_2$O$_4$ from an aqueous solution of cobalt(II) chloride and iron(III) chloride. CoFe$_2$O$_4$ precipitates directly or, at least, without any long-lived iron oxide hydroxide or cobalt hydroxide intermediates. The predominance of CoFe$_2$O$_4$ as a thermodynamically stable crystalline phase in coexistence with an aqueous solution of neutral–basic pH at 80 °C is evident in the E_H–pH diagram,[18] as shown in Figure 8.6.

Cobalt is a toxic element, and sodium hydroxide is hazardous to the eyes. Submicrometre particles of CoFe$_2$O$_4$ as produced in this work are potentially hazardous to the skin and lungs, and so the inhalation and contact with these particles must be avoided. Therefore, eye protection, rubber gloves and an appropriate dust mask, must be used when handling these chemicals. The work should be carried out in a fume cupboard because of the hazards associated with toxic dust.

Prepare 20 g of CoFe$_2$O$_4$ by the following method: Weigh appropriate[19] amounts of CoCl$_2$·6H$_2$O and FeCl$_3$·6H$_2$O to a precision of ±0.001 g. Dissolve these salts together in 200 ml of distilled water in a 500 ml glass beaker. In a separate 1-l glass beaker, slowly add an appropriate quantity of sodium hydroxide pellets to 400 ml of distilled water in order to obtain a 2 mol dm^{-3} NaOH(aq) solution. Heat this solution to 80 ± 5 °C on a magnetic hot plate with a magnetic stirrer bead. With the use of a glass burette (and a small glass funnel placed at its top) add the above salt solution drop-wise, into the hot sodium hydroxide solution whilst stirring magnetically. This process should be controlled manually so as to take about 45 min.

---

[18] E_H–pH diagrams are analogous to Pourbaix diagrams, and are useful for visualising the stability domain of solid phases in aqueous systems that are affected by variations in E_H and pH. The reader is referred to Pourbaix [27] for further details.
[19] The reader may wish to consider *Problem 1* before embarking on these calculations.

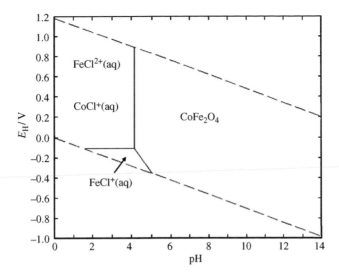

Figure 8.6 E_H–pH diagram for the Fe-Co-Cl aqueous system at 80 °C. Molalities of aqueous cobalt and iron species $= 10^{-3}$ mol kg^{-3}. Molalities of aqueous chlorine species $= 1$ mol kg^{-3}. The blue lines in dash indicate the upper and lower stability limits for H$_2$O. This illustration was drawn using Outokumpu HSC Chemistry$^{®}$ Software together with thermodynamic data from the database included in the software

Then, switch off the heater in the hot plate and place a large watch-glass over the mouth of the glass beaker and leave the contents to cool overnight, with the magnetic stirrer left on.

On the following morning, switch off the magnetic stirrer and remove the magnetic bead. Pour the contents of the beaker into a 1-l plastic beaker. Since a magnetic field should accelerate the settling of the precipitate, the physical separation of the CoFe$_2$O$_4$ precipitate from the aqueous phase can be enhanced by placing the plastic beaker (with its contents) on top of a large Nd$_2$Fe$_{14}$B (NEOMAX) magnet. You must avoid using a glass beaker for this purpose, since the sudden movements of the NEOMAX magnet can easily crack the glass beaker, on account of the strong attractive magnetic forces experienced here.

Measure the pH of the solution; which should be > 12 with pH indicator paper (range: 11–13) or a glass electrode. Decant the clear liquor carefully, using a small (~ 100 ml) glass beaker and a bulb pipette if necessary. Dispose of the liquor by carefully pouring it down the sink with cold running water. Add appropriate quantities (~ 200 ml) of distilled water to the precipitate; agitate gently, then leave to settle, and decant as before. It is important to wash the precipitate thoroughly in order to remove the undesirable contaminants, i.e., NaOH and NaCl. If these are not removed, then they will degrade the appearance of the ceramic magnet.[20]

Once the wash liquor has a pH $\leq$ 8, decant the remaining liquor, wash the precipitate finally with a small quantity ($\sim$ 100 ml) of acetone, then decant, and leave the precipitate to dry at ambient temperature in the plastic beaker at the back of the fume cupboard. Retain a small sample ($\sim$1 g) of the dried precipitate for analysis by powder XRD (scan: 10–70° 2θ). Compare your pattern with the PDF 22-1086 for CoFe$_2$O$_4$.

Place a sufficient quantity of this powder (4–5 g) into a silicon rubber mould (14 mm ID), and introduce this into a condom, eliminate the air and seal with a tight knot. If time permits, a second ceramic monolith should be prepared likewise, in order to produce a pair of magnets. Then place these in a hydraulic press and apply an isostatic pressure of 15 kbar for 10 min at ambient temperature.[21] Spread the remaining powdered product in a small alumina crucible (CC47 Almath Ltd) so as to form a sacrificial powder bed. Place the *green* monoliths on top of this powder bed. Place the alumina crucible in the pottery kiln and heat to 1000 °C at 200 °C/h. After 10 h at 1000 °C cool to ambient temperature (20 °C) with a slow cooling rate of 10 °C/h. This heating cycle will last (113 h) nearly five days.

Before you proceed any further, you should confirm that your CoFe$_2$O$_4$ ceramic monoliths can neither attract nor repel each other, before you attempt to magnetise them. In freshly prepared material, such as this, the magnetic domains are oriented randomly, therefore the magnetic field that is generated by the material, is effectively zero, or at best, extremely weak, and so the material is referred to as being in its *unmagnetised* state.

The CoFe$_2$O$_4$ monoliths can now be magnetised by exposing them to a strong external magnetic field. This is best achieved by introducing the material into a solenoid with an external field greater than 3 kOe [26]. Alternatively, the material can be more simply, but less effectively, magnetised by bringing it into contact with a Nd$_2$Fe$_{14}$B (NEOMAX) magnet. In either case, provided that the external field is sufficiently strong, then the net magnetic domains within the material will rotate so as to align with the external field. Once all the domains are fully aligned, then no further magnetisation is possible; thus corresponding to the saturation magnetisation, M_s (see Figures 8.4 and 8.5). This process is to some extent irreversible in

---

[20] The alternative approach is to filter the precipitate using a Buchner filter funnel, with a medium-speed filter paper (e.g., *Whatman No. 2 Qualitative*). The initial powder bed will soon become quite compacted, and so will act as a filter in its own right. Therefore, a reasonably high degree of suction and plenty of time (1–2 h) is necessary for this operation. The powder bed should be washed with distilled water, followed by acetone, and then left to dry overnight, at ambient temperature, whilst still attached to the filter paper.

[21] If a hydraulic press is unavailable, then use a 32-mm diameter Specac stainless steel die and a uniaxial press in order to form a disc of $\sim$15 g compacted CoFe$_2$O$_4$ powder under a load of $\sim$10 tonne.

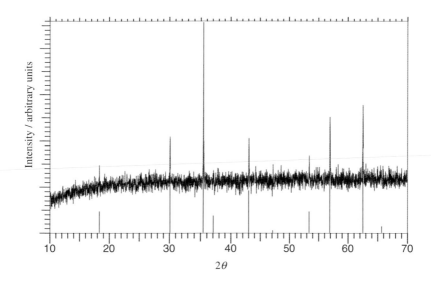

Figure 8.7 Powder XRD pattern (Cu-$K_{\alpha 1}$ radiation) of powdered ceramic $CoFe_2O_4$ as prepared by the method described in this chapter. PDF 22-1086 for $CoFe_2O_4$ is shown green for comparison

$CoFe_2O_4$, such that the material retains a significant amount of its magnetisation when it is removed from the external field; yielding a permanent magnet. A simple qualitative test of the remanence of your $CoFe_2O_4$ magnets is to see if they now have the ability to attract, and repel, each other.

Submit a powder sample ($\sim$1 g) of the sintered powder bed for analysis by powder XRD (scan: 10–70° 2θ). Compare your pattern with the PDF 22-1086 for $CoFe_2O_4$ (see Figure 8.7). The reader should bear in mind that powder X-ray diffractometry can not resolve the ordering of cobalt and iron among the tetrahedral and octahedral sites within $CoFe_2O_4$. This information can, in principle, be obtained through analysing the material with XANES, EXAFS or, [57]Fe-Mössbauer spectroscopy.

REFERENCES

1. M. Sugihara, H. Sato, A. Izaka, S. Kuroda and M. Saito, *Magnetic Ferrites*, Japanese Patent (1973) 48030758 19730922
2. F. K. Lotgering, P. H. G. M. Vromans and M. A. H. Huyberts, *Journal of Applied Physics* **51** (1980) 5913–5918.
3. G. Natta and L. Passerini, *Gazzetta Chimica Italiana* **59** (1929) 280–288.
4. S. M. Yunus, H. Yamauchi, A. K. M. Zakaria, N. Igawa, and A. Hoshikawa, Y. Ishii, *Journal of Alloys and Compounds* **454** (2008) 10–15.
5. A. R. West, *Basic Solid State Chemistry*, 2nd edn, 1999, Wiley, Chichester, 480 pp.
6. J. Smit and H. P. J. Wijn, *Ferrites*, Wiley, New York, 1959, 369 pp.

7. A. F. Orchard, *Magnetochemistry*, Oxford University Press, Oxford, 2003, 176 pp.

8. J. B. Goodenough *The Verwey Transition Revisited*, 413–425 in *Mixed-Valence Compounds: Theory and Applications in Chemistry, Physics, Geology, and Biology*. Proceedings of the NATO Advanced Study Institute held at Oxford, England, September 9–21, 1979, ed. D. B. Brown, D. Reidel Publishing Company, Boston Massachusetts.

9. J. Bostock and H. T. Riley (Trans. and eds.), *Natural History: Pliny the Elder*, Book XXXVI, Chapter 25, Taylor and Francis, London, 1855.

10. J. D. Livingston, *Driving Force: The Natural Magic of Magnets*, Harvard University Press, 1997, 334 pp.

11. W. L. Roth, *Physics and Chemistry of Solids* **25** (1964) 1–10.

12. T. A. S. Ferreira, J. C. Waerenborgh, M. H. R. M. Mendonça, M. R. Nunes and F. M. Costa, *Solid State Sciences* **5** (2003) 383–392.

13. C. M. B. Henderson, J. M. Charnock and D. A. Plant, *Journal of Physics: Condensed Matter* **19** (2007) 076214/1–076214/25.

14. G. A. Sawatzky, F. van der Woude, and A. H. Morrish, *Journal of Applied Physics* **39** (1968) 1204–1206.

15. G. A. Sawatzky, F. van der Woude and A. H. Morrish, *Physical Review* **187** (1969) 747–757.

16. C. Housecroft and A. G. Sharpe, *Inorganic Chemistry*, 3rd edn, Prentice Hall, London, 2008, 1098 pp.

17. W. H. Wang and X. Ren, *Journal of Crystal Growth* **289** (2006) 605–608.

18. X.-M. Liu, S.Y. Fu, H.M. Xiao and C.-J. Huang, *Physica B* **370** (2005) 14–21.

19. C. N. Chinnasamy, B. Jeyadevan, K. Shinoda, K. Tohji, D. J. Djayaprawira, M. Takahashi, R. Justin Joseyphus and A. Narayanasamy, *Applied Physics Letters* **83** (2003) 2862–2864.

20. K. E. Mooney, J. A. Nelson and M. J. Wagner, *Chemistry of Materials* **16** (2004) 3155–3161.

21. R. A. McCurrie, *Ferromagnetic Materials: Structure and Properties*, Academic Press, San Diego, 1994, 297 pp.

22. R. Valenzuela, *Magnetic Ceramics*, Cambridge University Press, Cambridge, 1994, 312 pp.

23. S. Holgersson, *Kungliga Fysiografiska Sällskapets Handlingar N. F.* **38** (1927) 1–112.

24. Y. Kato and T. Takei, *Journal of the Institute of Electrical Engineers of Japan* **53** (1933) 408–412.

25. Y. Kato and T. Takei, *Metallic Oxide Magnet*, Japanese Patent 110822 19350517 (1935).

26. H. E. Finke, *Materials and Methods* **25** (1947) 72–76.

27. M. Pourbaix, *Atlas of Electrochemical Equilibria in Aqueous Solutions*, Pergamon Press, Oxford, 1966, 644 pp.

PROBLEMS

1. Write a balanced chemical equation to describe the coprecipitation of CoFe$_2$O$_4$ from an aqueous solution of cobalt(II) chloride and iron(III) chloride as described in this chapter.

2. State the minimum number of moles of NaOH required for the above process in order to prepare 20 g of $CoFe_2O_4$.

3. Comment on the actual number of moles of NaOH used in your preparation.

4. In this preparative procedure, the salt solution was added to the sodium hydroxide solution. Would it make any difference if these were added the other way round?

5. Fe_3O_4 is unstable in air at 1000 °C with respect to hematite, Fe_2O_3. Therefore, offer an explanation as to why $CoFe_2O_4$ is apparently stable under these conditions.

6. Explain why a slow cooling rate was used during the annealing of the ceramic $CoFe_2O_4$.

7. What influence, if any, does the inversion parameter, i, (see Footnote 6) have on the remanence and coercivity of $CoFe_2O_4$.

8. $CoFe_2O_4$ exhibits magnetocrystalline anisotropy. Explain briefly, the meaning of this term, and search the literature for information regarding the preferred or 'easy' direction (with respect to the unit cell in Figure 8.1) for the magnetisation of $CoFe_2O_4$.

9. $CoFe_2O_4$ exhibits magnetostriction. Explain briefly, the meaning of this term, and search the literature for quantitative information regarding the magnetostriction of $CoFe_2O_4$.

10. Originally, lodestone was used to magnetise the steel needles as used in magnetic compasses. In this chapter, a $Nd_2Fe_{14}B$ (NEOMAX) magnet is used to magnetise ceramic $CoFe_2O_4$. Explain how it was possible in the years prior to the discovery of electromagnetism by Ørsted in 1820, for one to make a permanent magnet with a *higher* remanence than lodestone itself?

9

Lead Zirconate Titanate $PbZr_{0.52}Ti_{0.48}O_3$ by a Coprecipitation Method Followed by Calcination

The *Second World War* drove the United States, Japan, and the Soviet Union to discover independently, the ferroelectric[1] properties of ceramic barium titanate, $BaTiO_3$ for use as a piezoelectric[2] material in sonar,[3] and as a

---

[1] Ferroelectric materials are a subclass of piezoelectric materials that exhibit spontaneous electric polarisation and can remain polarised when the applied electric field is removed. Moreover, and by definition, the polarity of the polarisation can be reversed by applying an electric field of the opposite polarity. The expression *ferroelectric* is derived from the analogous expression, ferromagnetic (cf. Chapter 8). This analogy, however, should not be taken too literally, since ferromagnetism is due to magnetic dipoles on individual atoms, whereas the polarity of ferroelectric materials resides in the crystal structure as a whole; see Jaffe for further details [1].

[2] The expression *piezoelectric* is derived from the Greek word *piezo* for pressure, and was first suggested by W. Hankel in 1881. Piezoelectricity (or the piezoelectric effect) refers to the electric polarisation ($P/C\,m^{-2}$) arising in certain anisotropic crystals (i.e., crystals that do not possess a centre of symmetry) when subject to mechanical stress ($\sigma/N\,m^{-2}$). In the converse situation, an applied electric field will produce mechanical deformation within the crystal. Piezoelectric materials can therefore be considered as transducers for converting mechanical energy to electrical energy, and vice versa. The *piezoelectric effect* is also referred to as an *electromechanical response*. Piezoelectric materials are a subclass of dielectric materials.

[3] The expression *sonar* is an acronym from: 'sound, navigation and ranging'.

Synthesis, Properties and Mineralogy of Important Inorganic Materials   By Terence E. Warner
© 2011 John Wiley & Sons, Ltd

high-permittivity dielectric[4] medium in high farad-rated capacitors. This was followed in the early 1950s by a study of the phase relationships within the $PbZrO_3$–$PbTiO_3$ binary system, in a search for ceramic materials with enhanced dielectric, ferroelectric and piezoelectric properties. As a result of these investigations, $PbZr_{0.52}Ti_{0.48}O_3$ is generally considered to be the optimal composition for maximising the piezoelectric and dielectric properties within this binary system. Thus, $PbZr_{0.52}Ti_{0.48}O_3$ became a commercially important piezoelectric and ferroelectric material, and is known in the industry as PZT. In practice, PZT contains various dopants in order to tailor its properties for specific applications; and these materials have largely replaced $BaTiO_3$ as a piezoelectric and ferroelectric material. PZT has been exploited in film and ceramic form. Applications include piezoelectric transducers for the generation and detection of acoustic energy (e.g., sound); gramophone pick-ups; high-voltage generators, such as, gas igniters; and, ferroelectric memory devices.[5]

$PbZr_{0.52}Ti_{0.48}O_3$ is a member of the solid-solution series, $PbZr_{1-x}Ti_xO_3$ ($0 \leq x \leq 1$) and has a pale yellowish-brown coloration. This material crystallizes at high temperature with the cubic form of the perovskite structure, and undergoes tetragonal and monoclinic distortions upon cooling to room temperature. This chapter describes the preparation of a ceramic monolith of $PbZr_{0.52}Ti_{0.48}O_3$ through the coprecipitation of precursor substances from a solution of organometallic reagents (as sources for the PbO, ZrO_2 and TiO_2 components) at room temperature. This is followed by calcination, milling, compaction, and sintering processes. Finally, the ceramic monolith is polarised by applying a high electric field (typically, $60\,kV\,cm^{-1}$) across a thin slice of the monolith. But one should be aware; that it is only after this polarisation (a process known as 'poling'), that the ceramic material (a sintered aggregate of crystallites) becomes ferroelectric, and thereby, piezoelectric. These aspects will be discussed in detail later in this chapter.

The perovskite structure (including its derivatives and modifications) is one of the most important crystal structures within the subject of electroceramics. The mineral, perovskite was discovered at Yekaterinburg, in the Ural Mountains, Russia, by the German mineralogist Gustav Rose (1798–1873) in 1839, and named in honour of the Russian mineralogist Count L. A. Perovski (1792–1856) [5]. Perovskite has the ideal chemical composition, $CaTiO_3$, and crystallises with an orthorhombic unit cell. Naturally occurring material normally contains appreciable substitutions for both Ca^{2+} and Ti^{4+}, and is therefore more typically described as, (Ca,Na,

---

[4] *Dielectric* materials can withstand an applied electric field without the field being dissipated internally, i.e., they are good electrical insulators.

[5] The reader is referred to Jaffe *et al.* [1], Moulson and Herbert [2], Haertling [3] and Fujishima [4], for further information regarding the subject of piezoelectric and ferroelectric materials.

Figure 9.1   Perovskite, Gardiner Complex, Greenland. Field of view: $60 \times 90\,\text{mm}$ (Photograph used with permission from Ole Johnsen. Copyright (2010) Ole Johnsen)

Ce,Sr)(Ti,Nb)O_3 [6]. Perovskite occurs in basic igneous and contact-meta-morphic rocks, and is thought to be a major constituent of the earth's upper mantle [7] (see Figure 9.1). Although CaTiO_3 comprises Ca^{2+} and Ti^{4+} ions, the chemical bonding in this and other perovskite phases is not necessarily purely ionic. The reader is referred to the monograph by Mitchell [8] for a comprehensive survey of the enormous family of perovskite-related phases.

The high temperature PbZr$_{1-x}$Ti$_x$O$_3$ solid solution crystallises with the perovskite structure, and belongs to the cubic crystal system and space group $Pm\overline{3}m$ [9]; see Figure 9.2. The perovskite structure is very densely packed, such that calcium titanate, CaTiO_3 (which comprises comparatively light elements) has a relatively high density of $4.1\,\text{g\,cm}^{-3}$. Yet, having stated that, one way to visualise the ideal cubic ABO$_3$ perovskite structure is to consider it as a *framework* structure in which corner-sharing BO$_6$ octahedra are linked together in a regular cubic fashion. Thereby, the B-sites form a primitive cubic lattice, which in the case of PbZr$_{1-x}$Ti$_x$O$_3$, are occupied randomly by the Zr^{4+} and Ti^{4+} cations. Figure 9.2 shows a group of eight BO$_6$ octahedra, in which each octahedron is centred at one of the eight corners of the unit cell. The A-site is situated at the centre of this unit cell, within the relatively large interstice created by the surrounding octahedra. Consequently, the A-sites also form a primitive cubic lattice, and are occupied by the divalent Pb^{2+} cation with a high coordination number of 12 towards the oxide ions.

At temperatures below $387\,^{\circ}\text{C}$, PbZr$_{0.52}$Ti$_{0.48}$O$_3$ crystallises with the tetragonally distorted perovskite structure, as shown in Figure 9.3. The tetragonal cell constants are reported by Noheda *et al.* [10] as: $a = 4.0460\,\text{Å}$ and $c = 4.1394\,\text{Å}$ (at $52\,^{\circ}\text{C}$). In this tetragonal phase, the BO$_6$ octahedra are no longer regular, since the B-site is displaced from its central position (with

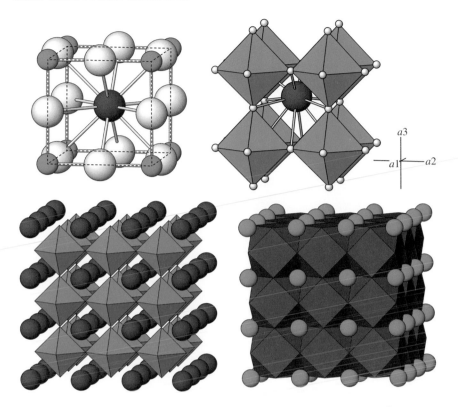

Figure 9.2 Crystal structure of cubic PbZr$_{0.52}$Ti$_{0.48}$O$_3$ at 850 K. The Pb^{2+} ions are located on the A-sites and are shown in purple-pink. The Zr^{4+} and Ti^{4+} ions occur randomly on octahedral B-sites and are shown in blue. The oxide ions are shown in white. A unit cell is outlined in the top left figure. The two figures at the bottom emphasise that the A-sites and the B-sites belong to separate primitive cubic lattices. This figure was drawn based upon data for cubic PbTiO$_3$ from Kuroiwa *et al.* [9] cf. ICSD 153406

respect to the oxygen substructure) and in the direction of one of the apical oxygen ions. For the sake of convention, this is taken towards the *c*-direction, as shown in Figure 9.3. Likewise, the Pb^{2+} ions on the A-site undergo a similar displacement in the *c*-direction. To accommodate these structural changes, there is a contraction within the basal $(a - b)$ plane resulting in a slight distortion of the oxygen substructure.

The net effect of these uniform ionic displacements gives rise to a permanent potential difference across the crystal. This may involve the entire crystal or, more commonly, a discrete part of it known as a domain; separated from an adjacent domain by a domain boundary (analogous to a crystallographic twin boundary). This phenomenon, referred to as ferroelectricity, is exploited in storage media in which the polarity of the crystal (or one of its domains) can be flipped over by an applied electric field, and can then be 'read' at a later time.

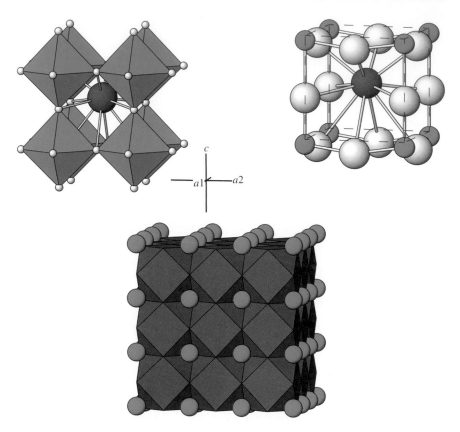

Figure 9.3 Crystal structure of tetragonal $PbZr_{0.52}Ti_{0.48}O_3$ at 325 K. The Zr^{4+} and Ti^{4+} ions occur randomly on distorted octahedral sites and are shown in blue. These ions are displaced (relative to the oxygen substructure) in the c-direction. Likewise, the Pb^{2+} ions are displaced in the c-direction, as shown in purple-pink. A unit cell is outlined in the top right figure, in which the Pb^{2+} ion lies at the centre. The figure at the bottom gives the opportunity to compare these subtle distortions with the cubic counterpart in Figure 9.2; and given the uniformity of these ionic displacements, can be regarded as a single domain. This figure was drawn using data from Noheda *et al.* [10], cf. ICSD 92059

A high-temperature equilibrium phase diagram of the $PbZrO_3$–$PbTiO_3$ binary system ($\geq 1200\,°C$) was reported by Fushimi and Ikeda in 1967 [11] and is shown in Figure 9.4. From this diagram, it can be seen that $PbZr_{1-x}Ti_xO_3$ forms a continuous solid solution below the solidus; which links the incongruent melting point of $PbZrO_3$ (1570 °C) with the congruent melting point of $PbTiO_3$ (1300 °C). Below 500 °C, however, this system is more complex, and involves several solid-solution phases with slightly dissimilar crystal structures that exhibit ferroelectric and antiferroelectric properties. In 1950 Shirane *et al.* [12, 13] reported the ferroelectric tetragonal to

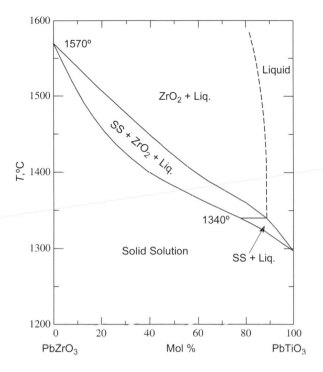

Figure 9.4 An equilibrium phase diagram of the PbTiO$_3$–PbZrO$_3$ binary system at high temperature (Reproduced with permission from Journal of the American Ceramic Society, Phase Equilibrium in the System PbO-TiO$_2$–ZrO$_2$ by S. Fushimi and T. Ikeda, 50, 3, 129–132 Copyright (1967) American Ceramic Society)

paraelectric[6] cubic phase transition in lead titanate, PbTiO$_3$ at 490 °C. In 1951 Sawaguchi *et al.* [14] reported the antiferroelectric[7] orthorhombic to para-electric cubic phase transition in lead zirconate, PbZrO$_3$ at 224 °C. Thereby establishing the antiferroelectric and ferroelectric Curie temperatures for the two end-members of the PbZrO$_3$–PbTiO$_3$ binary system, respectively.

    In 1952, Shirane and Suzuki [16] investigated the phase relationships within the PbZrO$_3$–PbTiO$_3$ binary system, and discovered the existence of a phase boundary between a rhombohedral ferroelectric phase and a tetragonal

---

[6] A *paraelectric* phase is a nonpolar phase into which ferroelectric and antiferroelectric phases transform above the Curie temperature, T_c. A material in its *paraelectric* state behaves similar to a normal dielectric material.

[7] An *antiferroelectric* material contains electric dipoles, but these dipoles are aligned in an antiparallel fashion, such that the net spontaneous polarisation is zero. The term has certain analogies to the term antiferromagnetism. It is interesting to note that although PbZrO$_3$ is antiferroelectric below the antiferroelectric Curie temperature, $T_c = 224$ °C, PbZrO$_3$ becomes ferroelectric when placed in a sufficiently strong electric field; especially at temperatures close to T_c [15].

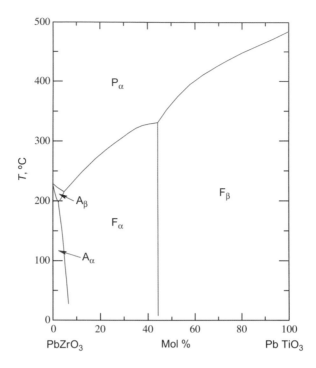

Figure 9.5 Equilibrium phase diagram of the $PbTiO_3$–$PbZrO_3$ binary system. $P\alpha$ = paraelectric, cubic phase; A_α = antiferroelectric, orthorhombic phase; A_β = antiferro-electric; F_α = ferroelectric, rhombohedral phase; F_β = ferroelectric, tetragonal phase (Reproduced with permission from Journal of the Physical Society of Japan, Ferro-electricity versus Antiferroelectricity in the Solid Solutions of $PbZrO_3$ and $PbTiO_3$ by E. Sawaguchi, 8, 5, 615–629 Copyright (1953) Physical Society of Japan)

ferroelectric phase, lying within the compositional limits, $0.40 < x < 0.45$ with respect to the formula, $PbZr_{1-x}Ti_xO_3$. In 1953, Sawaguchi [17] constructed an equilibrium phase diagram for the $PbTiO_3$–$PbZrO_3$ binary system; which is in agreement with the work by Shirane and Suzuki [16] (see Figure 9.5).[8]

Encouraged by these events in Japan, Jaffe *et al.* [18] in the United States made the important discovery in 1954 that compositions in close vicinity to the rhombohedral–tetragonal phase boundary exhibit a strong piezoelectric effect and a very high relative permittivity.[9] The principle figure of merit used to evaluate the piezoelectric properties of a ceramic disc is the radial coupling

---

[8] In order to construct this phase diagram, Sawaguchi [17] prepared various samples by reacting TiO_2, ZrO_2 and PbO together in a PbO vapour at 1200 °C. The reaction product was crushed to a fine powder compacted and sintered 1300 °C for 1 h. Volatilisation of PbO in the course of preparing specimens of $PbZr_{1-x}Ti_xO_3$ is a problem noted by most workers in this field, especially when adopting the route of using the binary metal oxides as the chemical reagents for the synthesis.

[9] The expression relative permittivity, $\varepsilon_r = \varepsilon / \varepsilon_0$ is also known as the dielectric constant.

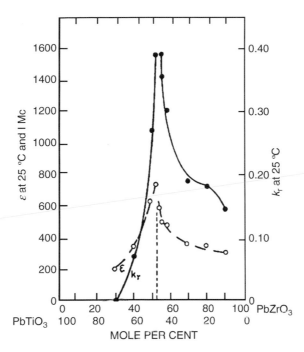

Figure 9.6 Relative permittivity, ε_r and radial coupling coefficient, k_r for PbZrO$_3$–PbTiO$_3$ solid solution ceramics at room temperature (Reproduced with permission from Journal of Research of the National Bureau of Standards, B. Jaffe, R. S. Roth, and S. Marzullo, 55, 239–254 Copyright (1955) National Bureau of Standards)

coefficient.[10] Figure 9.6 shows the dependence of the radial coupling coefficient, k_r and the relative permittivity, ε_r on composition, at room temperature. These values are enhanced in opposite directions from either side of the PbZrO$_3$–PbTiO$_3$ system, so as to converge as a maximum at the phase boundary. This discovery led eventually to the optimal composition, PbZr$_{0.52}$Ti$_{0.48}$O$_3$. In 1955, Jaffe *et al.* [19] coined the phrase, *morphotropic phase boundary* (MPB) to describe this compositional boundary, in which an abrupt change in crystal structure arises through the variation in composition of a solid solution.

Over the following years, the PbTiO$_3$–PbZrO$_3$ binary system was studied extensively by many workers, which resulted in certain improvements to the phase diagram. Gradually, it became apparent that the *morphotropic phase boundary* (MPB) is temperature dependent, and moreover, instead of being a sharp boundary, it was considered to be an immiscibility gap comprising a mixture of two phases. For many years these two phases were believed to be of rhombohedral and tetragonal symmetry; although in 1977, doubt was cast on

---

[10] The expression, *radial coupling coefficient* squared is a measure of the materials ability to convert electrical energy to mechanical energy, or, conversely, to convert mechanical energy to electrical energy [19]. With the 'radial' term referring to radial vibration of a thin disc.

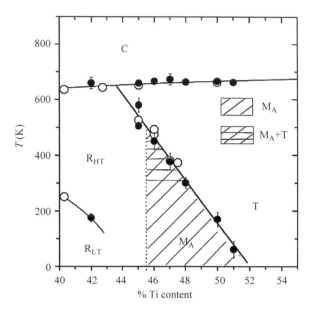

Figure 9.7 A more recent and detailed depiction of the PbTiO$_3$–PbZrO$_3$ binary system within the vicinity of the morphotrophic phase boundary. C = cubic phase; R$_{HT}$ = high-temperature rhombohedral phase; R$_{LT}$ = low-temperature rhombohedral phase; T = tetragonal phase; M$_A$ = monoclinic phase (Reproduced with permission from Physical Review B, Stability of the monoclinic phase in the ferroelectric perovskite PbZr$_{1-x}$Ti$_x$O$_3$ by B. Noheda, D. E. Cox, G. Shirane, R. Guo, B. Jones, and L. E. Cross, 63, 1, 014103 Copyright (2000) American Physical Society)

the existence of the immiscibility gap, by Kakegawa *et al.* [20]. Then, in 2000, Nohedra *et al.* [10] discovered the existence of a monoclinic phase within the MPB, which was found to lie within the composition range: $0.455 \leq x \leq 0.48$ with respect to the formula PbZr$_{1-x}$Ti$_x$O$_3$ (at 27 °C); i.e., intermediately, between the rhombohedral and tetragonal phases (see Figure 9.7).[11] However, knowledge of the boundaries between the rhombohedral, monoclinic and tetragonal phases, still remain somewhat uncertain within this part of the equilibrium phase diagram, because of the difficulty in distinguishing between these three phases.

The structure of the PbZr$_{0.52}$Ti$_{0.48}$O$_3$ monoclinic phase can be regarded as a slightly distorted modification of the parent tetragonal phase [21]; as shown in Figure 9.8. The monoclinic unit cell, however, is twice as massive as the tetragonal cell, and has the *b*-axis as the unique axis (oriented in the pseudocubic [1$\bar{1}$0] direction). The monoclinic angle, $\beta = 90.5°$ is only slightly

---

[11] Nohedra *et al.* [10] came to this conclusion after analysing various compositions using high-resolution synchrotron X-ray powder diffractometry.

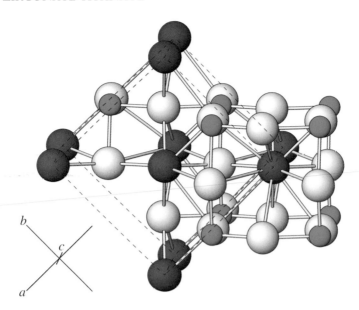

Figure 9.8   Crystal structure of monoclinic $PbZr_{0.52}Ti_{0.48}O_3$ at 20 K. The Zr^{4+} and Ti^{4+} ions are shown in blue. The Pb^{2+} ions are shown in purple-pink. The monoclinic unit cell with $Z = 2$ (shown in dash and corresponding to the axes) is shown in conjunction with an extended part of the structure toward the front-right that corresponds to the pseudocubic cell. This figure was drawn using data from Noheda et al. [10], cf. ICSD 92061

larger than that for an orthogonal system and is hardly discernible in illustrations of the unit cell. Very importantly, the ferroelectric polarisation in the monoclinic phase is no longer constrained by symmetry to lie along a symmetry axis (as it is in the tetragonal phase), but can be directed in any direction within the monoclinic $a - c$ plane; i.e., the polar axis is free to rotate within this plane [21]. The monoclinic phase, therefore, is instrumental for the high piezoelectric response in PZT [22]. According to data reported by Nohedra et al. [10] the composition of interest, $PbZr_{0.52}Ti_{0.48}O_3$ lies on the monoclinic–tetragonal phase boundary at 27 °C (see Figure 9.7). Therefore, at room temperature (i.e., 20 °C) the composition $PbZr_{0.52}Ti_{0.48}O_3$ is most probably monoclinic (see Figure 9.12).[12]

In summary, the composition $PbZr_{0.52}Ti_{0.48}O_3$ has an incongruent melting point of 1380 °C. Below 1380 °C, $PbZr_{0.52}Ti_{0.48}O_3$ crystallises with a cubic structure. The cubic symmetry prevents it from undergoing spontaneous polarisation, and hence this material is paraelectric. $PbZr_{0.52}Ti_{0.48}O_3$ undergoes a

---

[12] Interestingly, the powder X-ray diffraction pattern shown in Figure 9.12 gives the impression of an additional peak in between, $d_{011} = 30.88°\ 2\theta$ and $d_{110} = 31.24°\ 2\theta$. If true, this would be consistent with the monoclinic cell discussed above. Hence these peaks might perhaps be more appropriately indexed according to the monoclinic unit cell as: $d_{11-1} = 30.84°\ 2\theta$; $d_{111} = 31.03°\ 2\theta$; $d_{200} = 31.26°\ 2\theta$; and $d_{020} = 31.33°\ 2\theta$.

phase transition[13] from paraelectric cubic to ferroelectric tetragonal upon cooling below 387 °C; corresponding to the ferroelectric Curie temperature. The ferroelectric tetragonal $PbZr_{0.52}Ti_{0.48}O_3$ phase remains stable between 387 and 27 °C, and is completely miscible with the isostructural end-member, $PbTiO_3$ throughout this temperature range. Upon cooling below 27 °C, $PbZr_{0.52}Ti_{0.48}O_3$ undergoes a transition from a tetragonally distorted perovskite phase to a monoclinically distorted perovskite phase. The effects of pressure and temperature on the phase relationships within the $PbTiO_3$–$PbZrO_3$ binary system have been reported by Porat et al. [23]; although this diagram predates the discovery of the monoclinic phase (see Figure 9.9).

$PbZr_{0.52}Ti_{0.48}O_3$ has a very high relative permittivity, $\varepsilon_r > 2000$, and therefore, is used as the dielectric medium in ceramic capacitors. But the prime commercial interests in this material lie in its piezoelectric and ferroelectric properties. Each individual crystal of tetragonal or, monoclinic $PbZr_{0.52}Ti_{0.48}O_3$ is ferroelectric; by its very nature. For instance, in the case of the tetragonal $PbZr_{0.52}Ti_{0.48}O_3$ phase, there exists a spontaneous and orderly displacement of all the positive ions towards the c-direction of the crystal. But it is important to appreciate that a polycrystalline ceramic body of $PbZr_{0.52}Ti_{0.48}O_3$ comprises a sintered mass of randomly oriented ferroelectric crystallites (whether they be tetragonal or monoclinic), and therefore one should expect the ceramic body to be isotropic in its physical properties on a macroscopic scale, and therefore, completely useless as either a ferroelectric or piezoelectric material. Fortunately, a process called 'poling' allows for the as-near-as-possible parallel alignment of the polar axes associated with each of the crystallites (i.e., the c-axis in the case of the tetragonal phase) in the direction of the applied static electric field.[14] After poling, the ceramic body should now retain its induced anisotropic physical properties (i.e., the mass of crystallites should remain co-oriented), so that it can now be used as a ferroelectric, and thereby, piezoelectric material. For example, the piezoelectric response of poled ceramic PZT is sensitive enough that when used in a gramophone pick-up; the mere force of 0.01–0.05 N available within the record groove is sufficient to induce an electric potential across the PZT ceramic of between 0.1–0.5 V [1].

---

[13] This phase transition is generally considered to be first order at ambient pressure.

[14] The specimen of PZT that was used to obtain the ferroelectric hysteresis loop, as shown in Figure 9.10, was polarised by corona poling. During corona poling, a high electric field is induced across the ceramic specimen, by ensuring that one side of the ceramic specimen is earthed, and then bringing an array of needles very close to the opposite surface, and applying a potential in the order of 10 kV. Corona poling has an advantage in that it offers a lower risk of inducing an electrical breakdown within the ceramic specimen [2]. An alternative poling method involves first of all applying silver electrodes to the opposite faces of the ceramic specimen (using silver paint or by a silver sputtering technique). The ceramic specimen is then immersed in transformer oil and poled under a direct current field of 20 kV cm^{-1} at approximately 125 °C for 10 min, and then cooled to near room temperature whilst maintaining the applied electric field [2, 22].

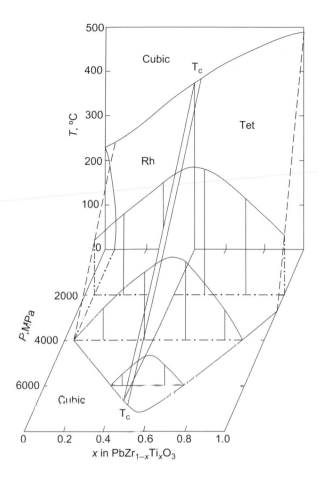

Figure 9.9 Calculated pressure and temperature dependence of the morphotrophic phase boundary in the PbTiO$_3$–PbZrO$_3$ binary system. Tet = tetragonal phase; Rh = rhombohedral phase. The depiction of the morphotrophic phase boundary in this diagram shares a slight resemblance to one of Escher's illusions (Reproduced with permission from Ferroelectrics, Concentration-pressure-temperature phase diagram of PZT by Y. Porat, Y. Imry, A. Aharony *et al.*, 37, 1, 591–594 Copyright (1981) Taylor & Francis.)

Guo *et al.* [22] have demonstrated that high-resolution synchrotron X-ray diffractometry can be used to distinguish poled from unpoled ceramic PbZr$_{0.52}$Ti$_{0.48}$O$_3$. For instance, the relative intensities of the (002) and (200) reflections in a *ceramic* specimen of tetragonal PbZr$_{0.52}$Ti$_{0.48}$O$_3$ switch very distinctively in magnitude from, 1 : 2 in the unpoled state to, 2 : 1 in the poled state, respectively.

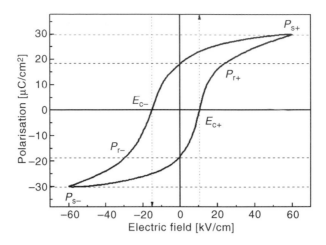

Figure 9.10 Ferroelectric hysteresis loop for a ceramic monolith of $PbZr_{0.52}Ti_{0.48}O_3$ (11 mm diameter and 600 μm thick) as prepared by the method described in this chapter (Reproduced courtesy of Professor Dr Mohammed Es-Souni)

The ferroelectric hysteresis loop for a poled ceramic monolith of $PbZr_{0.52}Ti_{0.48}O_3$ at room temperature is shown in Figure 9.10.[15] In general, the ferroelectric hysteresis loop gives useful information as to how effectively a material can be polarised; and how well it can retain its polarisation after removal, or reversing the polarity, of the applied electric field at a given temperature. The area inside the loop is a measure of the loss in energy density (ideally, in units of $J\,m^{-3}$) per cycle. The general features of the hysteresis loop are analogous, in many ways, to the ferromagnetic hysteresis loop for ferromagnetic (and ferrimagnetic) materials as discussed in Chapter 8.

From Figure 9.10 it can be seen that the saturation polarisation, $P_s = 30$ $\mu C\,cm^{-2}$; the remnant polarisation, $P_r = 18.5\,\mu C\,cm^{-2}$; and the average value[16] for the coercive field, $E_c = 12.8\,kV\,cm^{-1}$. These values are comparable to those reported by Sahoo et al. [24]: $P_s = 17.0\,\mu C\,cm^{-2}$; $P_r = 11.5\,\mu C\,cm^{-2}$; and $E_c = 12.6\,kV\,cm^{-1}$; for $PbZr_{0.52}Ti_{0.48}O_3$ at room temperature; and to those reported by Sharma et al. [25]: $P_s = 38\,\mu C\,cm^{-2}$; $P_r = 32\,\mu C\,cm^{-2}$; and $E_c = 12\,kV\,cm^{-1}$; for their specimen of PZT. It is generally known

---

[15] The present author is grateful to Professor Dr Mohammed Es-Souni, Kiel University of Applied Science, Germany, for kindly preparing the poled specimen and measuring these data. This sample was prepared by cutting a slice from the original specimen (cf. Figure 9.11). This was then ground to a thickness of 600 μm, and finished with 4000 abrasive paper. The diameter of the sample was 11 mm. Silver conductive paste was applied on both faces to make a planar capacitor. The measurement was performed using a TF Analyzer 2000 (aixACCT Systems GmbH, Germany), coupled with a High Voltage Amplifier (TREK 609E-6, TREK Inc., USA) at room temperature and 100 Hz. The sample was polarised prior to the measurement by corona poling.
[16] The slight shift to negative values for coercive fields in Figure 9.10 is most likely an artefact of the measurement setup.

that both the saturation polarisation, P_s and the remnant polarisation, P_r are dependent upon the microstructure of the material, in particular, the grain size, and therefore are influenced by the method of preparation. Presumably, the monoclinic and tetragonal modifications of $PbZr_{0.52}Ti_{0.48}O_3$ also exhibit differences in the values for their P_s and P_r.

All of these aspects are plausible contributory reasons for the disparties in the above values for the P_s and P_r as well. Furthermore, the ferroelectric parameters (and piezoelectric coefficients) typically decrease exponentially with time, due to inherent instability within the domain structure of the ceramic material; a process known as 'ageing'.

Single crystals of $PbZr_{0.52}Ti_{0.48}O_3$ are reported to have a reddish-brown coloration [26]; whereas, the material in ceramic form, is described as being yellow [27]; and in powder form, as being white [28]. The pale yellowish-brown coloration of the ceramic material prepared by the method described in this chapter is consistent with these descriptions (see Figure 9.11). Shimakawa and Kubo [28] mention that the reddish-yellow coloration of massicot, β-PbO is comparable with the orange-yellow tint in one of their specimens of tetragonal $PbZr_{0.48}Ti_{0.52}O_3$; implying that β-PbO contamination might be the cause of the coloration. But even so, the cause of the coloration in PZT has not been fully explained in the literature. Furthermore, it is interesting to note that $PbZr_{0.52}Ti_{0.48}O_3$ transforms into a black n-type semiconductor upon annealing in hydrogen [27].

Figure 9.11   Ceramic monolith (11 mm diameter) showing the pale yellowish-brown coloration of $PbZr_{0.52}Ti_{0.48}O_3$ as prepared by the method described in this chapter. This specimen was, cut, ground and polished, prior to recording the measurements as shown in Figure 9.10

PZT has been prepared by a variety of methods. The original synthesis of PZT was performed by Shirane and Suzuki [16] and Sawaguchi [17] in the early 1950s at the Tokyo Institute of Technology. Their method of synthesis involved a conventional solid-state reaction between the constituent binary metal oxides: PbO; ZrO_2; and TiO_2. These were mixed together (presumably as powders) in the desired ratio, compacted, and then calcined at $1200\,°C$ under an atmosphere of lead oxide vapour. These authors noted difficulties in procuring sufficient quantities of ZrO_2 as a high-purity reagent, and there-fore, were often obliged to use an inferior grade of ZrO_2; at 98%.[17] Nevertheless, the product was crushed to a fine powder, compacted into cylindrical monoliths (30 mm diameter and 5 mm thick) and sintered at $1300\,°C$ for one hour; and resulted in a dense ceramic. As a slight adaptation of this procedure, Shimakawa and Kubo [28] synthesised powder samples of $PbZr_{0.48}Ti_{0.52}O_3$, from a mixture of $PbZrO_3$ and $PbTiO_3$.

Preparing PZT by these solid-state methods proved to be unsatisfactory for obtaining single-phase material with a precise and homogenous compo-sition. So this inspired the development of alternative methods of synthesis. For example, in 1977, Kakegawa et al. [20] prepared specimens of $PbZr_{1-x}Ti_xO_3$ with compositions near the morphotropic phase boundary (MPB) by a 'wet and dry combination technique' involving aqueous solutions of $TiCl_4$ and $ZrOCl_2$ reacted with aqueous ammonia solution at room temperature. The dried precipitate was mixed with PbO powder and fired at $1100\,°C$. However, this technique was still dependent upon the use of PbO as one of the chemical reagents. The volatility of PbO at elevated temperature is a notorious problem that commonly results in the product being deficient in its lead content. To a certain degree, this can be circumvented by performing the synthesis under an atmosphere of PbO vapour, through the deployment of discrete quantities of $PbZrO_3$ in order to maintain a sufficiently high PbO vapour pressure.

More sophisticated techniques have been developed over the years, which include sol-gel techniques [29-31]; coprecipitation methods [32-34]; and a hydrothermal method [35]. The method described here is adopted from the work by Guiffard and Troccaz [36], whom in turn were inspired by the original work of Eyraud et al. [37]. This procedure involves a coprecipita-tion method using, lead acetate, tetra-n-butyl zirconate and tetra-n-butyl titanate as the chemical reagents. As a general comment, this method has the attraction in that the chemical reagents are mixed intimately on a molecular level at an early stage of the process. The first stage of the process involves the formation of a precipitate comprising the binary metal oxalates;

---

[17] Jaffe et al. [1] pointed out that high-grade zirconia only became readily available as a chemical reagent because of its requirement in the nuclear industry. HfO_2 is a common contaminant within zirconia, but the effect of this on the ferroelectric properties of PZT is unclear to the present author.

$PbZrO(C_2O_4)_2\cdot6H_2O$ and $PbTiO(C_2O_4)_2\cdot6H_2O$. Upon their subsequent reaction with aqueous ammonia solution, this concoction yields a complex precipitate comprising; lead hydroxide, $Pb(OH)_2$; lead oxalate, $Pb(C_2O_4)$; zirconium oxyhydroxide, $ZrO(OH)_2$; and, titanium oxyhydroxide, $TiO(OH)_2$. This blend of semiamorphous phases, constitute the reactive precursor powders required for the production of PZT. This procedure yields a ceramic material of high phase purity comprising the much desired monoclinic phase, and with a homogenous and precisely defined composition. This particular method has been exploited by other workers, for example, Sharma *et al.* [25].

Preparative Procedure

Prepare 30 g of $PbZr_{0.52}Ti_{0.48}O_3$ by the following method: Dissolve 50 g (to a precision of ±0.01 g) of oxalic acid dihydrate ($H_2C_2O_4\cdot2H_2O$) in 200 ml of distilled water contained in a 500 ml glass beaker, so as to produce ~ 2 mol dm^{-3} solution of oxalic acid. Heat gently ($\sim 50\,°C$) on a hot plate with a magnetic stirrer bead until the acid is completely dissolved.

Meanwhile, weigh[18] the appropriate amounts (to a precision of ±0.01 g) of tetra-*n*-butyl zirconate $Zr(OC_4H_9)_4$ (remembering that this is normally supplied as a 80% (wt/wt) solution in 1-butanol) and tetra-*n*-butyl titanate $Ti(OC_4H_9)_4$ (as liquid) directly into a 1-l conical glass flask using a bulb pipette. Insert a magnetic stirrer bead and stir gently at room temperature on a magnetic stirrer plate. Then pour the above oxalic acid solution slowly into this flask and continue to stir the reaction mixture for 1 h.

In the meantime, weigh the appropriate amount (to a precision of ±0.001 g) of lead acetate trihydrate ($CH_3COO)_2Pb\cdot3H_2O$ and place this in a 250-ml glass beaker. Add ~ 100 ml of distilled water so as to dissolve the lead acetate. Lead acetate and the lead compounds produced in this work are toxic, so take appropriate precautions when handling them.

After the one hour has elapsed, pour this acetate solution slowly into the above 1-l conical glass flask containing the reaction mixture of tetra-*n*-butyl zirconate, tetra-*n*-butyl titanate and oxalic acid solution and continue to stir magnetically. Rinse the 250-ml glass beaker with a small amount of distilled water and add this fluid to the conical flask. It is important to ensure that all of the lead acetate solution enters the reaction flask, or else the final product material will be deficient in lead oxide. Measure the pH of the reaction mixture with a glass electrode and pH meter (or with pH indicator paper in the range 0–2.5); the pH should be ~ 1. After stirring for 1 h, add 200 ml of 24% (wt/wt) aqueous ammonia solution to the reaction mixture. Measure the pH of the reaction mixture with pH indicator paper (range 7.5–14); the pH

---

[18] Using for example, a Mettler PM2000 balance. Or, more ideally, to a greater degree of precision using a balance that can tolerate the weight of the conical flask.

should be ∼ 10. Place a stretched sheet of *parafilm*™ over the mouth of the conical flask and leave the reaction mixture overnight with gentle stirring at room temperature. The reaction mixture can be left for longer periods, if the reader has other commitments to attend.

The contents of the conical flask should then be filtered using a Buchner filter funnel with *Advantec 5B* 110 mm filter paper (or a similar filter paper with a slow filtration rate). The filtration may take 1–2 h. The powder-cake should be washed several times with distilled water so as to remove the oxalate anions, and then finally with acetone. Place the powder-cake in an evaporating dish and leave on a hot plate (low-medium heat) for a few minutes in order to evaporate the acetone and produce a dry powder. If necessary, grind the dried material with a porcelain pestle and mortar so as to produce a loose and homogenous powder. Note the colour of the powder. Submit ∼ 3 g of this powder for analysis by powder XRD (scan: 10–60° 2θ). Place the rest of the powder in a large alumina crucible (CC62 Almath Ltd) along with its lid. Heat in a pottery furnace (or a chamber furnace) at 100 °C/h to 700 °C. After 10 h at 700 °C cool to room temperature (20 °C) with a cooling rate of 100 °C/h. This heating cycle will last ∼ 24 h.

After the calcination, grind the material with a porcelain pestle and mortar so as to produce a loose and homogenous powder. Note the colour of the powder. Place a sufficient amount (∼5 g per mould) of the calcined powder into two silicone rubber moulds and insert each of these moulds inside a separate condom; exclude most of the air and tie a knot. These should then be pressed isostatically in the hydraulic press under a pressure of ∼15 kbar for

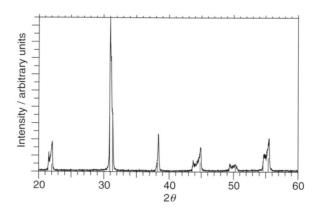

Figure 9.12   Powder XRD pattern (Cu-$K_{\alpha 1}$ radiation) of PbZr$_{0.52}$Ti$_{0.48}$O$_3$ as prepared by the method described in this chapter. This powder pattern corresponds closely to the PDF 33-784 for tetragonal PbZr$_{0.52}$Ti$_{0.48}$O$_3$ (shown in red), except for the red line at 53.3° 2θ; which is most likely incorrect. This red line should perhaps be moved to correspond with the d_{120} ∼ 50.43° 2θ; which is evidently unresolved from the d_{201} in the original PDF 33-784. The reader is advised to consult crystallographic data regarding the tetragonal phase (ICSD 92059) and the monoclinic phase (ICSD 92061) for a more accurate comparison

5 min, so as to form two compacted powder monoliths. Place a thin layer of lead zirconate ($PbZrO_3$) in the small conically tapered alumina crucible (BS67 Almath Ltd) – which is dedicated solely for this sintering process – and place the *green* monoliths on top of this sacrificial powder bed. Place this crucible with its alumina lid inside a larger one (CC62 Almath Ltd) also with its lid. Heat in a chamber furnace at 100 °C/h to 1200 °C. After 3 h at 1200 °C cool to ambient temperature (20 °C) with a cooling rate of 100 °C/h. This heating cycle will last ∼ 27 h.

After sintering, note the colour of the ceramic monolith. Crush one of the monoliths in a steel crusher and grind to a fine powder using a porcelain pestle and mortar. Submit this powder for analysis by powder X-ray diffraction (scan: 20–60° 2θ). Compare the powder pattern with the PDF 33-784 for $PbZr_{0.52}Ti_{0.48}O_3$ in conjunction with Figure 9.12.

In order to induce a permanent dipole, the $PbZr_{0.48}Ti_{0.52}O_3$ ceramic monolith needs to be poled. A summary of this process is given in the *footnote 14*. The reader should be aware of the dangers associated with high-voltage electric fields, and should only use equipment that is specifically designed and built for the purpose of poling ceramic materials of this kind.

REFERENCES

1. B. Jaffe, W. R. Cook and H. Jaffe, *Piezoelectric Ceramics*, Academic Press, London and New York, 1971.
2. A. J. Moulson and J. M. Herbert, *Electroceramics*, 2nd edn, 1993, Wiley, Chichester, 557pp.
3. G. H. Haertling, *Piezoelectric and Electrooptic Ceramics*, 129–205 in *Ceramic Materials for Electronics*, ed. R. C. Buchanan, Second Edition, Marcel Dekker, Inc., 1991, New York.
4. S. Fujishima, *IEEE Transactions on Ultrasonics, Ferroelectrics, and Frequency Control*, 47 (2000) 1–7.
5. J. W. Anthony, R. A. Bideaux, K. W. Bladh and M. C. Nichols, *Handbook of Mineralogy*: Volume III, *Halides, Hydroxides, Oxides*, Mineral Data Publishing, Tucsun, Arizona, 1997, 628pp.
6. W. A. Deer, R. A. Howie and J. Zussman, *An Introduction to the Rock-Forming Minerals*, 2nd edn, Longman, Harlow, Essex, 1992.
7. R. V. Gaines, H. C. W. Skinner, E. E. Foord, B. Mason and A. Rosenzweig, *Dana's New Mineralogy: The System of Mineralogy of James Dwight Dana and Edward Salisbury Dana*, 8th edn, 1997, Wiley-Blackwell, New York.
8. R. H. Mitchell, *Perovskites: Modern and Ancient*, Almaz Press, Ontario, 2002, pp 318.
9. Y. Kuroiwa, Y. Terado, S. J. Kim, A. Sawada, Y. Yamamura, S. Aoyagi, E. Nishibori, M. Sakata and M. Takata, *Japanese Journal of Applied Physics* 44 (2005) 7151–7155.
10. B. Noheda, J. A. Gonzalo, L. E. Cross, R. Guo, S.-E. Park, D. E. Cox and G. Shirane, *Physical Review B* 61 (2000) 8687–8695.
11. S. Fushimi and T. Ikeda, *Journal of the American Ceramic Society* 50 (1967) 129–132.
12. G. Shirane, S. Hoshino and K. Suzuki, *Physical Review* 80 (1950) 1105–1106.
13. G. Shirane, S. Hoshino and K. Suzuki, *Journal of the Physical Society of Japan* 5 (1950) 1105–1106.

14. E. Sawaguchi, G. Shirane and Y. Takagi, *Journal of the Physical Society of Japan* **6** (1951) 333–339.
15. A. R. West, *Basic Solid State Chemistry*, 2nd edn, 1999, Wiley, Chichester, 480pp.
16. G. Shirane and K. Suzuki, *Journal of the Physical Society of Japan* **7** (1952) 333.
17. E. Sawaguchi, *Journal of the Physical Society of Japan* **8** (1953) 615–629.
18. B. Jaffe, R. S. Roth and S. Marzullo, *Journal of Applied Physics* **25** (1954) 809–810.
19. B. Jaffe, R. S. Roth and S. Marzullo, *Journal of Research of the National Bureau of Standards* **55** (1955) 239–254.
20. K. Kakegawa, J. Mohri, T. Takahashi, H. Yamamura and S. Shirasaki, *Solid State Communications* **24** (1977) 769–772.
21. B. Noheda, D. E. Cox, G. Shirane, R. Guo, B. Jones and L. E. Cross, *Physical Review B* **63** (2000) 14103–1 to 9.
22. R. Guo, L. E. Cross, S-E. Park, B. Noheda, D. E. Cox and G. Shirane, *Physical Review Letters* **84** (2000) 5423–5426.
23. Y. Porat, Y. Imry, A. Aharony and I. Bransky, *Ferroelectrics* **37** (1981) 591–594.
24. B. Sahoo, V. A. Jaleel and P. K. Panda, *Materials Science and Engineering B* **126** (2006) 80–85.
25. P. K. Sharma, Z. Ounaies, V. V. Varadan and V. K. Varadan, *Smart Materials and Structures* **10** (2001) 878–883.
26. S. Fujii, Y. Sugie and H. Fujiwara, *Nippon Seramikkusu Kyokai Gakujutsu Ronbunshi* **99** (1991) 520–522.
27. H. Huang, W. Chu, Y. Su, K. Gao, L. Kewei and Q. L. Jinxu, *Jinshu Xuebao* **41** (2005) 1004–1008.
28. Y. Shimakawa and Y. Kubo, *Applied Physics Letters* **77** (2000) 2590–2592.
29. R. G. Dosch, *Materials Research Society Symposium Proceedings* **32** (1984) 199–204.
30. Z. Q. Zhuang, M. J. Haun, S. J. Jang and L. E. Cross, *Advanced Ceramic Material* **3** (1988) 485–490.
31. J. Schäfer, W. Sigmund, S. Roy and F. Aldinger, *Journal of Materials Research* **10** (1997) 2518–2521.
32. J. H. Choy, Y. S. Han and J. T. Kim, *Journal of Materials Chemistry* **5** (1995) 65–69.
33. J. H. Choy, Y. S. Han and J. T. Kim, *Journal of Materials Chemistry* **7** (1997) 1807–1813.
34. K. R. M. Rao, A. V. P. Rao and S. Komarneni, *Materials Letters* **28** (1996) 463–467.
35. H. Cheng, J. Ma, B. Zhu and Y. Cui, *Journal of the American Ceramic Society* **76** (1993) 625–629.
36. B. Guiffard and M. Troccaz, *Materials Research Bulletin* **33** (1998) 1759–1768.
37. L. Eyraud, P. Eyraud and F. Bauer, *Advanced Ceramic Materials* **1** (1986) 223–231.
38. M. T. Weller, *Inorganic Materials Chemistry*, Oxford University Press, Oxford, 1996.

PROBLEMS

1. (a) State the minimum number of moles of oxalic acid dihydrate ($H_2C_2O_4 \cdot 2H_2O$) required in order to prepare 30 g of $PbZr_{0.52}Ti_{0.48}O_3$ by the method described in this chapter.

   (b) Therefore, state the minimum mass of oxalic acid dihydrate required for a complete reaction.

   (c) State the mass of oxalic acid dihydrate that you used in the synthesis.

2. State the solubility (in terms of g dm^{-3}) of lead acetate trihydrate, $(CH_3COO)_2Pb \cdot 3H_2O$ in water at $\sim 25\,°C$.

3. State the concentration (in terms of g dm^{-3}) of the aqueous solution of lead acetate trihydrate that you prepared.

4. Comment on the following:
   (a) Which feature distinguishes a piezoelectric crystal from a normal dielectric crystal?
   (b) Which feature distinguishes a ferroelectric crystal from a normal piezoelectric crystal?
   (c) A piezoelectric single crystal is not necessarily also ferroelectric. But, a polycrystalline piezoelectric ceramic is by necessity also ferroelectric. Explain why this is so.
   (d) Where in this 'hierarchy' do pyroelectric crystals belong?

5. Predict what should happen to the dielectric properties, if you were to compress a previously poled ceramic monolith of $PbZr_{0.52}Ti_{0.48}O_3$ at $20\,°C$ under a pressure of:
   (a) 15 kbar
   (b) 80 kbar

6. The first six peaks in the powder XRD pattern for cubic $PbZr_{0.52}Ti_{0.48}O_3$ occur at: $d_{001} = 22.39°\ 2\theta$; $d_{011} = 31.88°\ 2\theta$; $d_{111} = 39.31°\ 2\theta$; $d_{002} = 45.70°\ 2\theta$; $d_{012} = 51.47°\ 2\theta$; and $d_{112} = 56.80°\ 2\theta$ (cf. ICSD 1613) Powder X-ray diffraction data for tetragonal $PbZr_{0.52}Ti_{0.48}O_3$ at 325 K corresponding to copper-$K_{\alpha 1}$ radiation with wavelength, $\lambda = 1.5405\ Å$, is given by Noheda et al. [10] as follows (cf. ICSD 92059):

2θ	Intensity
21.45	10
21.95	11
30.88	100
31.24	47
38.21	14
43.70	12
44.76	25
49.42	6
50.15	3
50.39	1
54.69	19
55.37	32

Determine the values for the corresponding lattice d-spacings in this powder pattern using the Bragg equation. Given that the unit cell is slightly elongated in the c-direction, deduce the multiplicity for each of these peaks, and use this information to index each of the peaks in terms of their Miller indices (hkl) Then use this information to determine the cell constants for tetragonal $PbZr_{0.52}Ti_{0.48}O_3$ at 325 K. The reader is referred to Weller [38] for further information regarding the indexing of tetragonally distorted cubic systems.

7. The Pb^{2+} ions in the monoclinic $PbZr_{0.52}Ti_{0.48}O_3$ phase belong to the c-centred monoclinic Bravais lattice. The body-centred monoclinic(I) and face-centred monoclinic (F) lattices do not appear among the list of fourteen Bravais lattices. State the Bravais lattice (or lattices) to which they are equivalent. Explain your reasoning for your choice of equivalent Bravais lattice in each case, using relevant illustrations were necessary.

10

Yttrium Barium Cuprate $YBa_2Cu_3O_{7-\delta}$ ($\delta \sim 0$) by a Solid-State Reaction Followed by Oxygen Intercalation

The $YBa_2Cu_3O_{7-\delta}$ ($0 \leq \delta \leq 1$) solid solution is a black material that crystallises with an oxygen-deficient perovskite-related structure and displays a large variation in oxygen stoichiometry. It is often referred to by the acronym, YBCO. At room temperature it undergoes the remarkable transition from an insulator to a metal in the vicinity of the oxygen composition, $\delta = 0.65$. At elevated temperature ($\geq 350\,^\circ\text{C}$) $YBa_2Cu_3O_{7-\delta}$ is a mixed (ionic/electronic) conductor and is an important oxygen intercalation compound that has been considered for use as a ceramic oxygen electrode.[1]

The oxygen-rich composition, $YBa_2Cu_3O_{7-\delta}$ ($0 \leq \delta < 0.2$) has the even more remarkable property of exhibiting high critical temperature (high-T_c) superconductivity, with the onset critical temperature, $T_c \sim 92\,\text{K}$; i.e., above the boiling point of liquid nitrogen (77 K). These features have made $YBa_2Cu_3O_{7-\delta}$ one of the most extensively studied solid solutions ever

---

[1] One can only ponder over how long it might have taken $YBa_2Cu_3O_{7-\delta}$ to have been discovered if it were not for the discovery of superconductivity within $(La,Ba)_2CuO_4$ in 1986. But one thing is fairly sure; if $YBa_2Cu_3O_{7-\delta}$ had been discovered before 1986, then the initial interest in this material would have focused on its mixed (ionic/electronic) conducting and intercalation properties at elevated temperature.

Synthesis, Properties and Mineralogy of Important Inorganic Materials   By Terence E. Warner
© 2011 John Wiley & Sons, Ltd

Figure 10.1 The Meissner effect showing a cubic rare-earth magnet levitated above a 32-mm diameter ceramic monolith of YBa$_2$Cu$_3$O$_{7-\delta}$ as prepared by the method described in this chapter, cooled with liquid nitrogen. The white rim is condensed ice (Photograph used with kind permission from Simon Svane. Copyright (2010) Simon Svane)

synthesised; revealing complex subsolidus phase relationships. Consequently, there are numerous methods reported in the literature, using a wide variety of chemical precursors, for preparing YBa$_2$Cu$_3$O$_{7-\delta}$. This chapter describes the preparation of a ceramic monolith of oxygen-rich YBa$_2$Cu$_3$O$_{7-\delta}$ through a solid-state chemical reaction between, Y$_2$O$_3$, BaCO$_3$ and CuO at 950 °C; followed by controlled annealing under a flowing atmosphere of oxygen. The YBa$_2$Cu$_3$O$_{7-\delta}$ ceramic monolith, as produced by this method, is suitable for demonstrating the Meissner effect (see Figure 10.1).

The first chemical compound to exhibit superconductivity was the sulfide mineral, covellite, CuS with the $T_c = 1.6$ K, as reported originally by Meissner in 1929 [1], cf. Chapter 3. A generation later, in 1964, Schooley *et al.* [2] discovered the first ternary oxide superconductor; the oxygen-deficient perovskite, strontium titanate,[2] SrTiO$_{3-\delta}$ with the modest $T_c = 0.28$ K. Three years later Remeika *et al.* [3] reported the occurrence of superconductivity in the potassium-bronze, K$_{0.33}$WO$_3$, with the $T_c = 6$ K. In 1973, attention shifted to the spinel structure, with the discovery of superconductivity in lithium titanate, LiTi$_2$O$_4$ with the $T_c = 12$ K [4]; and in the accompanying solid solution, Li[Ti$_{2-x}$Li$_x$]O$_4$ containing mixed Ti^{3+}/Ti^{4+} valency [5, 6]. Research interests at that time tended to focus on complex

---

[2] Stoichiometric SrTiO$_3$ is a colourless insulator with a bandgap ~ 3 eV. Upon reduction, for example, under an atmosphere of hydrogen, the material darkens to form the oxygen-deficient phase, SrTiO$_{3-\delta}$ [9].

metal oxides containing transition metals from the left-hand side of the d-block of the chemical elements; especially, titanium. Consequently, these phases were predominantly n-type superconductors. In 1975, the chemistry expanded to include some of the p-block metals, with the discovery of superconductivity in the perovskite, barium lead bismuthate, $BaPb_{0.75}Bi_{0.25}O_3$, with the $T_c = 13\,K$ [7].

By tampering with the chemistry of lanthanum cuprate,[3] $LaCuO_3$, a comparatively huge increase in the T_c came in 1986, when Bednorz and Müller [8] discovered superconductivity in an oxygen-deficient multiphase material with the global composition, '$La_{1-x}Ba_xCuO_{3-\delta}$' ($x = 0.2$ and 0.15; $\delta > 0$) with the onset $T_c \sim 30\,K$. For this discovery (which brought metal oxides to the forefront of superconductivity), Bednorz and Müller were promptly awarded the Nobel Prize for Physics in 1987. Their results were soon verified by other workers, and the phase responsible for the superconductivity was identified as, $La_{2-x}Ba_xCuO_4$ ($x \sim 0.15$); which crystallises with an orthorhombic distorted version of the tetragonal K_2NiF_4 structure [9]. It is interesting to note that this superconducting phase contains the element, copper, i.e., a transition metal from the right-hand side of the d-block of the chemical elements; and in the same (first) row as titanium.

In 1987, Chu et al. [11] observed that the T_c for $La_{2-x}Ba_xCuO_4$ increased under high pressure, at a rate of approximately $1\,K\,kbar^{-1}$, raising the onset T_c to $57\,K$ (with the 'zero-resistance' state reached at $40\,K$; i.e., the highest in any known superconductor at that time). This unusually large pressure effect on T_c tempted Wu et al. [12] to induce a contraction in the molar volume for this material at ambient pressure, by replacing La^{3+} with an ion of a higher charge/radius ratio, such as, Y^{3+}. Their attempts at preparing the analogous phase, $Y_{2-x}Ba_xCuO_{4-\delta}$, failed; in the sense that their product was comprised of multiphase material, with little indication from their powder X-ray diffraction patterns for the presence of a phase with the K_2NiF_4 structure, that they were originally hoping for. But to their surprise, superconductivity with an exceptionally high onset $T_c \sim 92\,K$ (and the 'zero-resistance' state at $80\,K$), was observed in multiphase material with the global composition, '$Y_{1.2}Ba_{0.8}CuO_{4-\delta}$'.[4] Their results were reproduced by other

---

[3] $LaCuO_3$ crystallises with the rhombohedral $NdAlO_3$ structure. Previously, $LaCuO_3$ was considered to contain trivalent Cu^{3+} ions; nowadays, it is generally accepted that only divalent Cu^{2+} are present in this phase [10]. Bednorz and Müller [8] assumed the presence of trivalent Cu^{3+} ions in $LaCuO_3$. They supposed that Cu^{2+}/Cu^{3+} mixed valency might be induced within this material through the partial substitution of Ba^{2+} for La^{3+}; such that the creation of oxygen-deficient phases within the Ba–La–Cu–O system might involve itinerant electronic states between the non-Jahn–Teller Cu^{3+} and the Jahn–Teller Cu^{2+}, which might offer possibilities for enhancing electron-phonon coupling and metallic conductivity.

[4] These particular values relate to the absence of a magnetic field. Interestingly, the detrimental affect of a magnetic field on the value for T_c was already evident from the measurements on the multiphase material as produced and recorded by Wu et al. [12] *before* the pertinent phase (i.e., $YBa_2Cu_3O_{7-\delta}$) was even identified.

workers, and the relevant phase was soon identified as YBa$_2$Cu$_3$O$_{7-\delta}$ [13, 14]. The discovery of high-T_c superconductivity in YBa$_2$Cu$_3$O$_{7-\delta}$ created the opportunity for developing superconducting devices that can operate using cheap[5] and readily available liquid nitrogen as the refrigerant, and thus, stimulated an immense technological and scientific interest in this material. Potential applications include; superconducting high-power electric cables, levitated trains, superconducting magnets for use in generators and motors, efficient computer circuits, and SQUIDS (superconducting quantum interference devices).

YBa$_2$Cu$_3$O$_{7-\delta}$ ($0 \leq \delta \leq 1$) crystallises with an oxygen-deficient perovskite structure that can be visualised as follows. It is constructive to imagine three perovskite ABO$_3$ cubic unit cells stacked one on top of the other, such that the B-sites are located at the corners (see Figure 10.2). The barium and yttrium ions are ordered among the A-sites (which lie at the centre of these cubes), whilst the copper ions are located on the B-sites; such that both sites are fully occupied. With regards to the oxygen-rich composition, YBa$_2$Cu$_3$O$_7$, two oxygen sites (per unit formula) are permanently unoccupied in this idealised structure; as emphasised by the empirical formula, (YBa$_2$)Cu$_3$O$_7\square_2$.[6] These intrinsic oxygen vacancies are ordered within the crystal structure, and give rise to eight-fold coordination for the Y^{3+} ion; ten-fold coordination for the larger Ba^{2+} ion and two crystallographically distinct, and chemically dissimilar, copper ions (see Figure 10.3). One-third of the copper ions, designated as Cu(1), have square planar coordination, which link together to form chains running parallel to the b-direction. The remaining copper ions, designated as Cu(2), have five-fold coordination with square-pyramidal symmetry. Interestingly, the Cu(2) ions are displaced ~ 0.3 Å from the base of the square pyramid, which gives rise to the characteristic buckling of the Cu(2)$-$O bonds within the (CuO$_2$)$_\infty$ plane (see Figure 10.5). Edwards et al. [15] suggest that this distortion may arise from the combined effects of the large Ba^{2+} ion and the small, highly charged Y^{3+} ion. YBa$_2$Cu$_3$O$_7$ belongs to the orthorhombic crystal system and space group $Pmmm$, with cell constants; $a = 3.8178$ Å, $b = 3.8839$ Å, and $c = 11.6828$ Å as determined by Chaplot et al. [16] using the method of single-crystal X-ray diffractometry. As a consequence of the above features, the unit cell for YBa$_2$Cu$_3$O$_7$ has a high c/a ratio, as emphasised by the complete absence of oxygen ions in the basal ($a-b$) plane surrounding the small Y^{3+} ion.

---

[5] Liquid nitrogen is comparatively inexpensive, and much easier to handle, than liquid helium.
[6] The squares ($\square$) represent here, the normally unoccupied sites in YBa$_2$Cu$_3$O$_7$ with respect to the oxygen lattice in the idealised perovskite structure, ABO$_3$.

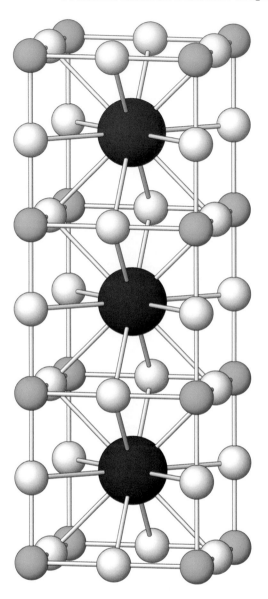

Figure 10.2   A stack of three idealised perovskite ABO$_3$ cubic unit cells, with the B-sites (light blue) situated at the corners. The larger A-sites are shown purple-pink, and the oxygen sites are shown white. The colour coding corresponds to that used in Chapter 9 for the perovskite, PbZr$_{1-x}$Ti$_x$O$_3$

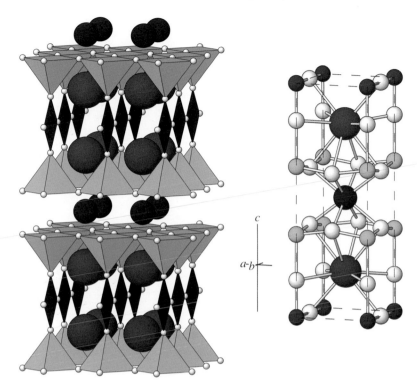

Figure 10.3   Crystal structure of YBa$_2$Cu$_3$O$_7$ showing the orthorhombic unit cell (on the right). The Cu(1) sites are shown dark blue and the Cu(2) sites are shown light blue. The Y^{3+} ions are shown red and the Ba^{2+} ions are shown purple-pink. The illustration on the left shows the square pyramidal coordination of the Cu(2) sites (light blue) and the square planar coordination of the Cu(1) sites (dark blue). This figure was drawn using data from Chaplot *et al.* [16], cf. ICSD 41646

Under reducing conditions (i.e., low oxygen partial pressure, P_{O_2}) or, at elevated temperature, YBa$_2$Cu$_3$O$_7$ undergoes a progressive loss of oxygen, resulting eventually in the formation of the oxygen-deficient end-member, YBa$_2$Cu$_3$O$_6$.[7] Figure 10.6 shows the oxygen content in YBa$_2$Cu$_3$O$_x$ as a logarithmic function of oxygen partial pressure for various temperatures. Since the oxygen composition, x is a function of both T and P_{O_2}, alternative temperatures result in correspondingly unique $x = f(\log P_{O_2})$ profiles [17]. This feature has a great bearing on the synthesis of this material, necessitating an annealing process for achieving the desired stoichiometry through the intercalation of oxygen. YBa$_2$Cu$_3$O$_6$ belongs to the tetragonal crystal system

---

[7] Extreme reducing conditions will result in the subsequent decomposition of YBa$_2$Cu$_3$O$_6$ to further reduced chemical phases.

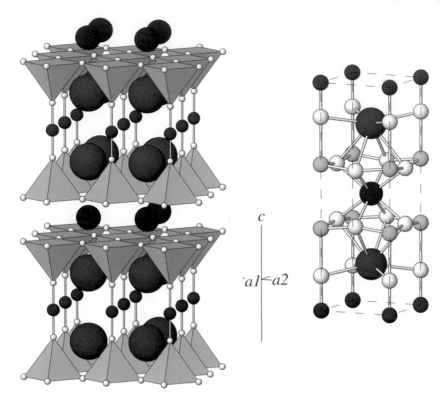

Figure 10.4   Crystal structure of YBa$_2$Cu$_3$O$_6$ showing the tetragonal unit cell (on the right). The Cu(1) sites (occupied by the Cu$^+$ ions) are shown dark blue, and the Cu(2) sites (occupied by the Cu^{2+} ions) are shown light blue. The Y^{3+} ions are shown red and the Ba^{2+} ions are shown purple-pink. The illustration on the left shows the square pyramidal coordination of the Cu^{2+} ions (light blue) and the dumb-bell (180°) coordination of the Cu$^+$ ions (dark blue). This figure was drawn using data from Parise and McCarron [18], cf. ICSD 203116

and space group $P4/mmm$, with cell constants; $a = 3.866$ Å, and $c = 11.777$ Å as determined by Parise and McCarron [18] (see Figure 10.4).

The intercalatory oxygen ions are extracted (and inserted) quasireversibly within the O(4) sites of the YBa$_2$Cu$_3$O$_{7-\delta}$ solid solution in a progressive and semi-ordered manner.[8] There are, however, some subtle differences in the degree of ordering within the intermediate compositions depending upon the temperature, and the direction by which the intercalation process is driven. But ultimately, all the O(4) sites become vacant in the oxygen-deficient

---

[8] The O(4) sites are located within the top (and, of course, the bottom) face of the unit cell as shown in Figures 10.3 and 10.4.

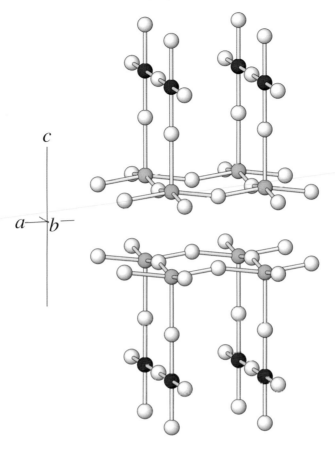

Figure 10.5   The copper–oxygen substructure in YBa$_2$Cu$_3$O$_7$. The Cu(1) ions in the charge reservoir layers are shown dark blue; Cu(2) ions are shown light blue; and the oxygen ions are shown white. The illustration emphasises the buckled (CuO$_2$)$_\infty$ planes in which the holes are delocalised in the normal metallic state (p-type metal) and form the Cooper pairs in the superconducting state (p-type superconductor). This figure was drawn using data from Chaplot *et al.* [16], cf. ICSD 41646

end-member, YBa$_2$Cu$_3$O$_6$.[9] The extraction of oxygen, has the direct effect of reducing the coordination for the Cu(1) ion from square-planar (four-fold) to linear (two-fold) coordination; and the coordination of the Ba^{2+} ions, from ten-fold, to an even more irregular, eight-fold coordination. Apparently, the

---

[9] Apparently, intermediate oxygen compositions are believed to form superstructures; at least at relatively low temperature. For example, Burns [19] describes that in the midintermediate composition, $\delta = 0.5$ (YBa$_2$Cu$_3$O$_{6.5}$), every other row of oxygen ions are missing along the *b*-direction. Moreover, in the vicinity of $\delta \sim 0.65$, YBa$_2$Cu$_3$O$_{7-\delta}$ undergoes a structural phase transition from orthorhombic to tetragonal symmetry. The demarcation (at, $\delta = 0.7$) between the orthorhombic and tetragonal regions in Figure 10.7 is a generalised simplification, since the composition, δ at which this transition occurs is also a function of P_{O_2}.

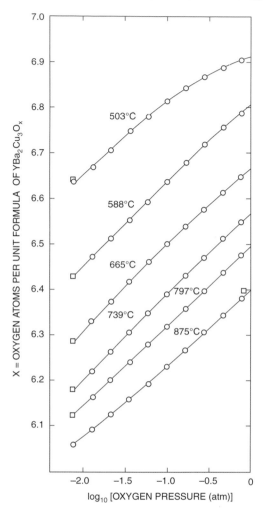

Figure 10.6 Oxygen content in YBa$_2$Cu$_3$O$_x$ at various temperatures and oxygen partial pressures (Reproduced with permission from Physical Review, Effect of oxygen pressure on the orthorhombic-tetragonal transition in the high-temperature superconductor YBa$_2$Cu$_3$O$_x$ by E. D. Specht, C. J. Sparks, A. G. Dhere, J. Brynestad, O. B. Cavin, D. M. Kroeger, and H. A. Oye, 37, 7426-7434 Copyright (1988) American Physical Society)

immediate coordination environment of the Cu(2) ion is only slightly affected. Therefore, the (CuO$_2$)$_\infty$ plane is in structural terms, essentially unaltered throughout the compositional range of the YBa$_2$Cu$_3$O$_{7-\delta}$ solid solution. Nevertheless, the effects that these changes have on the formal oxidation states, particularly regarding the oxygen and copper ions, are of fundamental interest.

Following the discovery of YBa$_2$Cu$_3$O$_{7-\delta}$, many authors assumed the conventional wisdom of ascribing the formal oxidation states for, Y(3+),

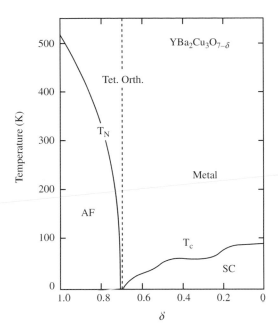

Figure 10.7 Temperature vs. composition, δ for YBa$_2$Cu$_3$O$_{7-\delta}$ showing the tetragonal and orthorhombic structure regions. AF = antiferromagnetic region; T$_N$ = Néel temperature; SC = superconductor region (Reproduced with permission from High-Temperature Superconductivity: An Introduction by G. Burns, Academic Press, San Diego 199pp Copyright (1992) Academic Press)

Ba(2 +), and O(2−) within YBa$_2$Cu$_3$O$_{7-\delta}$, thus leading to an average valence of 2.33 + for the Cu ions in the fully oxygenated material. This ionic model was commonly expressed as, Y^{3+}(Ba^{2+})$_2$(Cu^{2+})$_2$Cu^{3+}(O^{2-})$_7$ with discussion as to the proportioning of the Cu^{2+} and Cu^{3+} ions among the two copper sites; Cu(1) and Cu(2). But this model, however, falls short of reality. X-ray absorption spectroscopy (XAS) on metal cuprate superconductors such as, YBa$_2$Cu$_3$O$_7$, indicate no evidence for the presence of any considerable amount of trivalent Cu^{3+}; in contradiction to the above model [10, 20, 21]. Electron spin resonance (ESR) studies reveal an extensive delocalisation of the d-band electrons within YBa$_2$Cu$_3$O$_7$ in the metallic state [22], which is incompatible with an ionic model involving divalent Cu^{2+} ions with a *localised* electron configuration of $3d^9$.

Another model, whereby the extra holes (as generated, for example, by the intercalation of excess oxygen atoms beyond the composition, YBa$_2$Cu$_3$O$_{6.5}$) are considered to reside on some of the O^{2-} ions, with the formation of singly charged O$^-$ ions, has at least some precedence. For example, the soft Ba^{2+} ions within the barium peroxide, BaO$_2$ are known to bond with singly charged O$^-$ ions [23]. Whilst the theoretical calculations performed by Singh [24], regarding the electronic structure of the highly oxygenated sodium cuprate, NaCuO$_2$, indicate that the Cu ions are not trivalent, but have a surprisingly low valence; intermediate between divalent and monovalent! One common belief is that the highly mobile holes reside in the oxygen band [25]. Another popular belief (and perhaps the most realistic) is that they are resident within the partially oxidised (CuO$_2$)$_\infty$ planes; which, incidentally, avoids the problem of discriminating between Cu^{3+} and O$^-$. The Cu(1) ions on the other hand, are considered to function as a charge reservoir (i.e., a source and sink of electrons) for the (CuO$_2$)$_\infty$ planes [9] (see Figure 10.5).[10] It is generally agreed, however, that these excess holes form the Cooper pairs[11] within the superconducting phase; hence YBa$_2$Cu$_3$O$_7$ is a p-type Class II superconductor; although the mechanism for superconductivity in this class of superconductors is not fully understood.

The above model (involving highly mobile holes within the partially oxidised (CuO$_2$)$_\infty$ planes) is consistent with the distinct anisotropy regarding the resistivity of YBa$_2$Cu$_3$O$_7$ in its normal metallic state (> 92 K). Friedmann *et al.* [26] performed a series of resistivity measurements as a function of temperature on a single-crystal of YBa$_2$Cu$_3$O$_7$ for the three orthogonal directions. The following values measured at 275 K; $\rho_a = 146\,\mu\Omega$ cm; $\rho_b = 68\,\mu\Omega$ cm; and $\rho_c = 4800\,\mu\Omega$ cm; reveal appreciable metallic conductivity within the basal (a–b) plane, but with a notably lower metallic conductivity in the c-direction. This anisotropy prevails over the entire temperature range (92–275 K) as studied by these workers (see Figure 10.8). Furthermore, the resistivity of YBa$_2$Cu$_3$O$_7$ in its superconducting state (< 92 K) is zero for all three orthogonal directions!

A similar anisotropy manifests in the critical current density, J_c with regards to YBa$_2$Cu$_3$O$_7$ in its superconducting state (< 92 K). Generally, J_c (parallel to the c-direction) is significantly less than J_c (parallel to the a–b plane). Furthermore, in polycrystalline materials, the values for J_c are consistently lower compared with those for a single crystal, on account of the weak links across the grain boundaries. For instance, the fabrication of the sintered polycrystalline (ceramic) monolith of YBa$_2$Cu$_3$O$_7$ as described in this chapter is suitable for demonstrating the Meissner effect; but it is unsuitable for conveying a large current density. Hence, there are enormous challenges in the processing of this material (especially in controlling the microstructure) for use at high current density. High-quality long-length conductors of

---

[10] The extension of the metallic and superconducting regions to values of $\delta > 0.5$ as shown in Figure 11.7 is believed to be due to an increase in the concentration of the holes within the (CuO$_2$)$_\infty$ planes, due to the reduction of some of the Cu^{2+} ions to Cu$^+$ that are situated in the Cu(1) site [25]; Cu^{2+} + O^{2-} → Cu$^+$ + O$^-$. Furthermore, the reversible extraction of oxygen ions from YBa$_2$Cu$_3$O$_{6.5}$ leading ultimately to the oxygen-deficient end-member, YBa$_2$Cu$_3$O$_6$, is also believed to result in the reversible reduction of the Cu(1) ions from Cu^{2+} to Cu$^+$.

[11] A Cooper pair is the name given to electrons or, holes, that are bound together at low temperatures resulting in a lower energy than the Fermi energy within a metal and play an important role in the superconducting mechanism.

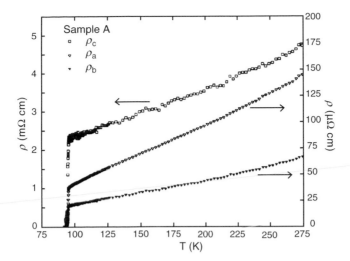

Figure 10.8 The specific resistivities for the orthogonal directions as a function of temperature for a single YBa$_2$Cu$_3$O$_7$ crystal (Reproduced with permission from Physical Review B., Direct measurement of the anisotropy of the resistivity in the a-b plane of twin-free, single-crystal, superconducting YBa$_2$Cu$_3$O$_{7-\delta}$ by T. A. Friedmann, M. W. Rabin, J. Giapintzakis, J. P. Rice, and D. M. Ginsberg, Phys. 42, 6217-6221 Copyright (1990) American Physical Society)

YBa$_2$Cu$_3$O$_7$ demand the conflicting requirements for an almost perfect alignment of the grains *and* an attachment to a flexible substrate; which accounts for why they have taken many years to be developed.[12] The reader is referred to the review article by MacManus-Driscoll [27] for details concerning this topic.[13]

Microstructure, on the other hand, has little influence on the value of the critical temperature. Besides the destructive effect of magnetic fields, T_c is strongly influenced by the oxygen composition, δ (see Figure 10.9). A high value of $T_c = 93.3$ K is reported by Kirby *et al.* [28] for a twinned crystal with an oxygen-rich composition, YBa$_2$Cu$_3$O$_{6.99}$ as grown in a BaZrO$_3$ crucible. From Figures 10.7 and 10.9 it can be seen that an oxygen composition, $\delta \leq 0.2$ is sufficient to allow superconductivity in YBa$_2$Cu$_3$O$_{7-\delta}$ when cooled in liquid nitrogen.

Over the past 23 years, a multitude of high-T_c superconductors have been discovered. The majority of these phases display exotic chemical compositions and complex structures, and many have a T_c substantially higher than that for YBa$_2$Cu$_3$O$_7$. The highest at present is HgBa$_2$Ca$_2$Cu$_3$O$_{8+\delta}$ with the

---

[12] The company American Superconductor has commercialised a second-generation (2G) high-temperature superconductor wire (so-called 344 superconductor) based on a 4-mm wide YBa$_2$Cu$_3$O$_{7-\delta}$ strip laminated between two metallic stabiliser strips and soldered at the edges [29].

[13] The author is grateful to Professor Judith L. Driscoll, University of Cambridge, for sharing information regarding YBa$_2$Cu$_3$O$_7$.

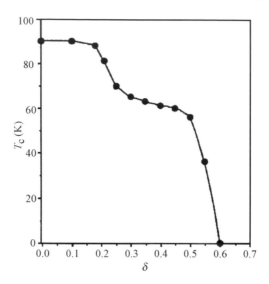

Figure 10.9 Variation in critical temperature, T_c with oxygen content, δ in $YBa_2Cu_3O_{7-\delta}$ (Reproduced with permission from Annual Reports Section C (Physical Chemistry) C, Chemistry of superconducting oxides by A. R. Armstrong and P. P. Edwards, 88 259-331 Copyright (1991) Royal Society of Chemistry)

$T_c = 135$ K at ambient pressure [30]. But at the time of writing, none of these phases can rival the performance of $YBa_2Cu_3O_7$ at 77 K within a magnetic field. This is actually quite peculiar, considering that $YBa_2Cu_3O_7$ happened to be the first of many superconductors to be discovered with a T_c above the boiling point of nitrogen. But putting aside any notions of fate, it is not just the value of T_c, but rather the features concerning magnetic flux pinning[14] that are the key issues favouring the use of $YBa_2Cu_3O_7$ above all others.

An isothermal section for the $BaO-YO_{1.5}-CuO$ ternary system in artificial air $(P_{O_2} = 0.21$ bar; the balance being argon) at 950 °C, was constructed by Osamura and Zhang [31, 32] in 1991; as shown in Figure 10.10. From this diagram it can be seen that under these conditions, $YBa_2Cu_3O_7$ can coexist with; Y_2BaCuO_5; $YBa_4Cu_3O_9$; $BaCuO_2$; liquid; and in all cases, $O_2(g)$, as in accordance with the phase rule. For example, compositions based on

---

[14] The following passage is taken from the review article by Foltyn *et al.* [33]: 'In the interior of Type-II superconductors such as $YBa_2Cu_3O_7$, magnetic field exists in the form of tubular structures called flux lines or vortices, each of them carrying one unit of magnetic flux, or flux quantum. The application of an electric current to the superconductor generates a lateral force on the vortices known as the Lorentz force, and the resulting vortex motion dissipates energy and causes electrical resistance to appear in the superconductor. Only if vortices can be immobilised by a counteracting 'pinning force' can a superconductor sustain high current density. The higher the current density, the greater the Lorentz force acting on the vortices – the critical current density is essentially the point at which the Lorentz force begins to exceed the maximum available pinning force.'

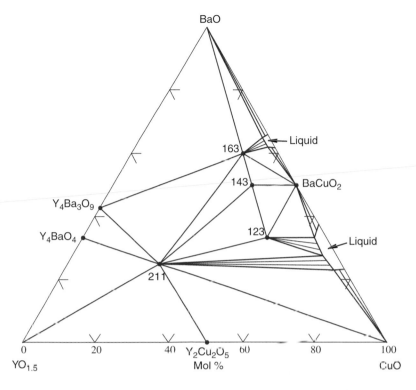

Figure 10.10 An isothermal section for the BaO–YO₁.₅–CuO ternary system in artificial air ($P_{O_2} = 0.21$ bar) at 950 °C. 123 = YBa₂Cu₃O₇; 211 = Y₂BaCuO₅ (green phase); 143 = YBa₄Cu₃O₉; 163 = YBa₆Cu₃O₁₁. After Osamura and Zhang [30], Zhang and K. Osamura [31], cf. ACerS-NIST Phase Equilibria Diagram SII-421

YBa₂Cu₃O₇ but with a slight excess of CuO at the expense of Y₂O₃, will have a tendency to form a small amount of liquid at 950 °C.

With regards to the synthesis of YBa₂Cu₃O₇, one of the notorious impurity phases commonly encountered within the product material, is the so-called, 'green phase', i.e., Y₂BaCuO₅.[15] Therefore, with reference to Figure 10.10, the reader may wish to draw a straight pencil-line originating at the composition of the 'green phase' (211), and drawn so as to intersect the composition, YBa₂Cu₃O₇ (123) and then finally crossing the BaO–CuO binary system at the composition, 37.5 mol. % BaO and 62.5 mol. % CuO (i.e., close to the boundary of the liquid phase). This pencil-line depicts the horizontal coordinate (at $T = 950$ °C) for the corresponding diagram in Figure 10.11; which shows the phase relationships within the Y₂BaCuO₅ −(3BaCuO₂ + 2CuO)

---

[15] Y₂BaCuO₅ exhibits a bright green coloration, which gave rise to the expression; 'the green phase'.

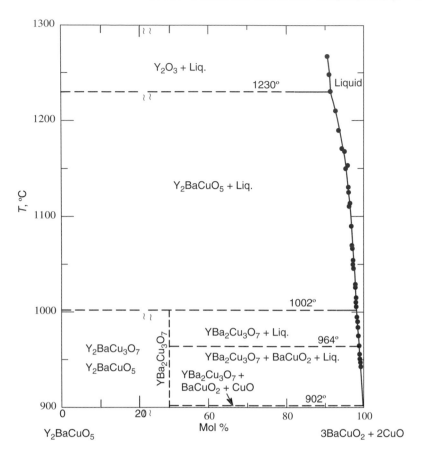

Figure 10.11   Phase relationships within the Y$_2$BaCuO$_5$ $-$(3BaCuO$_2$ + 2CuO) pseudobinary join in air ($P_{O_2} = 0.21$ bar). NB. The composition coordinate has a break in the vicinity of 20–50 mol. % (Reproduced with permission from Advanced Superconductors VI, Proc. Int. Symp. Supercond., 6th, Hiroshima, Japan, October 26–29, 1993, Vol. 2. Edt. T. Fujita and Y. Shiohara, Tokyo, Tokyo, Japan, Ch. Krauns, M. Tagami, M. Sumida, Y. Yamada, and Y. Shiohara, Liquidus compositions of Re-Ba-Cu-oxides (RE = Y, Sm), 767–769 Copyright (1994) Springer-Verlag)

pseudobinary join as a function of temperature, as constructed by Krauns *et al.* [34].[16]

From Figure 10.11, it can be seen that YBa$_2$Cu$_3$O$_{7-\delta}$ has an incongruent melting point at 1002 °C in air. Although not apparent from this diagram, the incongruent melting point varies depending on the oxygen composition, δ, which in turn is dependent on the oxygen partial pressure in the surrounding atmosphere. Nevertheless, YBa$_2$Cu$_3$O$_{7-\delta}$ decomposes in air at about 1002 °C, resulting in the

---

[16] Apparently, both of these phase diagrams (cf. Figures 10.10 and 10.11) ignore the deviations in oxygen stoichiometry, presumably, for the sake of simplicity. The information given in Figure 10.6, should imply that $\delta > 0.5$ in YBa$_2$Cu$_3$O$_{7-\delta}$ at 950 °C; and not YBa$_2$Cu$_3$O$_7$ as portrayed in Figure 10.9; and likewise, in Figure 10.10.

formation of the *green phase*, Y_2BaCuO_5, and a melt that is very deficient in Y_2O_3. The peritectic reaction can be expressed from Figure 10.11 as:

$$YBa_2Cu_3O_{6.5}(s) \rightleftharpoons 0.49\,Y_2BaCuO_5(s) + 0.51\,Y_{0.04}Ba_{2.96}Cu_{4.92}O_{7.94}(l)$$

This peritectic reaction is exploited in the principle method for growing single crystals of $YBa_2Cu_3O_{7-\delta}$; a technique known as, melt processing. The reader is referred to Campbell and Cardwell [35], and Lo *et al.* [36] for further details. Ceramic crucibles made of $BaZrO_3$ are apparently one of the most inert for containing the $BaCuO_2$–CuO melt [28]. Alumina crucibles are attacked by this melt and therefore are unsuitable for this particular purpose.

Preparative Procedure

The method described in this chapter for the preparation of a ceramic monolith of oxygen-rich $YBa_2Cu_3O_{7-\delta}$ involves a solid-state chemical reaction between, Y_2O_3, $BaCO_3$ and CuO at $950\,^\circ C$, followed by controlled annealing under a flowing atmosphere of oxygen. The temperature of $950\,^\circ C$ is sufficiently high to afford a reasonable rate of reaction between these chemical reagents, without encroaching upon the incongruent melting point of $YBa_2Cu_3O_{7-\delta}$. The use of alumina crucibles is just about acceptable here, since the amount of liquid phase produced in the course of the solid-state reaction is presumed to be minimal. But even so, for the production of a more refined product, the reader should consider using alternative ceramic crucibles, for example, $BaZrO_3$, and gold.

Prepare 35 g of $YBa_2Cu_3O_{7-\delta}$ (ideally with $\delta = 0$) by the following method: Weigh appropriate amounts of Y_2O_3, $BaCO_3$ and CuO, preferably as fine powders, and within the precision ± 0.001 g. Grind these powders together thoroughly, with a porcelain pestle and mortar.[16] It is important to avoid spillage of material at all stages of the preparation, since this will effect the final composition of the ceramic monolith. Barium carbonate and the barium compounds produced in this work are toxic, so appropriate care should be exercised when handling them.

Place the reaction mixture inside a small alumina crucible (CC47 Almath Ltd). Insert this crucible inside a larger one (CC62 Almath Ltd), which will act as a shield to help prevent any spillage of the contents onto the furnace floor. Then introduce these into the chamber furnace, and heat at $100\,^\circ C/h$ to $950\,^\circ C$ in air ($P_{O_2} = 0.21$ bar). After 72 h at $950\,^\circ C$, cool to ambient temperature ($20\,^\circ C$) with a cooling rate of $300\,^\circ C/h$. This heating cycle will last ~ 85 h ($3^1/_2$ days).

Remove as much of the product material from the alumina crucible as possible. This material should then be crushed using a steel crusher (or a similar device), and then ground it to a fine powder using a porcelain pestle and

---

[17] If available, the reader is advised to use a planetary mono mill (e.g., a *Fritsch Pulverisette 6* with zirconium oxide bowl and balls and cyclohexane as a liquid medium) for this part of the procedure.

mortar; whilst taking care to avoid any loss of material.[17] Retain ~ 3 g sample of this powder for analysis by powder X-ray diffractometry (scan: 5–60° 2θ). Compare your powder pattern with the PDF 40-159 for YBa$_2$Cu$_3$O$_7$; the PDF 39-1496 for YBa$_2$Cu$_3$O$_6$; and the PDF 38-1434 for the 'green phase', Y$_2$Ba-CuO$_5$; and compare with other phases if considered necessary.

Press ~ 20 g of the powdered product into a compacted monolith, using a 32 mm die and a uniaxial press with a load ~ 10 tonne (1.2 kbar). Spread the remaining powder, evenly, into an alumina boat (SRX110 Almath Ltd) so as to form a 'sacrificial' powder bed. Then gently place the compacted monolith on top of the powder bed. Introduce the alumina boat into the central zone of a tube furnace, and heat at 200 °C/h to 950 °C under an atmosphere of flowing oxygen, with a flow rate of ~ 0.5 l/min.[18] After sintering at 950 °C for 24 h, cool to 350 °C with a cooling rate of 20 °C/h. After 10 h at 350 °C, cool at 60 °C/h to ambient temperature (20 °C). This heating cycle will last ~ 70 h (3 days). Finally, remember to close the oxygen gas flow using the tap on the gas cylinder.

With the aid of a diamond saw or a chisel, cut or chip, a small piece (~ 250 mg) from the YBa$_2$Cu$_3$O$_7$ ceramic monolith for analysis by powder X-ray diffractometry. Grind this small piece of ceramic to a fine powder, using an agate pestle and mortar. Place your powder in a tightly sealed specimen bottle and submit it for analysis by powder X-ray diffraction (scan: 5–60° 2θ). Compare your powder pattern with the PDF 40-159 for YBa$_2$Cu$_3$O$_7$; the PDF 39-1496 for YBa$_2$Cu$_3$O$_6$; and the PDF 38-1434 for the 'green phase', Y$_2$BaCuO$_5$; see Figure 10.12.

YBa$_2$Cu$_3$O$_{7-\delta}$ is thermodynamically unstable under a moist atmosphere and will eventually degrade if left exposed to the air. To help prevent this, the YBa$_2$Cu$_3$O$_7$ monolith should be stored in a tightly sealed specimen jar. If the material is to be used for demonstration purposes, it is advantageous to also coat the monolith with a suitable lacquer. As part of a lecture theatre exercise, the YBa$_2$Cu$_3$O$_7$ monolith can be used at the boiling point of liquid nitrogen, in conjunction with a small rare-earth (e.g., NEOMAX) magnet, to demonstrate the Meissner effect (see Figure 10.1).

Aslan et al. [37] have reported a relatively simple hot-pressing technique for improving the quality of YBa$_2$Cu$_3$O$_{7-\delta}$ ceramic monoliths in terms of reducing the porosity, inducing preferential grain alignment, and thereby enhancing the critical current density, J_c. Several parameters were investigated by these workers, with satisfactory results obtained by the method as described in the following summary. Presintered compacts of YBa$_2$Cu$_3$O$_{7-\delta}$ were prepared by heating isostatically pressed cylinders (7.4 mm diameter, 4.7 mm length) of prereacted YBa$_2$Cu$_3$O$_{7-\delta}$ powder at 950 °C for 1 h. These were then deformed uniaxially whilst subjected to an applied initial load

---

[18] Gas flow meters are normally controlled with a needle valve. A needle valve is a very delicate device; it is not a tap! Therefore, a needle valve should *not* be used to close the gas supply. Instead, a ball valve should be inserted in the gas supply line for the intended use as a tap.

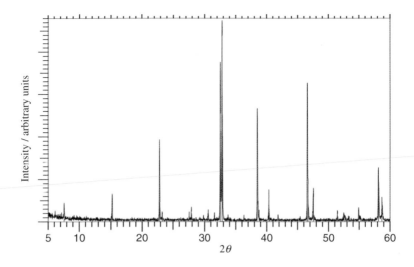

Figure 10.12　Powder XRD pattern (Cu-$K_{\alpha 1}$ radiation) of YBa$_2$Cu$_3$O$_{7-\delta}$ as prepared by the method described in this chapter. This powder pattern corresponds closely to the PDF 78-2463 for YBa$_2$Cu$_3$O$_{6.9}$ (shown in red)

of 110 N ($\sim$ 40 bar) at 950 °C for 4 h (presumably in air), before cooling to room temperature at a rate of 300 °C/h under reduced load. These deformed monoliths were then annealed in flowing oxygen inside a tube furnace at 400 °C for 10 h. The final ceramic density was claimed to be $> 98\%$. The

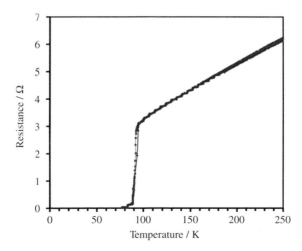

Figure 10.13　Resistance as a function of temperature for ceramic YBa$_2$Cu$_3$O$_{7-\delta}$ as prepared by the method described in this chapter corresponding to the ceramic monolith shown in Figure 10.1. The cooling profile is in blue and the heating profile is in red. T_c (on-set) = 92 K (Reproduced with kind permission from Simon Svane, U. Søndergaard and C. Jacobsen.)

microstructure was studied by reflected light microscopy under polarised light. This hot-pressing technique should give some ideas for improving the quality of the $YBa_2Cu_3O_{7-\delta}$ ceramic monoliths as prepared by the method described in this chapter.

REFERENCES

1. W. Meissner, *Zeitschrift für Physik* **58** (1929) 570–572.
2. J. F. Schooley, W. R. Hosler and M. L. Cohen, *Physical Review Letters* **12** (1964) 474–475.
3. J. P. Remeika, T. H. Geballe, B. T. Matthias, A. S. Cooper, G. W. Hull and E. M. Kelly, *Physics Letters A* **24** (1967) 565–566.
4. D. C. Johnston, H. Prakash, W. H. Zachariasen and B. Vishvanathan, *Materials Research Bulletin* **8** (1973) 777–784.
5. D. C. Johnston, *Journal of Low Temperature Physics* **25** (1976) 145–175.
6. M. R. Harrison, P. P. Edwards and J. B. Goodenough, *Philosophical Magazine B* **52** (1985) 679–699.
7. A. W. Sleight, J. L. Gillson and P. E. Bierstedt, *Solid State Communications* **17** (1975) 27–28.
8. J. G. Bednorz and K. A. Müller, *Zeitschrift für Physik B: Condensed Matter* **64** (1986) 189–193.
9. A. R. Armstrong and P. P. Edwards, *Annual Reports on the Progress of Chemistry, Section C: Physical Chemistry* **88** (1991) 259–331.
10. A. W. Webb, K. H. Kim and C. Bouldin, *Solid State Communications* **79** (1991) 507–508.
11. C. W. Chu, P. H. Hor, R. L. Meng, L. Gao, Z. J. Huang and Y. Q. Wang, *Physical Review Letters* **58** (1987) 405–407.
12. M. K. Wu, J. R. Ashburn, C. J. Torng, P. H. Hor, R. L. Meng, L. Gao, Z. J. Huang, Y. Q. Wang and C. W. Chu, *Physical Review Letters* **58** (1987) 908–910.
13. T. Siegrist, S. A. Sunshine, D. W. Murphy, R. J. Cava and S. M. Zahurak, *Physical Review B* **35** (1987) 7137–7139.
14. P. M. Grant, R. B. Beyers, E. M. Engler, G. Lim, S. S. P. Parkin, M. L. Ramirez, V. Y. Lee, A. Nazzal, J. E. Vazquez and R. J. Savoy, *Physical Review B* **35** (1987) 7242–7244.
15. P. P. Edwards, M. R. Harrison and R. Jones, *Chemistry in Britain* **23** (1987) 962–964.
16. S. L. Chaplot, W. Reichardt, L. Pintschovius and N. Pyka, *Journal of Physics and Chemistry of Solids* **53** (1992) 761–770.
17. E. D. Specht, C. J. Sparks, A. G. Dhere, J. Brynestad, O. B. Cavin, D. M. Kroeger and H. A. Øye, *Physical Review* **37** (1988) 7426–7434.
18. J. B. Parise and E. M. McCarron, *Journal of Solid State Chemistry* **83** (1989) 188–197.
19. G. Burns, *High-Temperature Superconductivity: An Introduction*, Academic Press, San Diego 1992, 199pp.
20. G. van der Laan, R. A. D. Pattrick, C. M. B. Hendersen and D. J. Vaughan, *Journal of Physics and Chemistry of Solids* **53** (1992) 1185–1190.

21. K. C. Hass, *Electronic structure of copper-oxide superconductors*, in *Solid State Physics*, eds. H. Ehrenreich and D. Turnbull, 42 (1989) Academic Press, New York, 213–270.

22. R. Jones, R. Janes, R. Armstrong, N. C. Pyper, P. P. Edwards, D. J. Keeble and M. R. Harrison, *Journal of the Chemical Society, Faraday Transactions* 86 (1990) 675–682.

23. H. H. Otto, e-Print Archive, *Condensed Matter* (2008) 1–8.

24. D. J. Singh, *Physical Review B* 49 (1994) 1580–1585.

25. V. J. Emery, *Materials Research Science Bulletin* 14 (1989) 67–71.

26. T. A. Friedmann, M. W. Rabin, J. Giapintzakis, J. P. Rice and D. M. Ginsberg, *Physical Review B* 42 (1990) 6217–6221.

27. J. L. MacManus-Driscoll, *Annual Review of Materials Science* 28 (1998) 421–462.

28. N. M. Kirkby, A. Trang, A. van Riessen, C. E. Buckley, V. W. Wittorff, J. R. Cooper and C. Panagopoulos, *Superconductor Science and Technology* 18 (2005) 648–657.

29. M. W. Rupich *et al. IEEE Transactions on Applied Superconductivity* 17 (2007) 3379–3382.

30. C. W. Chu, L. Gao, F. Chen, Z. J. Huang, R. L. Meng and Y. Y. Xue, *Nature* 365 (1993) 323–325.

31. K. Osamura and W. Zhang, *Zeitschrift für Metallkunde* 82 (1991) 408–415.

32. W. Zhang and K. Osamura, *Phase diagram of Y-Ba-Cu-O system*, 437–440 in: *Advances in Superconductivity III: Proceedings of the 3rd International Symposium on Superconductivity*, Sendai, Japan, November 6–9, 1990. eds. K. Kajimura and H. Hayakawa, Springer-Verlag Tokyo, Tokyo, Japan, 1991.

33. S. R. Foltyn, L. Civale, J. L. MacManus-Driscoll, Q. X. Jia, B. Maiorov, H. Wang and M. Malcy, *Nature Materials* 6 (2007) 631–642.

34. Ch. Krauns, M. Tagami, M. Sumida, Y. Yamada and Y. Shiohara, *Liquidus compositions of Re-Ba-Cu oxides (RE = Y, Sm)*, 767–769 In: *Advances in Superconductivity VI: Proceedings of the 6th International Symposium on Superconductivity*, Hiroshima, Japan, October 26–29, 1993, Vol. 2. ed. T. Fujita and Y. Shiohara, Springer-Verlag Tokyo, Tokyo, Japan, 1994.

35. A. M. Campbell and D. A. Cardwell, *Cryogenics* 37 (1997) 567–575.

36. W. Lo, D. A. Cardwell, C. D. Dewhurst and S. L. Dung, *Journal of Materials Research* 11 (1996) 786–794.

37. M. Aslan, H. Jaeger, P. Majewski, K. Schulze, J. Kunesch and G. Petzow, *Proceedings of the ICMC '90 Topical-Conference on Materials Aspects of High-Temperature Superconductors Materials Aspects*, eds. H. C. Freyhardt, R. Flükiger and M. Peuckert; DGM Informations GmbH, Garmisch-Partenkirchen, Germany, May 9–11, Vol. 2 (1990) 1045–1050.

PROBLEMS

1. Barium carbonate, $BaCO_3$ is used here as one of the chemical reagents in the synthesis of $YBa_2Cu_3O_{7-\delta}$.
   (a) State the mineralogical name for $BaCO_3$.
   (b) Can you suggest a more suitable alternative barium compound to $BaCO_3$ for the synthesis of $YBa_2Cu_3O_{7-\delta}$.

2. Explain the reason for sintering the $YBa_2Cu_3O_{7-\delta}$ monolith on a bed of $YBa_2Cu_3O_{7-\delta}$ powder?

3. In Figure 10.10, various lines can be seen radiating from the '211' phase so as to join with the liquid phase. State the name given to describe these lines, and explain what they signify.

4. State the magnetic and electrical properties of the following phases at the temperature quoted:
   (a) $YBa_2Cu_3O_6$ at 550 K.
   (b) $YBa_2Cu_3O_6$ at 298 K.
   (c) $YBa_2Cu_3O_6$ at 77 K.
   (d) $YBa_2Cu_3O_7$ at 298 K.
   (e) $YBa_2Cu_3O_7$ at 77 K.
   (f) $YBa_2Cu_3O_{6.5}$ at 900 K.

5. Comment briefly, on the mobility of the oxide ions in $YBa_2Cu_3O_{7-\delta}$ at high temperature.

   How might their mobility be influenced by the oxygen composition, δ in $YBa_2Cu_3O_{7-\delta}$?

6. Why is the $YBa_2Cu_3O_{7-\delta}$ monolith cooled slowly under an atmosphere of flowing oxygen?

7. Comment on why the measurement of T_c is perfectly acceptable on a twinned crystal of $YBa_2Cu_3O_{7-\delta}$ but the measurement of J_c is perhaps less meaningful.

11

Single Crystals of Ordered Zinc–Tin Phosphide ZnSnP$_2$ by a Solution-Growth Technique Using Molten Tin as the Solvent

Zinc–tin phosphide, ZnSnP$_2$, is a noncommercialised dark grey semiconductor that crystallises with a diamond-like structure. It is an example of a stoichiometric compound that undergoes a progressive order–disorder phase transition with respect to the distribution of the metal atoms. The ordered, partially ordered, and fully disordered phases are considered to be direct bandgap p-type semiconductors. ZnSnP$_2$ was first prepared by Goodman in England in 1957 [1] and was characterised fairly extensively during the 1960s and 1970s; particularly in the Soviet Union for potential use in the manufacture of heterojunctions [2].[1]

In 1985, Goodman [3] published a profound article shortly before he died, in which he drew attention to the challenge offered by the less-conventional semiconductors. He emphasised that these materials could offer unusual and highly desirable *combinations* of properties, but concluded that the industry will not be investigating novel semiconductors until they have been demonstrated to possess valuable properties that are unobtainable with more conventional materials. This places the onus on academic research; for which ZnSnP$_2$ is just one of many promising candidates within

---

[1] A heterojunction is the interface that occurs between two layers or regions of dissimilar crystalline semiconductors that exhibit unequal bandgaps.

Synthesis, Properties and Mineralogy of Important Inorganic Materials   By Terence E. Warner
© 2011 John Wiley & Sons, Ltd

this field. ZnSnP$_2$ is particularly attractive, because it comprises relatively nontoxic, cheap and abundant chemical elements.[2] Recently, there has been a renewed interest in ZnSnP$_2$ as a photovoltaic material (photoconductor) with potential applications as a light absorber in solar cells, and in nonlinear optics.[3] This has stimulated research into the epitaxial growth of ZnSnP$_2$ on GaAs. The majority of the preparative methods described in the earlier literature, however, utilise molten tin as a high-temperature solvent (or flux) for growing single crystals of ZnSnP$_2$, and this approach is adopted in the present chapter.

The discovery of ZnSnP$_2$ can be traced to the year, 1956, when Austin *et al.* [4], described how various compounds based upon the chalcopyrite structure (CuFeS$_2$) can be derived from more familiar compounds that crystallise with the sphalerite structure (ZnS), such as, gallium arsenide, GaAs. In 1957, Goodman [1] demonstrated that zinc–germanium arsenide, ZnGeAs$_2$, can be derived from gallium arsenide, GaAs, through the simple transposition of a pair of gallium atoms (group 13 of the periodic table) for one zinc atom (group 12) and one germanium atom (group 14), with respect to the unit formula. Consequently, ZnGeAs$_2$ and GaAs are isoelectronic. After preparing ZnGeAs$_2$, Goodman went on to show that ZnSnP$_2$ can be derived from ZnGeAs$_2$, through the transposition of one germanium atom (group 14) for one tin atom (group 14), and one arsenic atom (group 15) for one phosphorus atom (group 15). The additional zinc–tin pnictides,[4] ZnSnAs$_2$ [5] and ZnSnSb$_2$ [6] were prepared subsequently by other workers; whilst ZnSnN$_2$ and ZnSnBi$_2$ are yet to be prepared. ZnSnP$_2$ has not yet been found in nature, although a few metal phosphides have been discovered in meteorites, for example, schreibersite, (Fe,Ni)$_3$P and florenskyite, (Fe,Ni)TiP.

The only major study of the phase relationships in the Zn–Sn–P ternary system was reported by Borshchevskii and Vysotina [2] in 1976, and is discussed below. Prior to this, the information is fragmentary and conflicting. For example, in 1968, Rubenstein and Ure [7] reported that ZnSnP$_2$ has a congruent melting point at $968 \pm 2\,°C$ [7]; whilst in the following year Mughal *et al.* [8] reported that ZnSnP$_2$ has an incongruent melting point at $930\,°C$ (corresponding to the solidus), with the liquidus at $970\,°C$.

---

[2] Cadmium, gallium, indium, germanium, arsenic, selenium and tellurium, are rare and/or toxic elements; although they are the essential components in several commercial semiconductors. Interestingly, both their poisonous and semiconducting properties are a consequence of the high polarisability of these atoms.

[3] Nonlinear optics describes the behaviour of light in nonlinear media, i.e., media in which the dielectric polarisation responds nonlinearly to the electric field. This branch of optics is particularly relevant to lasers.

[4] *pnictide* is a compound of a pnicogen with one or more elements of greater electropositivity; in which a pnicogen is any of the elements in group 15 of the periodic table, comprising nitrogen, phosphorus, arsenic, antimony, and bismuth.

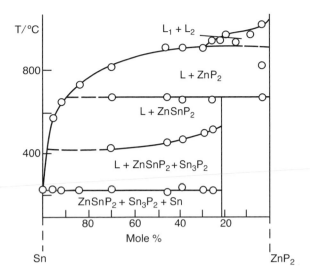

Figure 11.1   An equilibrium phase diagram of the ZnP$_2$–Sn pseudobinary system as constructed by Borshchevskii and Vysotina [2]. Perhaps, the composition coordinate should be expressed in terms of atomic percentage of the tin content (rather than in mole percentage as shown) in order for the ternary phase, ZnSnP$_2$ to be positioned more meaningfully . Note that ZnP$_2$ has a melting point of 1040 °C (Reproduced with permission from Neorganicheskie Materialy, A. S. Borshchevskii and M. G. Vysotina, Izvestiya Nauka Russian Academy of Sciences, 12, 615 618 Copyright (1976) Academizdatcenter Nauka Publishers, Russian Academy of Sciences)

Borshchevskii and Vysotina [2] constructed an equilibrium phase diagram of the ZnP$_2$–Sn pseudobinary system, as illustrated in Figure 11.1. From this diagram[5] it can be seen that the composition 'ZnSnP$_2$' exists as a homogenous liquid > 970 °C. Upon cooling ≤970 °C, this homogenous liquid segregates into two immiscible liquids, whilst ZnP$_2$ begins to crystallise from the melt. Upon cooling ≤920 °C, these two liquid phases recombine to form a homogeneous liquid, whilst ZnP$_2$ continues to crystallise progressively from this melt. Upon cooling further, the ternary phase, ZnSnP$_2$ crystallises completely at 667 °C through the following peritectic reaction:

$$ZnP_2(s) + Sn(l) \rightleftharpoons ZnSnP_2(s)$$

Crystalline ZnSnP$_2$ remains stable down to ambient temperature as a stoichiometric phase, and is considered to be the only ternary compound within the Zn–Sn–P system at ambient pressure. Borshchevskii and Vysotina [2] also described the liquidus in other regions within the ternary field; but these

---

[5] In this diagram the ternary phase ZnSnP$_2$ is shown to lie incorrectly at 21 mol.% Sn on the Sn–ZnP$_2$ coordinate. It should lie at 50 mol.% Sn (or 25 at.% Sn).

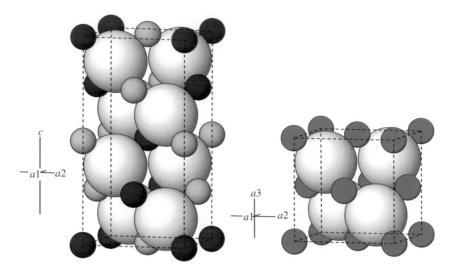

Figure 11.2   The illustration on the left shows the tetragonal unit cell for the ordered ZnSnP$_2$ phase corresponding to the chalcopyrite structure with $Z = 4$. Zinc atoms are shown blue, tin atoms are shown yellow, and phosphorus atoms are shown white. Cell constants, $a = 5.651$ Å and $c = 11.302$ Å [9]. The illustration on the right shows the cubic unit cell for the disordered ZnSnP$_2$ phase corresponding to the sphalerite structure with $Z = 2$. In this phase, the randomised positions of the zinc and tin atoms are shown green, and phosphorus atoms are shown white. Cell constant, $a = 5.651$ Å [26]. This figure was drawn using data from Vaipolin *et al.* [9, 26], cf. ICSD 648170 and 77804, respectively

aspects are not discussed here. The phase relationships in Figure 11.1 have important consequences for growing single crystals of ZnSnP$_2$ from molten tin, as will be shown below.

Concerning the crystal structure, Goodman [1] merely presumed that ZnSnP$_2$ adopted the chalcopyrite structure (CuFeS$_2$), on account of the analogies made in his original work in 1957. In 1968, Vaipolin *et al.* [9] used X-ray diffractometry to show that ZnSnP$_2$ can be obtained as two structural modifications; an ordered phase that crystallises with the chalcopyrite structure, and a disordered phase that crystallises with the sphalerite structure (see Figures 11.2 and 11.3).[6] In both of these phases, the phosphorus substructure comprises a highly ordered cubic close packed (*ccp*) structure of phosphide ions. Within this substructure lie a set of

---

[6] Both of these structures can be derived from the diamond structure. The diamond structure can be described as a three-dimensional network of tetrahedrally coordinated carbon atoms. Or alternatively, as a *ccp* structure of carbon atoms, in which half the tetrahedral interstices are occupied by carbon atoms. In other words, if one considers the zinc, tin and phosphorus atoms to be indistinguishable; the result is the diamond lattice.

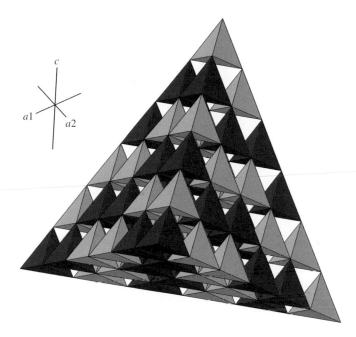

Figure 11.3   The chalcopyrite structure emphasising the {112} face on the tetragonal tetrahedron

tetrahedral interstices; such that half of these are occupied by the metal cations, whilst the other half remain unoccupied.[7] One important point is that the zinc and tin cations can be ordered among these tetrahedral sites, as in the chalcopyrite structure, or disordered, as in the sphalerite structure. It is interesting to note that the ordered ZnSnP$_2$ phase has no measurable tetragonal distortion, i.e., $c/a = 2.000$, and that both phases share the same density; i.e., $4.528 \, \mathrm{g \, cm^{-3}}$.

Vaipolin *et al.* [9] showed that crystals with various degrees of ordering can be obtained by varying the cooling rate of the melt during the crystal growth process. A low cooling rate of 5 °C/h resulted in a material with a high degree of order. Whereas, cooling rates greater than this produced increasing amounts of the disordered phase within the material. A small amount of the ordered phase was detected in material that had been quenched from 800 °C, which suggests that the fully disordered high-temperature phase is difficult to prepare by this route. A common hypothesis, is that these partially ordered crystals are considered to have a mosaic structure and consist of microscopic blocks

---

[7] The relatively modest density of ZnSnP$_2$ $(4.528 \, \mathrm{g \, cm^{-3}})$ is consistent with this structural arrangement, rather than the converse situation in which the phosphide ions are regarded as interstitial species within the zinc/tin *ccp* substructure; the latter would necessitate a much greater density.

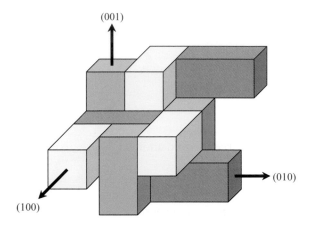

Figure 11.4   Schematic illustration of the relative orientation of the chalcopyrite unit cells that give rise to a mosaic structure

(or domains) with the respective chalcopyrite and sphalerite structures.[8] The x-, y- and z-axes of these domains are all aligned in a parallel, yet otherwise, random way to each other [10]; see Figure 11.4. Ryan *et al.* [11] proposed a model involving zinc and tin site exchange to account for the partially disordered material; as based on the results of ^{119}Sn Mössbauer spectroscopy and ^{31}P nuclear magnetic resonance (NMR) spectroscopy. Vaipolin *et al.* [9] concluded that this order–disorder phase transition occurs over a wide temperature range, and that it depends on several factors, such as the cooling rate and the concentration of the solution. Borshchevskii *et al.* [12] studied the morphology and microstructure of these crystals, and deduced that their crystallisation from molten tin solution was complex, and that the growth of the ordered phase from this solution was particularly slow.

Ryan *et al.* [11] attempted to match the different habits of their ZnSnP$_2$ crystals to the ordered and disordered phases. They concluded that their platy crystals were more disordered than their elongate-columnar crystals.[9] Elyukhin *et al.* [13] noted that since the ordered and disordered phases have identical densities, then the phase transition is very likely to be continuous or, at most, only weakly discontinuous. In other words, the transition is accompanied by an exceedingly large number of negligibly small incremental steps in the degree of ordering, and therefore need not be merely kinetically

---

[8] The order of magnitude at which an intimate mechanical mixture of chalcopyrite- and sphalerite-type phases become sufficiently minuscule, so as to be considered as just a single fully disordered phase, is somewhat semantic, but lies ultimately at the dimensions of the unit cell.

[9] The platy crystals of ZnSnP$_2$ as prepared by the method described in this chapter are dark grey and show a substantial amount of ordering as revealed by powder X-ray diffractometry (see Figure 11.9); which conflict with the disordered nature of the platy crystals prepared by Ryan *et al.* [11]. The occasional silver-grey acicular crystals are considered by the present author to be SnP$_2$.

sluggish. Furthermore, the change in symmetry from $F43m$ (for the disordered phase) to $I42d$ (for the ordered phase) is consistent with a second-order phase transition, in which the changes in enthalpy and entropy are continuous with respect to temperature throughout the transition.

In 1969, Mughal et al. [8] reported a thermal anomaly for $ZnSnP_2$ at 720 °C, but presented no further details, besides expressing their uncertainty as to its cause.[10] Therefore, it is surprising that other workers have been able to ascribe this particular anomaly to the order–disorder phase transition in $ZnSnP_2$. If, for instance, the thermal anomaly relates to enthalpy (rather than heat capacity) then this would tend to contradict the nature of a second-order phase transition, at least under reversible conditions.[11] One way to resolve this problem might be to perform powder X-ray diffractometry on a sample of ordered single-phase $ZnSnP_2$ at progressively higher temperatures in the vicinity of the alleged thermal anomaly (i.e., 720 °C). The disappearance of diffraction peaks relating to $ZnSnP_2$, and the appearance of ZnP_2 together with an increase in the background level due to the presence of a liquid phase, might then confirm the peritectic temperature, 667 °C, as reported by Borshchevskii and Vysotina [2]; see Figure 11.1.

The nature of the chemical bonding in $ZnSnP_2$ is considered to be covalent, as is typical for compounds with a diamond like structure. $ZnSnP_2$ forms a band structure similar to that in GaAs. Details of this band structure are given by Goryunova et al. [14], and in the monograph by Shay and Wernick [10]. Vaipolin et al. [9] performed Hall measurements on $ZnSnP_2$ crystals prepared with different cooling rates. All of their materials showed p-type conductivity with a carrier concentration averaging 7×10^{16} cm^{-3}. Their measurements indicated a progressive decrease in the hole mobility associated with an increase in disorder; ranging from, 68 cm^2 V^{-1} s^{-1} in material that had been cooled at 5 °C/h; to, 4.5 cm^2 V^{-1} s^{-1} in material that had been cooled fairly rapidly at 1200 °C/h. Rubenstein and Ure [7] and Ryan et al. [11] are in agreement that $ZnSnP_2$ is a p-type semiconductor. Ryan et al. [11] performed Hall measurements on materials cooled at 13 °C/h, and obtained a carrier concentration $= 1.4 \times 10^{17}$ cm^{-3}, with a carrier mobility, $\mu_p = 46$ cm^2 V^{-1} s^{-1}. Berkovskii et al. [15] report that $ZnSnP_2$ has an electronic conductivity, σ (300 K) $= 10$ S m^{-1}. Shin and

---

[10] Besides the actual temperature (720 °C) no other details of the thermal anomaly are given by Mughal et al. [8]. Neither is it clear if the thermal anomaly relates to enthalpy or, to heat capacity.

[11] For a transfer of heat during a second-order phase transition to be sufficiently large so as to be detected as a thermal anomaly in a differential thermal analysis (DTA), the transition needs to occur under irreversible conditions. In this case, the structural changes in the system become retarded during the cooling (or heating) process, thereby resulting in a delayed exothermic (or endothermic) reaction, corresponding to the more abrupt changes that will then arise within the crystal structure, at a displaced temperature. But, very high cooling rates can lead to irreversible conditions in which the system remains in a state of metastability; i.e., with no significant transfer of heat taking place at all.

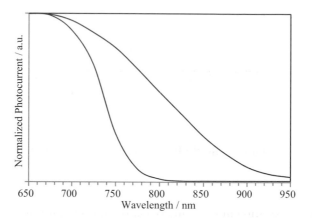

Figure 11.5 Photocurrent spectra for ZnSnP$_2$ samples prepared from melts at different cooling rate: 5 °C/h (red) and 50 °C/h (blue) (Reproduced with permission from Journal of Materials Research, Metal site disorder in zinc tin phosphide by M. A. Ryan, M. W. Peterson, D. L. Williamson, J. S. Frey, G. E. Maciel, B. A. Parkinson, 2(4) 528–537 Copyright (1987) Materials Research Society)

Ajmera [16] found that the resistivity of ZnSnP$_2$ follows an Arrhenius-type temperature dependence, which is consistent with being an intrinsic semiconductor.

The energy of the direct bandgap in these materials has been determined by different methods. Ryan *et al.* [11] recorded the photocurrent spectra for ordered and disordered ZnSnP$_2$ at ambient temperature, as reproduced here in Figure 11.5. From these spectra the values for the direct bandgap are found to be; 1.53 eV (805 nm) for the ordered phase and 1.33 eV (930 nm) for the disordered phase. Goryunova *et al.* [14] determined the fundamental absorption edge for ordered and disordered ZnSnP$_2$ at 298 K (from plots of the absorption coefficient as a function of energy) and ascribed these to direct allowed transitions at; 1.45 eV and 1.32 eV respectively. One consistent feature is that the bandgap in the ordered phase is greater than that in the disordered phase. The slight disparity in the values obtained by these two groups of workers most likely reflects the variation in the degree of ordering within the material.

Direct bandgap semiconductors have a fundamental advantage, as photovoltaic materials, over indirect bandgap semiconductors, such as, silicon, in regards to the efficiency of the excitation process. In a direct bandgap semiconductor, the energy associated with a photon can promote a valence electron directly into the conduction band. But in an indirect bandgap semiconductor, the valence electron requires a fortuitous boost from vibrations within the crystal lattice before the photon can promote it into the conduction band; which makes this class of semiconductor less efficient as a light absorber [17]. Since the direct bandgaps for the ordered and disordered phases are in the near-infrared, attention has been drawn to $ZnSnP_2$ as a photovoltaic material, with possible applications including solar cells and nonlinear optical devices, such as lasers [11, 18].

Several workers, including, Goodman [1], Mughal et al. [8], Rubenstein and Ure [7], Loshakova et al. [19], Borshchevskii et al. [12], Borshchevskii and Vysotina [2] have reported the preparation of $ZnSnP_2$. Loshakova et al. [19] were the first to successfully deploy a crystal-growth technique in which ZnP_2 is reacted at high temperature with a large excess of molten tin inside an evacuated quartz-glass ampoule, such that the tin is not only part of the desired compound but, when present in excess, also acts as the solvent for zinc and phosphorus at high temperature. Ryan et al. [11] prepared single crystals of $ZnSnP_2$ by a similar method, but through the direct chemical reaction of the constituent elements (zinc, tin and red phosphorus powders) together with an excess of molten tin.

From the literature quoted above, it is apparent that single crystals of ordered $ZnSnP_2$ are best prepared from a relatively dilute solution (i.e., 5 mol. % of $ZnSnP_2$ in Sn) at 800–1000 °C, followed by a low cooling rate of 5 °C/h down to at least the melting point of tin (232 °C). These conditions can be considered in reference to the phase diagram in Figure 11.1; in which it can be seen that the crystallisation of $ZnSnP_2$ from such a dilute melt does not commence until the temperature falls below 575 °C, and therefore should, in principle, avoid the formation of ZnP_2.[12] Furthermore, the more dilute the solution, the lower the corresponding temperature of the liquidus; which should enhance the prospect of growing the ordered phase *directly* from the melt, and thus avoid the need for a subsequent cation ordering process to take place within the solid state.

More recently, in 2008, Nakatani et al. [20] prepared single crystals of $ZnSnP_2$ by this tin flux method, but with the ampoule placed inside a two-zone vertical furnace with a temperature gradient (parallel to the ampoule) of $700 \, °C \, m^{-1}$. The temperature at the bottom of the ampoule was set initially at 700 °C, and then cooled (presumably to ambient temperature) at 3 °C/h. Their product comprised single crystals of ordered $ZnSnP_2$ and was devoid of binary phases. Shirakata and Isomura [21] prepared a polycrystalline ingot of the cadmium analogue, $CdSnP_2$ with a grain size 2–4 mm by a similar method.

Many workers have employed the precursor compound, zinc diphosphide, ZnP_2, as a source of zinc and phosphorus for the synthesis. This has certain advantages in that the phosphorus is already chemically combined with the zinc in the desired ratio of the target compound. This also avoids the possibility of generating an excessive vapour pressure of phosphorus at

---

[12] Inclusions of ZnP_2 are reported to occur in nearly all the alloys prepared by Borshchevskii and Vysotina [2] within the Zn–Sn–P system; as well as in many of the materials prepared by the present author's students during the preparation of this chapter. Ryan et al. [11] also report inclusion of ZnP_2 in their product material. Inadequate mixing of the reactants during the course of the reaction (inside the glass ampoule) is likely to be the cause of this nonequilibrium.

elevated temperature during the synthesis, which can cause the glass ampoule to explode. Unfortunately, ZnP$_2$ is unavailable commercially, and therefore needs to be prepared in-house.

Finally, the single crystals of ZnSnP$_2$ need to be isolated from the tin matrix. Rubenstein and Ure [7] removed the excess tin from their product material with mercury at $\sim 100\,°C$, and thereby obtained ZnSnP$_2$ as black platelets with dimensions averaging, $4 \times 4 \times 0.3\,\text{mm}^3$. Loshakova *et al.* [19] report that ZnSnP$_2$ is insoluble in mineral acids (except hydrofluoric acid), and therefore used concentrated nitric acid to remove the excess tin, to yield isolated dark grey ZnSnP$_2$ platy crystals, $3 \times 1.5 \times 0.5\,\text{mm}^3$ in size. Mughal *et al.* [8] used a mixture of nitric acid and hydrochloric acid for this purpose. Their crystals, however, were minute.

Rubenstein *et al.* [7] reported that the back-reflection Laue photographs taken on their ZnSnP$_2$ platelets indicated that the predominant faces are perpendicular to a three-fold axis. This symmetry is consistent with the predominant {111} face on crystals of naturally occurring sphalerite, ZnS (see Figure 11.6), and also with the predominant {112} face on crystals of naturally occurring chalcopyrite, CuFeS$_2$. It is therefore, very likely, that the platy ZnSnP$_2$ 'crystal' as shown in Figure 11.7 is a truncated tetragonal tetrahedron with a well-developed {112} face, corresponding to the ordered phase. But, in reality, this 'crystal' is probably a mosaic of various ordered and disordered domains as discussed above. Although twinning is not reported for ZnSnP$_2$, it is fairly common in naturally occurring chalcopyrite and sphaler-ite; and is a topic worthy of investigation particularly in the light of the hypothesis for a domain structure.

Since ZnSnP$_2$ has the same unit cell constant as GaAs, this feature lends itself to epitaxial growth on various orientations of GaAs substrates. Davis

Figure 11.6   Sphalerite, Binntal, Switzerland, showing the predominant {111} face. Field of view: $22 \times 33\,\text{mm}$ (Photograph used with permission from Ole Johnsen. Copyright (2010) Ole Johnsen)

Figure 11.7   Crystals of ZnSnP$_2$ as prepared by the method described in this chapter. Width of field: left = 1.64 mm, right = 410 μm. The image in the right photograph is an oblique view of the predominant face seen in the left photograph (see outline), and is taken at a higher magnification

and Wolfe [22] found from back reflection Laue patterns that their ZnSnP$_2$ layer grown on {111}-oriented GaAs has the sphalcrite structure; whilst their ZnSnP$_2$ layer grown on {110}-oriented GaAs has the chalcopyrite structure. Furthermore, the latter exhibits antiphase domain boundaries as a consequence of mixed orientation epitaxy, where three orthogonal orientations of the unit cell c-axis are present; see Figure 11.4. Seryogin *et al.* [23] reported that their ZnSnP$_2$ layer grown on {100} oriented GaAs by gas source molecular beam epitaxy in the temperature range 300–360 °C, has the chalcopyrite structure. Francoeur *et al* [24] concluded that modern epitaxial growth techniques offer the possibility to prepare heterostructures of ordered and disordered layers corresponding to different bandgaps, yet with identical chemical composition and density; thus bringing the pioneering work conducted in the Soviet Union, nearer to fruition.

On a rather different note, Goel *et al.* [25] in 1993, reported an organometallic synthesis of ZnSnP$_2$. Appropriate amounts of Zn[P(Si(CH$_3$)$_3$)$_2$]$_2$ and SnCl$_4$ in a 1:2 molar ratio were dissolved together in toluene at ambient temperature. An exothermic reaction occurred that formed a viscous sol. The mixture was refluxed for 24 h, resulting in an air-sensitive black precipitate. This was separated from the solution, and heated at 350 °C for 7 h whilst under dynamic vacuum. The product was then encapsulated in an evacuated quartz ampoule and heated at 600 °C for 12 h. The final product material was analysed by powder X-ray diffractometry; the results of which indicate the presence of disordered ZnSnP$_2$, together with Sn$_4$P$_3$ as an impurity phase. This approach might be an alternative way to prepare fully disordered ZnSnP$_2$.

Preparative Procedure

The following passage describes a method to prepare single euhedral crystals of ordered ZnSnP$_2$ by the direct chemical reaction of the constituent elements (i.e., red phosphorus, zinc, and a large excess of tin) at high temperature, encapsulated in an evacuated quartz-glass ampoule. The homogenous melt is then cooled slowly in order to promote the crystal growth of the ordered phase directly from the melt. The product material should comprise glistening dark grey platy crystals up to ~3 mm in diameter.

Prepare 4 g of zinc–tin phosphide (ZnSnP$_2$) by the following method: Calculate the amount (± 0.001 g) of red phosphorus required in order to obtain a 5 mol.% solution of ZnSnP$_2$ (solute) in Sn (solvent). First, weigh the empty quartz (silica) glass tube (~11 mm ID, ~14 mm OD, 250 mm long, and closed at one end) and then carefully introduce the red phosphorus powder directly into the bottom of this tube using a suitable funnel,[13] so as to avoid any powder sticking to the tube walls. Now reweigh the glass tube (plus contents) in order to obtain the precise mass of the phosphorus. Then based on this mass, calculate the appropriate amounts (± 0.001 g) of zinc shot and tin shot required for the preparation. Carefully place the zinc shot into the glass tube, followed by the tin shot on top of this mixture.

Keep the tube in a vertical position during handling. Attach the tube to a vacuum line and very slowly apply a vacuum; so as to avoid sucking the phosphorus powder out of the tube. Whilst under vacuum, seal the neck of the tube with an oxy-acetylene torch such that the ampoule has a maximum length of 200 mm (so that it can fit inside the chamber furnace). Place the sealed glass ampoule in a vertical position inside the 107-mm tall alumina crucible (CC63 Almath Ltd) that will act as a support, as well as, serving to protect the furnace floor from any accidental spillage of the contents in the event of failure or fracture of the glass ampoule. Then place this in a chamber furnace and heat at 60 °C/h to 1000 °C. After 48 h at 1000 °C, cool to 200 °C with a cooling rate of 5 °C/h. After 0.1 h at 200 °C, continue cooling with a rate of 100 °C/h to ambient temperature (20 °C). This heating cycle will last ~227 h ($9^1/_2$ days).

Whilst wearing goggles, carefully break open the end of the glass ampoule using a glass-cutting tool. Remove the product material taking care to keep the tin ingot intact (see Figure 11.8). Observe and note the morphology of the ingot. Cut a cross section (~ 5 mm thick) of the ingot for analysis by reflected light microscopy. The ZnSnP$_2$ crystals can be isolated from the remaining ingot by dissolving the tin matrix, using the mixed acid solution: 500 ml of 4 mol dm^{-3} hydrochloric acid solution + 500 ml of 4 mol dm^{-3} nitric acid solution (that is to say, the total volume of this mixed acid solution is 1 l). It is important to note that if Zn$_3$P$_2$ is present in the material, then this will form phosphine (PH$_3$) on contact with water or acid. Phosphine is a highly toxic gas, and therefore adequate ventilation is essential when handling these materials.

---

[13] A suitable funnel for this purpose can be made from a sheet of *baking-paper*, rolled up like a large drinking straw, which should be long enough to reach the bottom of the glass tube.

Figure 11.8 Tin ingot ($\sim$ 100 mm long) containing dark grey grains of ZnSnP$_2$ as prepared by the method described in this chapter (Photograph used with permission from Jan Hutzen Andersen. Copyright (2010) Jan Hutzen Andersen)

Place your product material as a single lump (i.e., without breaking it) in a suitably sized glass beaker that must be kept, from now on, in a fume cupboard under adequate ventilation. Add the above acid solution and leave this to stand until the excess tin has completely dissolved ($\sim$ 7 days). Separate the crystals carefully from the acid solution. This can be done by decanting the excess liquid, followed by filtering the remaining contents of the beaker using a Buchner filter funnel with a medium speed filter paper (e.g., *Whatman No. 2 Qualitative*); finally washing with distilled water and then acetone as appropriate. Handle and store your crystals carefully, so as not to damage them.

Observe and note the colour and crystal habit of your product material. For example, it should be interesting to record images of your crystals using

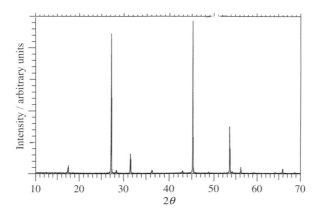

Figure 11.9 Powder XRD pattern (Cu-$K_{\alpha 1}$ radiation) of tetragonal phase ZnSnP$_2$ as prepared by the method described in this chapter. The PDF 72-341 (cf. ICSD 77803 by Vanderah and Nissan [27]) for tetragonal ZnSnP$_2$ is shown red for comparison

scanning electron microscopy (SEM), so keep your crystals safe and *do not* grind them; yet. With the aid of a binocular microscope, or hand lens, carefully select a few individual crystals for analysis by powder X-ray diffraction (scan: 10–70° 2θ). If you happen to observe two different crystal habits among your product material, then separate them and analyse each habit separately (see Figure 11.9). These powder patterns can be compared with the PDF 72-341 for tetragonal ZnSnP$_2$ and with the PDF 71-6473 for cubic ZnSnP$_2$.

The reader may wish to explore various modifications and improvements to this process. One obvious example, based upon Figure 11.1, would be to cool the fused mixture at 1000 °C to 700 °C at a relatively high rate of 100 °C/h, before embarking on the much lower cooling rate of 5 °C/h for the actual crystallisation process. If successful, this should reduce the synthesis time by two days! Another aspect worthy of consideration is to create a thermal gradient along the ampoule during the crystal growth process; cf. Nakatani *et al*. [20].

REFERENCES

1. C. H. L. Goodman, *Nature* **179** (1957) 828–829.
2. A. S. Borshchevskii and M. G. Vysotina, *Izvestiya Akademii Nauk SSSR, Neorganicheskie Materialy* **12** (1976) 615–618.
3. C. H. L. Goodman, *Materials Research Bulletin* **20** (1985) 237–250.
4. I. G. Austin, C. H. L. Goodman and A. E. Pengelly, *Journal of the Electrochemical Society* **103** (1956) 609–610.
5. O. G. Folberth and H. Pfister, *Acta Crystallographica* **13** (1960) 199–201.
6. N. A. Goryunova, B. V. Baranov, V. S. Grigor'eva, L. V. Kradinova, V. A. Maksimova and V. D. Prochukhan, *Izvestiya Akademii Nauk SSSR, Neorganicheskie Materialy* **4** (1968) 1060–1063.
7. M. Rubenstein and R. W. Ure, *Journal of Physics and Chemistry of Solids* **29** (1968) 551–555.
8. S. A. Mughal, A. J. Payne and B. Ray, *Journal of Materials Science* **4** (1969) 895–901.
9. A. A. Vaipolin, N. A. Goryunova, L. I. Kleshchinskii, G. V. Loshakova and E. O. Osmanov, *Physica Status Solidi* **29** (1968) 435–442.
10. J. L. Shay and J. H. Wernick, *Ternary Chalcopyrite Semiconductors: Growth, Electronic Properties, and Applications*, Pergamon Press, Oxford, 1975, 244 pp.
11. M. A. Ryan, M. W. Peterson, D. L. Williamson, J. S. Frey, G. E. Maciel and B. A. Parkinson, *Journal of Materials Research* **2** (1987) 528–537.
12. A. S. Borshchevskii, A. A. Vaipolin, N. A. Goryunova and G. V. Loshakova, *Izvestiya Akademii Nauk SSSR, Neorganicheskie Materialy* **4** (1968) 878–880.
13. V. A. Elyukhin, S. A. Nikishin and H. Temkin, *Crystal Growth and Design* **3** (2003) 733–775.
14. N. A. Goryunova, M. L. Belle, L. B. Zlatkin, G. V. Loshakova, A. S. Poplavnoi and V. A. Chaldyshev, *Fizika i Tekhnika Poluprovodnikov* **2** (1968) 1344–1351.

15. F. M. Berkovskii, V. Z. Garbuzov, N. A. Goryunova, G. V. Loshakova, S. M. Ryvkin and G. P. Shpenkov, *Fizika i Tekhnika Poluprovodnikov* 2 (1968) 744–747.
16. H. Y. Shin and P. K. Ajmera, *Materials Letters* 8 (1989) 464–467.
17. J. M. Crow, *Chemistry World* 5 (2008) 15–17.
18. B. Lita, M. Beck, R. S. Goldman, G. A. Seryogin, S. A. Nikishin and H. Temkin, *Applied Physics Letters* 77 (2000) 2894–2896.
19. G. V. Loshakova, R. L. Plechko, A. A. Vaipolin, B. V. Pavlov, Yu. A. Valov and N. A. Goryunova, *Izvestiya Akademii Nauk SSSR, Neorganicheskie Materialy* 2 (1966) 1966–1969.
20. K. Nakatani, T. Minemura, K. Miyauchi, K. Fukabori, H. Nakanishi, M. Sugiyams and S. Shirakata, *Japanese Journal of Applied Physics* 47 (2008) 5342–5344.
21. S. Shirakata and S. Isomura, *Journal of Crystal Growth* 99 (1990) 781–784.
22. G. A. Davis and C. M. Wolfe, *Journal of the Electrochemical Society* 130 (1983) 1408–1983.
23. G. A. Seryogin, S. A. Nikishin, H. Temkin, A. M. Mintairov, J. L. Merz and M. Holtz *Applied Physics Letters* 74 (1999) 2128–2130.
24. S. Francoeur, G. A. Seryogin, S. A. Nikishin and H. Temkin, *Applied Physics Letters* 74 (1999) 3678–3680.
25. S. C. Goel, W. E. Buhro, N. L. Adolphi and M. S. Conradi, *Journal of Organometallic Chemistry* 449 (1993) 9–18.
26. A. A. Vaipolin, E. O. Osmanov and V. D. Prochukhan, *Izvestiya Akademii Nauk SSSR, Neorganicheskie Materialy* 8 (1972) 947–949.
27. T. A. Vanderah and R. A. Nissan, *Journal of Physics and Chemistry of Solids* 49 (1988) 1335–1338.

PROBLEMS

1. Which particular feature of the 'Pilkington float-glass process' as commonly used for the manufacture of glass plate, gives you confidence that the molten tin used in this preparative method will not react with the quartz-glass ampoule?

2. In this method, the excess elemental tin is removed from the product material ($ZnSnP_2$) using the mixed acid solution; $500\,ml$ of $4\,mol\,dm^{-3}$ hydrochloric acid solution and $500\,ml$ of $4\,mol\,dm^{-3}$ nitric acid solution.
   (a) Write a chemical equation to describe the dissolution reaction.
   (b) Explain what would happen if you attempted to use $1\,l$ of $4\,mol\,dm^{-3}$ hydrochloric acid solution for this purpose instead of the mixed acid as used in this work.

3. Zn_3P_2 undergoes a chemical reaction with water. Write a chemical equation to describe this reaction.

4. State, using the correct nomenclature, the crystal habit(s) of your product material.

5. Use the crystallographic information given in Figure 11.2 in order to calculate the powder diffraction pattern for the cubic $ZnSnP_2$ phase corresponding to Cu-$K_{\alpha 1}$ radiation ($\lambda = 1.5405$ Å) within the range: 0–70° 2θ.

6. (a) Compare and contrast the powder pattern for ordered $ZnSnP_2$ (cf. Figure 11.9) with that for disordered $ZnSnP_2$ (as derived in *Problem 5*)

(b) Comment on the molar volume change (ΔV mol.%) for the order–disorder phase transition.

(c) Use the data to calculate the volume (in units of cm^3) that corresponds to 4 g of ZnSnP$_2$

7. Draw a speculative subsolidus phase diagram for the Zn–Sn–P ternary system (in mol.%)

   (a) Mark clearly, the ternary phase ZnSnP$_2$.

   (b) Mark clearly, the net composition of your starting reaction mixture (Zn + 2P + xSn)

   (c) Comment on any problems that you encounter upon drawing the tie-lines.

8. Describe a method to prepare 10 g of ZnP$_2$.

12

Artificial Kieftite CoSb₃ by an Antimony Self-Flux Method

The cobalt antimonide, $CoSb_3$ is a grey and brittle intermetallic compound that crystallises with the skutterudite structure. It was first prepared by the Norwegian metallurgist Terkel Rosenqvist in 1953 [1], and is considered to be a stoichiometric binary phase. $CoSb_3$ occurs in nature as the rare mineral kieftite, and was discovered by Dobbe *et al.* in 1994 [2] among the Tubaberg copper-cobalt sulfide bearing skarn ores, as collected from dumps and exposed rocks at the abandoned mine, south-eastern Bergslagen ($\sim$20 km south of Nyköping), Sweden [1] The mineralisation occurs in a meta-tuffite[2] formation (upper part of the Paleoproterozoic) that has been metamorphosed to the upper amphibolite facies: $T = 500$–$700\,°C$ and $p \sim 10$ kbar [2]. Another location for the occurrence of kieftite is yet to be identified.

$CoSb_3$ is a diamagnetic semiconductor that exhibits important thermo-electric properties and is a candidate material for use in the generation of

---

[1] Kieftite appears tin-white and highly reflective in polished section [2]. The mineral was named after the Dutch mineralogist, Cornelis Kieft.
[2] Meta-tuffite is a rock consisting of consolidated volcanic ash (ejected during a volcanic eruption) that has been subjected to metamorphism.

Synthesis, Properties and Mineralogy of Important Inorganic Materials   By Terence E. Warner
© 2011 John Wiley & Sons, Ltd

thermoelectric power.[3] The dimensionless figure of merit used to evaluate the thermoelectric properties of this class of material is the parameter ZT, where T is the temperature and $Z = \alpha^2\sigma/\kappa$ (α is the Seebeck coefficient[4], σ is the electrical conductivity, and κ is the thermal conductivity). As a rough guide, the ideal thermoelectric material should have a large Seebeck coefficient ($|\alpha| \geq 150\,\mu V\,K^{-1}$), a high electrical conductivity ($\sigma \geq 10^3\,\Omega^{-1}\,cm^{-1}$) and a low thermal conductivity ($\kappa \leq 2\,W\,m^{-1}\,K^{-1}$); in other words, it should have the combined attributes of conducting electricity like a metallic crystal and conducting heat like a glass. Ordinary metals are generally excluded because the Seebeck coefficient is far too low (typically, $|\alpha| < 10\,\mu V\,K^{-1}$).[5] On the other hand, the vast majority of glasses and semiconductors display an electrical conductivity to thermal conductivity ratio (σ/κ) that is too small; but there are some notable exceptions, such as, bismuth telluride, Bi_2Te_3, which is used commercially in thermoelectric refrigeration devices [4].

In the case of CoSb₃, ZT peaks at a temperature near 350 °C with values[6] of ZT approaching 0.2. Although 0.2 is at the low end of the scale, CoSb₃ is an important model compound in the search for related phases with more favourable thermoelectric properties.[7] For example, the partial substitution

---

[3] In 1823 Seebeck discovered that an electrostatic potential difference appeared across the junction of two dissimilar electrically conducting materials whilst under the influence of a temperature gradient. This phenomenon is known as the Seebeck effect, and is exploited for the production of thermocouples, and for the generation of thermoelectric power [3]. For example, radioisotope thermoelectric generators (using a SiGe alloy as the thermoelectric material) have been used in deep space probes, such as, NASA's *Voyager* missions for interplanetary exploration. The process can run in reverse, whereby an applied electric current produces heating and cooling at opposite ends of the material. In other words, the process can be considered as a transportation of entropy from one end of the material to the other. This phenomenon is known as the Peltier effect, and is exploited for thermoelectric refrigeration, particularly for cooling electronic components. These devices have no moving parts and therefore are maintenance free and silent. The reader is referred to the review articles on thermoelectric materials by: Sales [3]; Tritt and Subramanian [4]; and Snyder and Toberer [5], for further information.

[4] The Seebeck coefficient can be thought of as the entropy per charge carrier; and is also known by the alternative expression, 'thermopower' [4]. For the sake of comparison, the quantity $k_B/e \sim 87\,\mu V\,K^{-1}$ (in which k_B is the Boltzmann constant and e is the charge of the electron) is a constant that represents the thermopower of a classical electron gas [4].

[5] For example, the Pauli-paramagnetic metal, $FeNi_2S_4$ (see Chapter 13) has a Seebeck coefficient, $\alpha = -5\,\mu V\,K^{-1}$ at room temperature [6]. Superconducting phases have, characteristically, a Seebeck coefficient of zero, and therefore a ZT rating of zero, whilst in their superconducting state.

[6] For instance, the following peak values of ZT for CoSb₃ have been reported as: 0.095 at 673 K [7]; 0.11 at 650 K [8]; 0.17 at 550 K [9].

[7] Values of $ZT \geq 1$ would be more ideal. Although CoSb₃ has a relatively high Seebeck coefficient ($\sim 250\,\mu V\,K^{-1}$ at 300 °C) it has, unfortunately, a considerable thermal conductivity ($\sim 3\,W\,m^{-1}\,K^{-1}$ at 300 °C). Slack [10] has estimated that thermoelectric materials might exist with values of ZT as high as 4. It is interesting to note that the properties for ideal thermoelectric materials are opposite to those in diamond; which is a dielectric cubic polymorph of carbon with a very high thermal conductivity ($\sim 2000\,W\,m^{-1}\,K^{-1}$ at 25 °C).

Figure 12.1 A 20-g polycrystalline CoSb$_3$ ingot as prepared by the method described in this chapter (Photograph used with permission from Ngo Van Nong. Copyright (2010) Ngo Van Nong)

of iron for cobalt, together with the insertion of lanthanide metal atoms into the CoSb$_3$ structure, results in filled skutterudite phases such as LaFe$_3$CoSb$_{12}$ and CeFe$_3$CoSb$_{12}$ that have large values of ZT. Unfortunately, quaternary intermetallic phases are notoriously difficult to synthesise as single-phase materials. Therefore, as an introduction to this subject, this chapter describes the preparation of a polycrystalline CoSb$_3$ ingot (see Figure 12.1). This involves the direct chemical reaction of its constituent elements within an evacuated glass ampoule at high temperature This preparative method exploits the peritectic phase relationships within the Co–Sb binary system, and is referred to as the so-called, 'antimony self-flux' method. This can be contrasted against the tin-flux method for preparing ZnSnP$_2$, in which an excess of molten tin is used as the flux (see Chapter 11).

CoSb$_3$ crystallises with the skutterudite, CoAs$_3$ structure, as shown in Figure 12.2. It belongs to the cubic crystal system and space group $Im\overline{3}$, with cell constant $a = 9.0411$ Å, and has a calculated density of 7.63 g cm^{-3} [2]. The crystal structure was first determined in 1953 by Rosenqvist, using single-crystal X-ray diffractometry [1], and refined by Dobbe et $al.$ in 1994 using the same technique [2]. The skutterudite structure can be visualised as a distorted modification of the ReO$_3$ structure (as shown in Figure 12.3), which in turn is derived from the ideal cubic perovskite structure in which the A-sites are all vacant (cf. Figure 9.2). More specifically, the CoSb$_3$ structure consists of a framework of tilted corner-sharing trigonally distorted [CoSb$_6$] octahedra, with a tilt angle, $\phi = 34.79°$ [11]. This distortion brings certain antimony atoms into close proximity with one another, resulting in the formation of four-membered Sb$_4$ rings. The distribution and orientation of these rectangular (almost square) Sb$_4$ rings is somewhat complex; as revealed in Figure 12.4. This unit cell is drawn so as to emphasise the cobalt substructure (which belongs to the primitive cubic lattice) through a set of cobalt primitive

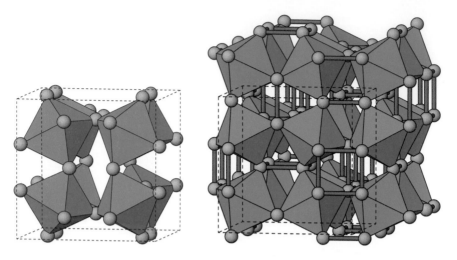

Figure 12.2   Left: The crystal structure of $CoSb_3$ with the outline of the unit cell, with $Z = 8$. The $[CoSb_6]$ octahedra are shown blue, and the Sb atoms are shown grey. Right: An extension of the unit cell showing the Sb−Sb bonds in orange; thus highlighting the Sb_4 rings (or parts of the rings). This figure was drawn using data from Dobbe *et al.* [2], cf. ICSD 79137

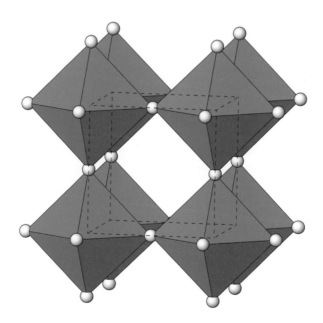

Figure 12.3   The ReO_3 structure with the outline of the unit cell. The $[ReO_6]$ octahedra are shown in blue and the oxygen ions are shown in white

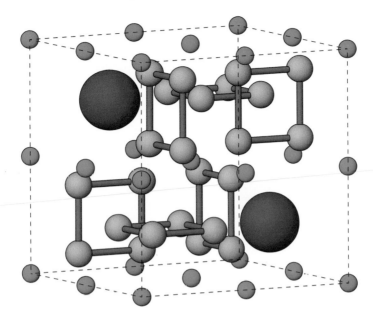

Figure 12.4 It is interesting to view the $CoSb_3$ structure in terms of the filled skutterudite prototype structure $LaFe_4P_{12}$ with the unit cell corresponding to $Z = 2$. The Fe atoms are shown blue. The P atoms are shown grey, with the P–P bonds in orange so as to emphasise the P_4 rectangular (almost square) rings. The Fe and P atoms are shown with very much reduced radii in order to see into the structure. The La sites are shown purple-pink; which happen to be vacant in the case of $CoSb_3$. This figure was drawn using data from Jeitschko and Braun [36], cf. ICSD 1286

cubic octants. These octants accommodate collectively the six Sb_4 rings at their centres; leaving two octants unoccupied. The plane of these Sb_4 rings is orientated with respect to the orthogonal axes according to the alternating sequence: x-, y-, z-, and vacant; thereby maintaining cubic symmetry [12]. The coordination structure of $CoSb_3$ can therefore be described as, $Co_4(\square[Sb_4]_3)$ in which the square represents an unoccupied octant within the cobalt substructure [13].

Cobalt and antimony have similar electonegativities, and so the chemical bonding between the Co and Sb atoms in $CoSb_3$ is considered to be strongly covalent [5]; a feature that is consistent with its semiconducting properties. $CoSb_3$ is described as a narrow-bandgap semiconductor with a highly nonparabolic band structure [14]. The reported values for the bandgap range from 0–0.56 eV [15, 16]. The values for the thermoelectric parameters also vary quite considerably.[6] This is most likely a consequence of certain differences in the microstructure and in the nature of the crystal defects, including chemical impurities, within the specimens of $CoSb_3$ as studied by the various authors.

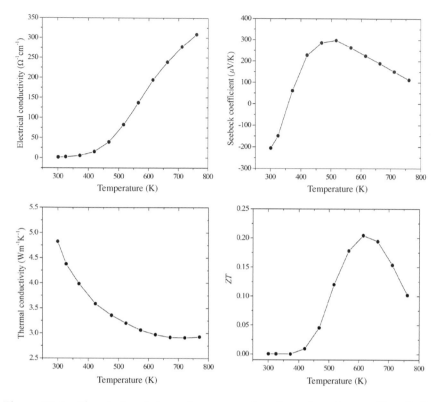

Figure 12.5 Electrical and thermal transport data from the polycrystalline CoSb$_3$ ingot as shown in Figure 12.1. NB *ZT* does not depend on sample porosity. The author is grateful to Dr Ngo Van Nong, Risø DTU National Laboratory for Sustainable Energy, Denmark, for recording these data

The electrical and thermal transport properties regarding the polycrystalline CoSb$_3$ ingot as prepared here are shown as a function of temperature in Figure 12.5. In this specimen, the Seebeck coefficient switches from negative to positive above 87 °C, indicating a transition from predominantly n-type to p-type conduction, with the relatively high value of 300 μV K^{-1} at 200 °C. A specimen of p-type CoSb$_3$ as prepared by Morelli *et al.* [14] has a hole concentration of ~3 × 10^{18} cm^{-3}, together with an exceptionally high hole mobility of ~2300 cm^2 V^{-1} s^{-1} at room temperature. Although these values are encouraging,[8] they fall short of the requirements for CoSb$_3$ to become commercially useful. Nevertheless, CoSb$_3$ is still attractive, since it offers the

---

[8] These values can be compared with the p-type ordered ZnSnP$_2$ photoconductor that has a hole concentration of 1.4 × 10^{17} cm^{-3}, and a hole mobility of 46 cm^2 V^{-1} s^{-1} at room temperature (as described in Chapter 11). ZnSnP$_2$ has a calculated thermal conductivity of 12.4 W m^{-1} K^{-1} at 298 K [17].

possibility of preparing both n-type and p-type material, which promise good compatibility in terms of thermal and chemical stability, and thermal expansion; essential requirements for use in thermoelectric devices.

Unfortunately, the thermal conductivity of CoSb$_3$ is not low enough (see Figure 12.5), and this aspect in particular limits its effectiveness as a thermoelectric material. However, the thermal conductivity can be reduced to a more acceptable level through the insertion of heavy atoms into the two vacant octants (as described above), resulting in the filled skutterudites, LaFe$_3$CoSb$_{12}$ and CeFe$_3$CoSb$_{12}$, and the semifilled skutterudite, Yb$_{0.2}$-Co$_4$Sb$_{12}$ [18], without adversely affecting the other thermoelectric parameters.[9] One of the ideas to account for the reduction in the thermal conductivity was the so-called 'rattling' model proposed by Slack [10]. This model was based on the premise that the large thermal motion or, 'rattling', of the heavy, but weakly bound, lanthanide atom inside the oversized cavity within the skutterudite structure, would serve to scatter the acoustic phonons and thereby reduce the lattice component of the thermal conductivity. Feldman et al. [19] have challenged the applicability of this model to the filled skutterudites, and suggested that anharmonic scattering of phonons by harmonic motions of the lanthanide atoms is a more plausible mechanism.

It is informative to compare CoSb$_3$ with other skutterudite phases. For instance, artificial skutterudite, CoAs$_3$ was shown to be a semiconductor with a Seebeck coefficient, $\alpha(298\,\mathrm{K}) = 100\,\mu\mathrm{V\,K}^{-1}$ by Hulliger, in 1961 [20].[10] The nickel-rich solid solution (Ni,Co)As$_3$ has been indentified as a mineralogical phase by Rudashevskii et al. [21]. The partial substitution of nickel for cobalt within CoSb$_3$ results in the solid solution series: Co$_{1-x}$Ni$_x$Sb$_3$ ($0 \leq x \leq 0.25$) and retains its semiconducting behaviour [22]. The metal antimonides, FeSb$_3$ and NiSb$_3$ have been prepared as metastable phases [23–25], but their thermoelectric properties have yet to be characterised; whereas RhSb$_3$ and IrSb$_3$ are apparently stable and exhibit promising thermoelectric properties [26]. Bismuth does not substitute for antimony in CoSb$_3$ [27], and the cobalt bismuthide, CoBi$_3$ is yet to be prepared. Knowledge of these phases is important, since the controlled alloying of some of them with CoSb$_3$ has shown significant enhancements in the values of ZT. For example, the pentenary alloy, Co$_{0.97}$Ir$_{0.03}$Sb$_{2.81}$-Te$_{0.04}$As$_{0.15}$ has $ZT \sim 0.6$ at 400 °C [27].

An equilibrium phase diagram for the Co−Sb binary system was constructed by Feschotte and Lorin in 1989 [31], as shown in Figure 12.6. From this diagram it can be seen that CoSb$_3$ has an incongruent melting point at

---

[9] The thermal conductivity (at 25 °C) for LaFe$_3$CoSb$_{12}$ = 1.6 W m^{-1} K^{-1} [28]; CeFe$_3$CoSb$_{12}$, = 1.5 W m^{-1} K^{-1} [29]; and Yb$_{0.2}$Co$_4$Sb$_{12}$ = 2.6 W m^{-1} K^{-1} [30].

[10] CoAs$_3$ is generally regarded as an inferior thermoelectric material in comparison to CoSb$_3$. The differences most probably lie in the different localisation of the valence As $4p$ and Sb $5p$ state [15].

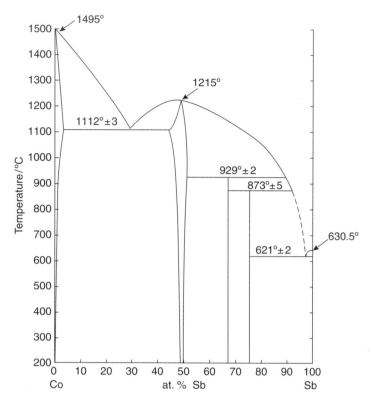

Figure 12.6 An equilibrium phase diagram of the Co–Sb binary system (Reproduced with permission from Journal of the Less-Common Metals, Les systemes binaires Fe-Sb, Co-Sb et Ni-Sb by P. Feschotte and D. Lorin, 1989, 155, 255–269 Copyright (1989) Elsevier Ltd)

873 °C, corresponding to the peritectic reaction:

$$CoSb_2(s) + L \rightleftharpoons CoSb_3(s)$$

and forms a binary eutectic with antimony at 621 °C. This creates a temperature 'window' between 621–873 °C in which CoSb$_3$ can coexist with a binary melt at ambient pressure; a feature that is exploited during its synthesis.

CoSb$_3$ was first prepared by Rosenqvist [1] in 1953 by heating and annealing a calculated mixture of cobalt and antimony in an evacuated and sealed quartz tube. Even though Dudkin and Abrikosov [32, 33] studied the thermoelectric properties of CoSb$_3$ in the Soviet Union during the mid 1950's, it was not until the early 1990s that CoSb$_3$ was identified as a potentially useful thermoelectric material [26]; which prompted the development of improved methods for its synthesis.

Souma and Ohtaki [34, 35] have performed a systematic study of the synthesis of CoSb$_3$ by the so-called, 'antimony self-flux' method. They devised an optimal procedure for preparing single-phase material with the composition, CoSb$_3$. The ideal conditions involved reacting cobalt powder (100 mesh) with antimony lump (3–10 mm) weighed in stoichiometric quantities corresponding to the composition, CoSb$_3$. These particle (or lump) sizes were apparently necessary for a successful reaction. These were encapsulated inside an evacuated quartz-glass ampoule, and introduced into an electric furnace in the vertical position, and then heated to 750 °C for 10 h. Both heating and cooling rates were set at 100 °C/h. During the process, the antimony melted and reacted with the cobalt to form single phase CoSb$_3$. The phase purity of their material was confirmed by powder X-ray diffractometry. Their preparative method is adopted here.

Preparative Procedure

Prepare a 20-g ingot of polycrystalline CoSb$_3$ by the following method: Weigh the appropriate amounts of cobalt powder[11] ($-100/+325$ mesh, 99.8 %, Alfa Aesar) and antimony lump (≤ 12.5 mm, 99.5 %, Alfa Aesar) to a precision of ± 0.0005 g corresponding to the empirical formula, CoSb$_3$. Particularly large lumps of antimony may require crushing to smaller millimetre-sized pieces in order to fit into the glass ampoule; but do not grind the antimony to a powder! Introduce the cobalt powder into the bottom of a quartz-glass tube sealed at one end ($\sim$12 mm ID; $\sim$14 mm OD; $\sim$200 mm long) with the aid of a small funnel, so as to avoid spillage. Then introduce the antimony lump, on top of the cobalt powder bed,

Keep the tube in a vertical position during handling. Attach the tube to a vacuum line and apply a vacuum very slowly; so as to avoid sucking the powders out of the tube. Whilst under vacuum, seal the neck of the tube with an oxy-acetylene torch such that the ampoule has a length of $\sim$130 mm. It is sensible to restrict the amount of vapour space within the ampoule. Place the sealed glass ampoule in a vertical position inside the 107 mm tall alumina crucible (CC63 Almath Ltd) that will act as a support, as well as, serving to protect the furnace floor from any accidental spillage of the contents in the event of a fracture to the glass ampoule. Then place this in a chamber furnace (or a pottery furnace) and heat at 100 °C/h to 750 °C. After 10 h at 750 °C, cool to room temperature (20 °C) with a cooling rate of 100 °C/h. This heating cycle will last $\sim$25 h (1 day).

Whilst wearing goggles, carefully break open the end of the glass ampoule using a glass-cutting tool. Remove the product material taking care to keep the CoSb$_3$ ingot intact; bearing in mind that CoSb$_3$ is a brittle material. Observe and note the morphology of the ingot (see Figure 12.1). Cut a small section (1–2 g) from the ingot using a diamond saw; alternatively, chip a piece

---

[11] The reader should be aware that cobalt powder is toxic, and therefore it must be handled with due care and attention.

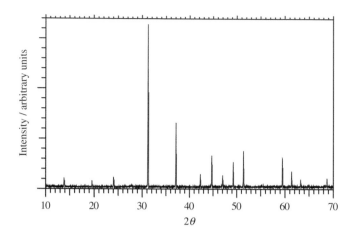

Figure 12.7 Powder X-ray diffraction pattern (Cu-$K_{\alpha 1}$ radiation) of CoSb₃ as prepared by the method described in this chapter. This powder pattern corresponds closely to the PDF 83-55 (cf. ICSD 79137) for the mineral kieftite, CoSb₃ (shown in red)

from the end of the ingot with a hammer and chisel. Grind this to a fine powder using an agate pestle and mortar. Submit the powder for analysis by powder X-ray diffractometry (scan: 10–70° 2θ). Compare your powder pattern with the PDF 83-55 (cf. ICSD 79137) for the mineral kieftite, CoSb₃,

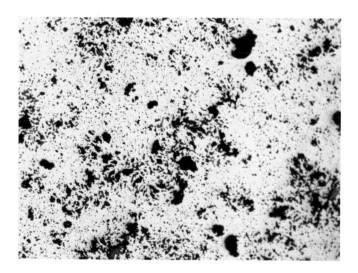

Figure 12.8 A polished section of the polycrystalline CoSb₃ ingot under reflected polarised light; appearing tin-white and highly reflective. The black regions are voids within the ingot. Width of field = 1.2 mm

and with Figure 12.7. Keep the remaining ingot in an appropriately labelled specimen jar.[12]

A polished section of the $CoSb_3$ ingot as prepared here is shown in Figure 12.8. The high porosity of this material is consistent with the results of Souma and Ohtaki [35]. If a more dense material is required, then the reader should crush the ingot using a hardened steel percussion impact piston, then pulverise the fragments using a planetary mono mill, and form a compacted powder monolith. This monolith should be encapsulated in an evacuated quartz-glass ampoule and sintered just below the incongruent melting point of 873 °C.

REFERENCES

1. T. Rosenqvist, *Acta Metallurgica* 1 (1953) 761–763.
2. R. T. M. Dobbe, W. J. Lustenhouwer, M. A. Zakrzewski, K. Goubitz, J. Fraanje and H. Schenk, *Canadian Mineralogist* 32 (1994) 179–183.
3. B. C. Sales, *Current Opinion in Solid State and Materials Science* 2 (1997) 284–289.
4. T. M. Tritt and M. A. Subramanian, *Materials Research Science Bulletin* 31 (2006) 188–198.
5. G. J. Snyder and E. S. Toberer, *Nature Materials* 7 (2008) 105–114.
6. M. G. Townsend, J. R. Gosselin, J. L. Horwood, L.G. Ripley and R. J. Tremblay, *Physica Status Solidi* A 40 (1977) K25–K29.
7. J. X. Zhang, Q. M. Lu, K. G. Liu, L. Zhang and M. L. Zhou, *Materials Letters* 58 (2004) 1981–1984.
8. Z. He, C. Stiewe, D. Platzek, G. Karpinski, E. Muller, S. Li, M. Toprak and M. Muhammed, *Journal of Applied Physics* 101 (2007) 53713.
9. N. Dong, X. Jia, T. C. Su, F. R. Yu, Y. J. Tian, Y. P. Jiang, L. Deng and H. A. Ma, *Journal of Alloys and Compounds* 480 (2009) 882–884.
10. G. A. Slack, *New Materials and Performance Limits for Thermoelectric Cooling*, 407–440, in CRC Handbook of Thermoelectrics, ed. D. M. Rowe, 1995, CRC Press, Boca Raton, Florida, 701 pp.
11. R. H. Mitchell, *Perovskites: Modern and Ancient*, Almaz Press, Ontario, 2002, pp318.
12. J. L. Feldman and D. J. Singh, *Physical Review* B 53 (1996) 6273–6282.
13. J.-P. Fleurial, T. Caillat and A. Borshchevsky, *Proceedings of the 16thInternational Conference on Thermoelectrics* (1997) 1–11.
14. D. T. Morelli, T. Caillat, J.-P. Fleurial, A. Borshchevsky, J. Vandersande, B. Chen and C. Uher, *Physical Review* B 51 (1995) 9622–9628.
15. E. Z. Kurmaev, A. Moewes, I. R. Shein, L. D. Finkelstein, A. L. Ivanovskii and H. Anno, *Journal of Physics: Condensed Matter* 16 (2004) 979–987.

---

[12] Thermoelectric measurements can be made by cutting (or grinding) sections from the $CoSb_3$ ingot with the dimensions (e.g., $10 \times 10 \times 3\,mm^3$) as required by the following specialist instruments used for the thermal and electrical transport measurements: electrical conductivity and Seebeck coefficient (ULVACZEM-3); carrier concentration measurements (Hall probe using PPMS); and the thermal conductivity measurements (Laser Flash instrument LFA457, NETZSCH).

16. T. Caillat, J.-P. Fleurial and A. Borshchevsky, *Journal of Crystal Growth* **166** (1996) 722–726.
17. S. M. Wasim, *Physica Status Solidi* A **51** (1979) K35–K40.
18. G. S. Nolas, M. Kaeser, R. T. Littleton and T. M. Tritt, *Applied Physics Letters* **77** (2000) 1855–1857.
19. J. L. Feldman, D. J. Singh, I. I. Mazin, D. Mandrus and B. C. Sales, *Physical Review B* **61** (2000) 9209–9212.
20. F. Hulliger, *Helvetica Physica Acta* **34** (1961) 782–786.
21. N. S. Rudashevskii, N. N. Shishkin, I. A. Bud'ko, A. F. Sidorov and G. V. Spiridonov, *Zapiski Vsesoyuznogo Mineralogicheskogo Obshchestva* **104** (1975) 209–216.
22. M. Toprak, Y. Zhang, M. Muhammed, A. A. Zakhidov, R. H. Baughman and I. Khayrullin, *Proceedings of the 18th International Conference on Thermoelectrics* (1999) 382–385.
23. M. D. Hornbostel, E. J. Hyer, J. Thiel and D. C. Johnson, *Journal of the American Chemical Society* **119** (1997) 2665–2668.
24. J. R. Williams, D. C. Johnson, and G. Nolas, *Book of Abstracts*, 219th ACS National Meeting, San Francisco, CA, March 26–30, (2000).
25. J. R. Williams and D. C. Johnson, *Inorganic Chemistry* **41** (2002) 4127–4130.
26. T. Caillat, A. Borshchevsky and J. P. Fleurial, *Proceedings of the 28th International Society Energy Conversion Engineering Conference*, Volume 1 (1993) 245–248.
27. J. W. Sharp, E. C. Jones, R. K. Williams, P. M. Martin and B. C. Sales, *Journal of Applied Physics* **78** (1995) 1013–1018.
28. B. C. Sales, D. Mandrus and R. K. Williams, *Science* **272** (1996) 1325–1328.
29. K. Matsubara, *Journal of Advanced Science* **9** (1997) 171–176.
30. H. Li, X. Tang, X. Su and Q. Zhang, *Applied Physics Letters* **92** (2008) 202114.
31. P. Feschotte and D. Lorin, *Journal of the Less-Common Metals* **155** (1989) 255–269.
32. L. D. Dudkin and N. Kh. Abrikosov, *Zhurnal Neorganicheskoi Khimii* **1** (1956) 2096–2105.
33. L. D. Dudkin and N. Kh. Abrikosov, *Seriya Fizicheskaya* **21** (1957) 141–145.
34. T. Souma and M. Ohtaki, *The 24th International Conference on Thermoelectrics* (2005) 121–124.
35. T. Souma and M. Ohtaki, *The 25th International Conference on Thermoelectrics* (2006) 598–602.
36. W. Jeitschko and D. Braun, *Acta Crystallographica B* **33** (1977) 3401–3406.

PROBLEMS

1. Comment on the oxidation states of cobalt and antimony in CoSb$_3$.

13

Artificial Violarite FeNi$_2$S$_4$ by a Hydrothermal Method Using DL-Penicillamine as the Sulfiding Reagent

FeNi$_2$S$_4$ is commonly considered to be the iron-rich end member of the solid-solution series, Fe$_{1-x}$Ni$_{2+x}$S$_4$ $(0 \leq x \leq 1)$ that crystallises with the spinel structure. This solid solution occurs naturally as the thiospinel mineral, violarite, and is mined for its nickel values.[1] Within the mining and extractive metallurgical industry, there is an interest in acquiring knowledge of the properties and behaviour of ore minerals in order to optimise the mineral processing and metal extraction processes[2] and violarite falls into this category. Petrological material is not always suited for scientific investigations, since practical-sized specimens of violarite are invariably accompanied by other mineralogical phases, whilst the violarite per se, displays a variable iron to nickel ratio, and often contains minor, yet valuable, amounts of cobalt

---

[1] Violarite is a common ore mineral in the nickel-sulfide deposits of Western Australia [1]. At the time of writing (27 May 2010) nickel commanded a price of US$ 21 580 per tonne on the London Metal Exchange, compared with copper at US$ 6 875 per tonne. Furthermore, these ores typically contain other valuables, especially cobalt and platinum.

[2] The term, *mineral processing*, refers to the physical separation of the valuable minerals from the gangue (i.e., the unwanted minerals) in order to form an ore concentrate; the reader is referred to the textbook by Napier-Munn and Wills [2] for more details. For information regarding the extraction of nickel metal from nickel ore concentrates, the reader is referred to Burkin [3], and Boldt and Queneau [4].

Synthesis, Properties and Mineralogy of Important Inorganic Materials   By Terence E. Warner
© 2011 John Wiley & Sons, Ltd

and copper. Consequently, there is a need to prepare violarite artificially as a phase pure material and with a definite composition. The most common method of synthesis, involves a lengthy two-stage process of reacting elemental iron and nickel together at high temperature with consecutive amounts of sulfur within a sealed evacuated glass ampoule. The product, however, is often inhomogeneous and is typically contaminated with pyrite, FeS$_2$, vaesite, NiS$_2$, and nickel-iron monosulfide solid solution, (Ni,Fe)$_{1-x}$S. This chapter describes the preparation of artificial violarite, with the idealised composition, FeNi$_2$S$_4$, by a novel hydrothermal process conducted in an autoclave at 130 °C using DL-penicillamine as the sulfiding reagent.[3]

Violarite was described originally by Lindgren and Davy in 1924 [5], as a violet grey nickel sulfide mineral, from which its name (in Latin; *violaris*) was derived. Its composition was determined as, (Ni,Fe)$_3$S$_4$, by Short and Shanon in 1930 [6]. In 1942, Harcourt [7] published the first set of powder X-ray diffraction data (No. 156) for violarite, using a specimen from Vermilion mine, Sudbury, Ontario, Canada [7].[4] Craig [8] used powder X-ray diffractometry to show that violarite belongs to the cubic crystal system with the space group $Fd\bar{3}m$, and has a cell constant, $a = 9.465$ Å (see Figure 13.1). Most recently, Tenailleau *et al.* [9] studied various specimens of artificial violarite using powder neutron diffractometry over the temperature range 100–498 K, and showed that FeNi$_2$S$_4$ crystallises with the inverse spinel structure, (Ni)^{tet}[FeNi]^{oct}S$_4$, with a cell constant, $a = 9.442$ Å at 298 K.

A tentative isothermal section of the equilibrium phase diagram for the Fe−Ni−S ternary system at 200 °C, was constructed by Lundqvist in 1947 [10], and is shown in Figure 13.2. The principle features of this diagram remain valid today. The violarite solid solution, Fe$_{1-x}$Ni$_{2+x}$S$_4$, is considered to be continuous between the limits, $0 \leq x \leq 1$, and lies towards the right of the diagram at ∼57 at.% sulfur, merging with polydymite, Ni$_3$S$_4$, in the Ni−S binary system. At the far left of the diagram, the pure iron analogue, greigite, Fe$_3$S$_4$, is considered to be an unstable phase at 200 °C with respect to a two-phase mixture of pyrite, FeS$_2$ and pyrrhotite, Fe$_{1-x}$S.

In 1971, Craig [8] published a phase diagram of the subsolidus relationships within the FeNi$_2$S$_4$−Ni$_3$S$_4$ pseudobinary join (see Figure 13.3). From

---

[3] Acknowledgment is given gratefully to the late Professor H. Toftlund, for perceiving the original idea of utilising DL-penicillamine as the sulfiding reagent for the preparation of Fe$_{1-x}$Ni$_{2+x}$S$_4$ ($0 \leq x \leq 1$) by the hydrothermal method as described in this chapter; and to, Miss W. Hagedorn Jørgensen, who performed the original syntheses of certain compounds within the solid-solution series Fe$_{1-x}$Ni$_{2+x}$S$_4$ ($0 \leq x \leq 1$), including, FeNi$_2$S$_4$ and Ni$_3$S$_4$ [11].
[4] The Sudbury Basin is one of the largest nickel-copper sulfide and platinum group metal ore deposits in the world. The ore mineralisation processes were accentuated by a huge meteorite impact 1850 million years ago in the Paleoproterozoic era. The reader is referred to Rousell *et al.* [12] for further details.

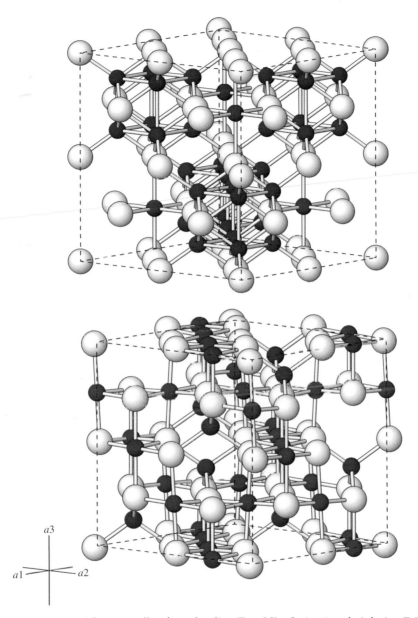

Figure 13.1   The unit cells of pentlandite, $Fe_{4.5}Ni_{4.5}S_8$ (top) and violarite, $FeNi_2S_4$ (bottom); illustrating their close structural relationship through their common face centred cubic (*fcc*) sulfur substructure (shown as white spheres). The tetrahedral cation sites are shown red, and the octahedral cation sites are shown blue. The nickel and iron cations are presumed to be disordered in pentlandite. Violarite has a metal atom content equal to $^2/_3$ of the metal content of pentlandite, corresponding to $\sim$18% contraction in the volume of the unit cell. The sulfur atoms in violarite lie in slightly off-centred positions compared with the idealised spinel structure as shown in the previous chapter. This figure was drawn using data for pentlandite from Pearson and Buerger [47], cf. ICSD 22304; and for violarite from Vaughan and Craig [17], cf. ICSD 42590

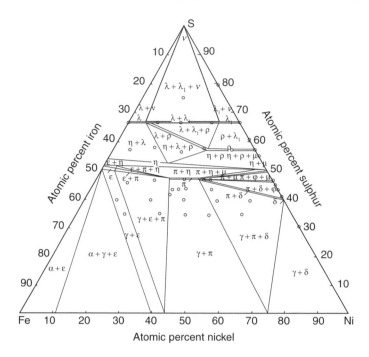

Figure 13.2   A tentative isothermal section of the equilibrium phase diagram for the Fe−Ni−S ternary system at 200 °C, as constructed by Lundqvist in 1947 [10]. The letters denote the following phases: α, body-centred cubic (Fe,Ni); γ, face-centred cubic (Fe,Ni); δ, Ni$_3$S$_2$; ε, the superstructure phase of FeS; π, pentlandite; φ, Ni$_7$S$_6$; μ, millerite; ρ, violarite; λ, pyrite; λ_1, NiS$_2$; ν, liquid sulfur; η, nickel-iron monosulfide solid solution (○ denotes the composition of the alloys studied by Lundqvist)

this diagram it can be seen that violarite with the composition, Fe$_{0.92}$Ni$_{2.08}$S$_4$, is the most thermally stable, and exists up to 461 ± 3 °C. Above this temperature, Fe$_{0.92}$Ni$_{2.08}$S$_4$ decomposes into the assemblage: nickel-iron monosulfide solid solution, (Ni,Fe)$_{1-x}$S; pyrite, FeS$_2$; and vaesite, NiS$_2$. The composition, FeNi$_2$S$_4$, is a stable phase up to at least 300 °C, then at a certain temperature in between 300–400 °C it becomes unstable with respect to the assemblage: pyrite, FeS$_2$; nickel-iron monosulfide solid solution, (Ni,Fe)$_{1-x}$S; and nickel-rich violarite, Fe$_{1-x}$Ni$_{2+x}$S$_4$. This imposes an upper limit of 300 °C for the preparation of FeNi$_2$S$_4$, in order to be confident that this compound will form and remain stable.

There is some uncertainty regarding the composition of the iron-rich end-member of the violarite solid solution; at least, so far as there are many examples of natural [13,14] and artificial violarite [9, 15, 16] containing a higher Fe:Ni ratio than portrayed by the formula, FeNi$_2$S$_4$. Vaughan and Craig [17], conclude that naturally occurring compositions between FeNi$_2$S$_4$ and Fe$_3$S$_4$ are very likely to be metastable, on account of the kinetic factors

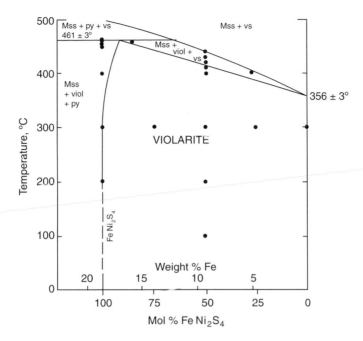

Figure 13.3   A phase diagram of the subsolidus relationships within the FeNi$_2$S$_4$–Ni$_3$S$_4$ pseudobinary join (● denotes the composition of the alloys studied by Craig) (Reproduced with permission from American Mineralogist, Violarite stability relations by J. R. Craig, 56, 1303–1311 Copyright (1971) Mineralogical Society of America)

relating to their paragenesis.[5] Likewise, Warner *et al.* [16] consider artificially prepared Fe-rich violarite, $\sim$Fe$_{1.2}$Ni$_{1.8}$S$_4$, to be metastable, since this composition occurs only within the grains of inhomogeneous violarite, in which it is located at the core and in intimate contact with a rim of Ni-rich violarite, $\sim$Fe$_{0.5}$Ni$_{2.5}$S$_4$; consequently, the material is presumed to be in a state of nonequilibrium.[6]

^{57}Fe-Mössbauer spectroscopy indicates the presence of low-spin Fe^{2+} in the octahedral sites of natural and artificial violarite [15,17,18]; which is consistent with violarite being an inverse spinel. The formal oxidation state of

---

[5] *Paragenesis* refers to the association of different minerals within the same deposit; and so reflects their origin (i.e., the conditions for their primary crystallisation and subsequent history) either collectively, or in reference to a specific mineral.

[6] The counter argument that FeNi$_2$S$_4$ is unstable with respect to a binary mixture of $\sim$Fe$_{1.2}$Ni$_{1.8}$S$_4$ and $\sim$Fe$_{0.5}$Ni$_{2.5}$S$_4$, is unlikely but not implausible. Interestingly, the radial distribution of nickel and iron within these grains indicate an abrupt peripheral change between these two compositions, rather than a continuous variation in solid-solution composition; this texture is yet to be explained. Furthermore, the disruption in local composition, caused by the presence of impurity phases, such as (Ni,Fe)$_{1-x}$S and FeS$_2$/NiS$_2$ within the material (see below), is likely to have an influence here as well.

the nickel content is generally believed to be Ni^{3+}, thus leading to the expression, $(Ni^{III})^{tet}[Fe^{II}Ni^{III}]^{oct}S_4$. Townsend et al. [15] studied the magnetic susceptibility and thermoelectric power of violarite, and concluded that violarite is a Pauli-paramagnetic metal; thus indicating that the valence electrons are extensively delocalised in this material. Vaughan and Tossell [18,19] constructed a qualitative molecular orbital model in an attempt to rationalise certain properties of violarite, and thereby attributed the metallic conduction of this material to a partially occupied sigma antibonding band.

The pure nickel end member, polydymite, Ni$_3$S$_4$, is thermally stable $< 356\,°C$, and is a Pauli-paramagnetic metal that exhibits ferrimagnetic ordering $< 20\,K$ [20], [21]. Polydymite has a steel-grey colour [5] and a high reflectivity [22], which distinguishes it from violarite.

Greigite, Fe$_3$S$_4$, adopts the inverse spinel structure, $(Fe^{III})^{tet}[Fe^{II}Fe^{III}]^{oct}S_4$, and has a relatively large cell constant, $a = 9.876$ Å [23], in which all the Fe^{2+} and Fe^{3+} ions are high-spin [24].[7] These unpaired electrons are coupled in a fashion similar to those in magnetite, Fe$_3$O$_4$; such that, Fe$_3$S$_4$ is a ferrimagnetic semiconductor, with a saturation magnetisation, M_s (at 300 K) $= 59$ A m^2 kg^{-1} [25], [26]. Greigite is considered by several workers to be metastable, but nonetheless, it has been prepared successfully in aqueous systems, for example, by Uda [27], Chen et al. [28] and, Chang et al. [25]. Attempts at preparing it in the dry Fe−Ni−S ternary system, have so far failed and by all accounts, Fe$_3$S$_4$ decomposes $> 250\,°C$. The notable immiscibility gap between Fe$_3$S$_4$ and FeNi$_2$S$_4$ is attributed to the fundamental differences in the electronic structure of these two phases [24].

Concerning the paragenesis of violarite, the subsolidus phase relationships as illustrated in Figure 13.3, indicate that violarite can form naturally as a hypogene[8] phase through the peritectoid reaction between; nickel-iron monosulfide solid solution, $(Ni,Fe)_{1-x}S$; vaesite, $(Ni,Fe)S_2$; and/or, pyrite, $(Fe,Ni)S_2$; upon cooling below 461 °C [1,8], and is expressed here more simply as:

$$2NiS(s) + FeS_2(s) \rightleftharpoons FeNi_2S_4(s)$$

From laboratory experiments, this reaction is found to be very slow, but under geological conditions, there should be ample time to enable the formation of violarite by this process. Nevertheless, natural occurrences of hypogene violarite are uncommon, although one notable example is within the serpentinised[9] ultramafic rock, at Black Swan, Western Australia [29].

---

[7] cf. low-spin Fe^{2+} in violarite [17].

[8] Hypogene; formed or occurring beneath the surface of the earth.

[9] Serpentisation refers to the hydrothermal alteration of the magnesium-rich minerals (e.g., olivine and enstatite) within mafic and ultramafic rocks, to the hydrous magnesium silicates; lizardite, antigorite, and chrysotile.

Violarite occurs more commonly as a secondary mineral through the super-gene[10] alteration of pentlandite, Fe$_{4.5}$Ni$_{4.5}$S$_8$, and to a lesser extent through the alteration of pyrrhotite, Fe$_7$S$_8$, and millerite, NiS, [1,8]. Supergene violarite is particularly abundant within the transition zone of the massive pyrrhotite–pentlandite ore body at Kambalda, Western Australia [13].[11]

In 1944, Michener and Yates [30] stumbled across a unique opportunity to investigate the weathering of pentlandite through the discovery of a collection of abandoned drill cores of primary sulfide ore from the Sudbury Basin, Canada. The core had been left in boxes on the ground, such that the boxes had rotted away, leaving the core exposed to atmospheric weathering for a period of about 25 years. A study of the core specimens showed them to be at various stages of oxidation. The first stage involved the formation of violarite along the cleavage cracks in the pentlandite. Once most of the pentlandite had been consumed, the oxidation progressed to pyrrhotite; and eventually, to chalcopyrite, CuFeS$_2$. This corrosion sequence is indicative of galvanic coupling between the respective sulfide phases during these processes.

The alteration of pentlandite to violarite is accompanied by ~18% con-traction in volume, resulting in characteristic shrinkage cracks within the supergene violarite. These cracks run parallel with the octahedral cleavage plane {111} of the pentlandite precursor, and are interpreted by many workers as evidence for pseudomorphic replacement [29]. A clear depiction of this process is revealed in the micrograph of a specimen from Virginia, by Craig and Higgins [32]; see Figure 13.4. Misra and Fleet [14] studied a fragment of altered pentlandite by X-ray diffractometry. Their precession photograph showed that violarite is formed topotactically from pentlandite.[12] In struc-tural terms, it should be quite feasible for pentlandite to transform into violarite by a process of deintercalation, given sufficient time; since violarite and pentlandite share an almost identical sulfur (*ccp*) substructure, as can be seen in Figure 13.1. This process would involve the removal of $\frac{1}{3}$ of the metal ions through coupled oxidative and solid state diffusion reactions, with a preferential loss of iron over nickel, and an ordering of the remaining Fe^{2+} ions into octahedral sites within the resultant violarite. The principle metastable equilibria between violarite and pentlandite are

---

[10] Supergene; involving enrichment or deposition by downward-moving solution; (of an ore or mineral) so enriched or deposited.

[11] The primary massive monoclinic pyrrhotite–pentlandite ore body at Kambalda, occur in Archean rocks of the Western Australia Shield; i.e., > 2500 million years old [31].

[12] The term, *topotactic*, refers to a solid-state reaction or process whereby crystals are produced having the same orientation as those in the original substance. Misra and Fleet do not actually use the term, *topotactic*, in their article, but their description implies precisely this. The reader should be aware that certain workers, for example, Tenailleau *et al*. [33] dispute the occurrence of a topotactic reaction for the supergene alteration of pentlandite to violarite, in favour of a complete dissolution of pentlandite followed by the in situ precipitation of violarite in close proximity to the original pentlandite. For a discussion of topotaxy in the context of extractive hydrometallurgy the reader is referred to Burkin [34,35].

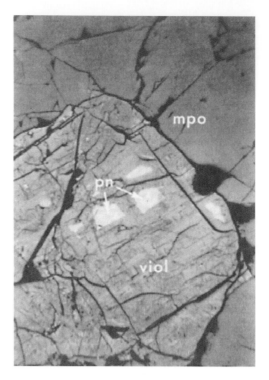

Figure 13.4 Photomicrograph under polarised-light showing a grain of violarite (viol) with remnant pentlandite cores (pn) in monoclinic pyrrhotite (mpo). Width of field = 40 μm (Reproduced with permission from American Mineralogist, Cobalt- and iron-rich violarites from Virginia by J. R. Craig and J. B. Higgins, 60, 36–38 Copyright (1975) Mineralogical Society of America)

portrayed graphically, in the E_H–pH diagram in Figure 13.5, where it can be seen that the transformation of pentlandite to violarite occurs under mild oxidising conditions, at neutral pH and ambient temperature.

In 1972, Thornber proposed that the supergene alteration of pentlandite to violarite occurs via an electrochemical process within certain ore bodies such as the one at Kambalda, Western Australia [36,37]. In Thornber's model, the sulfide ore body acts as a giant corrosion cell with anodic reactions occurring at depth, in which pentlandite transforms into violarite with a loss of metal cations to the ground water. The overall process is driven electrochemically by the cathodic reduction of atmospheric oxygen dissolved in surface water. This takes place on the part of the ore body that is exposed at the earth's surface, with the formation of aqueous hydroxide ions that percolate down-wards through the ground water and meet the aqueous iron and nickel ions percolating upwards, to form gossan (an encrusting mass of iron and nickel oxides and hydroxides). The ground water acts as the aqueous electrolyte, whilst the sulfide ore body acts as the electronic conductor. This geological

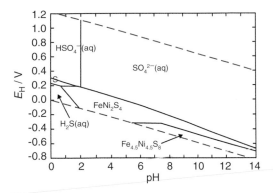

Figure 13.5 E_H–pH diagrams for the Ni-Fe-S aqueous system at 25 °C. The dashed blue lines indicate the upper and lower stability limits for H_2O. Molalities of aqueous nickel, iron and sulfur species $= 10^{-3}$ mol kg^{-3}. Only, aqueous sulfate and sulfide, together with pentlandite, violarite and orthorhombic sulfur are shown. These equilibria relate to $Ni(OH)_2$ and Fe_2O_3 as the solid oxidised phases. This illustration was drawn using Outokumpu HSC Chemistry® Software together with thermodynamic data given by Warner *et al.* [48] and from the references therein

profile, not surprisingly, bears a close resemblance to the E_H–pH diagram in Figure 13.5, in terms of oxidation potential versus depth underground.

The first reference to artificial violarite is by Lundqvist in 1947 [10,38], in connection with the construction of the phase diagram for the Fe–Ni–S system at 200 °C, as discussed above. Iron, nickel and sulfur powders[13] were mixed together and fused inside an evacuated and sealed quartz ampoule at 900 °C. The heating and cooling rates were not stated. The product material was ground to a fine powder, and once again enclosed in an evacuated sealed quartz ampoule and heated at 200 °C for 2 months. The product material was analysed by X-ray powder photograms, and shown to comprise the cubic phase, $FeNi_2S_4$, with a cell constant $a = 9.445$ Å.

In 1962, Kullerud and Yund [39] prepared artificial polydymite, Ni_3S_4, by reacting nickel and sulfur together at 300 °C in an evacuated silica-glass ampoule, in which the product was reground three times during the heating process. Finally, after 8 months, the product contained ~80% Ni_3S_4, the remainder being, NiS_2 and β-$Ni_{1-x}S$. These difficulties are consistent with those encountered earlier by Lundqvist [38].

In 1968, Craig [40] attempted to prepare artificial violarite by a direct reaction of the elements (iron, nickel and sulfur) at 300 °C and 400 °C, for a period of 5 months with repeated grinding. The products yielded only 10–20% violarite, together with pyrite, FeS_2, vaesite, NiS_2, and nickel-iron monosulfide solid solution, $(Ni,Fe)_{1-x}S$. From Lundqvist's phase diagram

---

[13] Iron and nickel powders were first prepared by reducing their respective oxides in hydrogen at 800 °C, whilst the sulfur was freshly distilled prior to use [10,38].

(see Figure 13.2), it can be seen that another plausible route to prepare violarite, is through the sulfiding of a previously prepared monosulfide solid solution, i.e., FeNi$_2$S$_3$ + S → FeNi$_2$S$_4$. This two-step process was employed successfully by Craig [8,40], in his preparation of artificial violarite, and is paraphrased here as follows.

The first stage of the synthesis involved the preparation of nickel-iron monosulfide solid solution, FeNi$_2$S$_{3+x}$ (with Fe:Ni = 1:2 molar proportions) by a direct reaction of the elements in an evacuated sealed quartz ampoule at 500 °C for 2 days. The intermediate product material was ground, resealed and reheated at 500 °C for a further 17 days. The second stage of the synthesis involved reacting the finely ground FeNi$_2$S$_{3+x}$ with an appropriate amount of additional sulfur at 300 °C[14] for 67 days, in order to yield single phase FeNi$_2$S$_4$. The best results were achieved by segregating the sulfur powder from the FeNi$_2$S$_{3+x}$ powder with the aid of a plug of silica wool; thus limiting the reaction to sulfur transported in the vapour phase (cf. the preparation of artificial covellite, CuS, as described in Chapter 4). The main draw back of this method is that the reaction time is about three months! Nevertheless, Craig's original method has been adopted, with slight modifications, by Townsend et al. [15], Warner et al. [16], and Tenailleau et al. [9]. In general, these product materials comprise an inhomogeneous violarite, including a Fe-rich violarite, contaminated with a certain amount of unreacted nickel-iron monosulfide solid solution, (Ni,Fe)$_{1-x}$S, and pyrite FeS$_2$ (or, vaesite, NiS$_2$ [15]). It is interesting to note, that none of the above workers have been able to prepare homogenous violarite as a single-phase material, with the idealised composition, FeNi$_2$S$_4$.

An alternative approach to preparing artificial violarite is through the precipitation of an aqueous solution containing the appropriate metal salts with the aid of a sulfiding reagent. In 1944, Michener and Yates [30] attempted to prepare violarite, with the composition, Fe$_{1.5}$Ni$_{1.5}$S$_4$, by treating an aqueous solution containing an equimolar mixture[15] of NiSO$_4$ and FeSO$_4$ with hydrogen sulfide. The precipitate was analysed by powder X-ray diffractometry and concluded to be amorphous; the matter was not perused any further.

In 1947 Lundqvist [41] prepared artificial polydymite, Ni$_3$S$_4$, by adding, drop-wise (within the span of ∼ 5 min), a rather concentrated aqueous solution of sodium thiosulfate (Na$_2$S$_2$O$_3$) to a boiling dilute neutral aqueous solution of NiSO$_4$ in an open vessel with rigorous stirring.[16] The reaction mixture was allowed to boil for 1 h with continuous stirring, whilst air was allowed to enter. Consequently, the wall of the vessel became coated

---

[14] If the second stage is conducted > 300 °C there is a high risk of forming pyrite.

[15] With a total metal concentration = 5 g dm^{-3}.

[16] Unfortunately, Lundqvist did not report the values for the concentrations in his article.

with a coherent precipitation that displayed a metallic gray lustre. X-ray photograms showed that this precipitate was essentially Ni$_3$S$_4$. No mention is made, however, of attempting to apply this method to the preparation of artificial violarite; even though artificial bravoite, (Fe$_{0.6}$Ni$_{0.4}$)S$_2$ was prepared by a hydrothermal synthesis at 300 °C in dilute aqueous sulfuric acid solution, as described in the same article.

Manthiram and Jeong [20] prepared a specimen of Ni$_3$S$_4$ in 1999, by a method similar to that devised originally by Lundqvist (as described above). 150 ml of 1 mol dm^{-3} sodium dithionite (Na$_2$S$_2$O$_4$) solution was added to a 50 ml of 0.25 mol dm^{-3} nickel chloride (NiCl$_2$) solution, with constant stirring and maintaining a pH $= 3$ through the addition of drops of hydrochloric acid solution. The precipitate was filtered and washed with deionised water, then left to dry at ambient temperature. The formation of artificial polydymite, Ni$_3$S$_4$, was confirmed by powder X-ray diffractometry.

In 2001, Shimizu and Yano [42] prepared artificial polydymite, Ni$_3$S$_4$, by a pH-controlled homogeneous precipitation method. 15 mmol of NiCl$_2 \cdot$ 6H$_2$O was dissolved in 200 ml distilled water with 20 mmol NH$_3$(aq)/ NH$_4$Cl(aq) buffered solution (pH $= 9.9$). This solution was then added drop-wise to a solution containing 20 mmol thioacetamide in 200 ml distilled water and refluxed within the temperature range 60–70 °C for 12 h. The precipitate was filtered, washed with distilled water and dried under vacuum at ambient temperature. Although the authors claim to have prepared CoNi$_2$S$_4$ and NiCo$_2$S$_4$ by this method, no reference is made to the preparation of FeNi$_2$S$_4$.

In summary, this chapter has described some of the practical problems associated with preparing violarite, FeNi$_2$S$_4$, through the reaction of its constituent chemical elements in the laboratory. In contrast to this, the formation of violarite in nature, most commonly involves the supergene alteration of pre-existing iron and nickel sulfide minerals (e.g., pentlandite) at near-ambient temperature. The following section describes a novel hydrothermal process for preparing artificial violarite, FeNi$_2$S$_4$, from an aqueous solution of iron(II) acetate and nickel(II) acetate tetrahydrate, using DL-penicillamine as the sulfiding reagent, as conducted in a Teflon®-lined stainless steel autoclave at 130 °C.[3]

Preparative Procedure

Penicillamine is a metabolite of penicillin. The D-enantiomer is manufactured as the immunosuppressant drug, Cuprimine™ (Merck) and Depen™, for the treatment of rheumatoid arthritis [43]. It is also used as a chelating agent in the treatment Wilson's disease (a rare genetic disorder of copper metabolism), whereby D-penicillamine binds to the Cu^{2+} ions in the body, and acts as a vehicle for their removal via the urine [44,45]. The other enantiomer, L-penicillamine, is toxic. A mixture of both enantiomers, DL-penicillamine, is available commercially at a reduced price (compared with the separate enantiomers) and is used here as the sulfiding reagent.

DL-penicillamine

Penicillamine is particularly attractive as a sulfiding reagent since certain Lewis acids, such as, copper(II) and nickel(II) aqueous ions are capable of breaking the bond between carbon and sulfur in molecular species similar to penicillamine at elevated temperature, with the possibility of forming the corresponding metal sulfide. The reader is referred to the article by Becher *et al.* [46] for further details. Therefore, it was tempting, whilst writing this book, to explore this approach as a method to prepare violarite and polydymite.

Prepare 1.5 g of FeNi$_2$S$_4$ by the following method: Calculate the mass corresponding to the molar quantities of the following chemical reagents: 5 mmol iron(II) acetate, Fe(OOCCH$_3$)$_2$ (Aldrich, 95%); 10 mmol nickel(II) acetate tetrahydrate, Ni(OOCCH$_3$)$_2 \cdot$ 4H$_2$O (Aldrich, 98%); and 25 mmol DL-penicillamine, (CH$_3$)$_2$C(SH)CH(NH$_2$)COOH (Aldrich, 97%).[17]

Weigh the appropriate amounts of these three reagents (to a precision of ± 0.001 g) directly into a 300-ml glass beaker.[18] Pour 100 ml of distilled water carefully from a measuring cylinder into this beaker in order to dissolve these reagents. Add a magnetic stirrer bead and stir magnetically for $\sim$ 5 min or, until all the solids have dissolved. It may be necessary to use a glass rod to help the DL-penicillamine to dissolve, since this compound has a tendency to float temporarily on the surface of the water.

Transfer the deep reddish-purple solution into a 180-ml Teflon® container (i.e., the liner for the autoclave; see Figure 2.3). Rinse the glass beaker with 10 ml of distilled water and add this to the solution already in the Teflon® container; thereby leaving $\sim$ 70 ml of headspace. Attach the Teflon® lid accordingly, and place this inside the stainless steel autoclave. Attach the lid of the autoclave which must be equipped with a safety-valve with a release pressure of $\sim$ 15 bar (218 psi), and seal tightly. Introduce this into a preheated

---

[17] The chemical reagent for the sulfur component was used in a slight excess in the original work by Hagedorn Jørgensen [11], from the premise that a certain amount of this reagent may be lost during handling. The fact that the preparative method works, is the reason that this aspect is retained here.

[18] The reader should wear protective gloves and goggles when handling these chemical reagents; especially persons with an allergy to nickel.

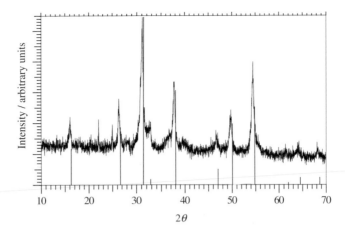

Figure 13.6 Powder XRD pattern (Cu-$K_{\alpha1}$ radiation) of artificial violarite, FeNi₂S₄ precipitate, as prepared by the method described in this chapter. Cell constant, $a = 9.524$ Å. PDF 47-1740 for FeNi₂S₄ ($a = 9.458$ Å) is shown in red for comparison

electric oven at a stable temperature of 130 °C. The reader should be aware that the Teflon® container (which is made of PTFE) begins to decompose above 260 °C, with the formation of highly toxic fluorocarbon gases. These gases can cause influenza-like symptoms in humans and may be lethal.

After 45 h at 130 °C switch the oven off, and allow the autoclave to cool to ambient temperature. Then remove the autoclave from the oven and retrieve the Teflon® container. Filter the precipitate using a normal 120-mm diameter glass funnel with a medium speed filter paper.[19] It may be necessary to use a spatula in order to scrape all the precipitate from the walls of the Teflon® container. Then wash the Teflon® container with a small portion (~10 ml) of distilled water, and add the wash liquor to the precipitate being collected in the filter funnel. Once most of the material has been filtered, wash the precipitate with ~20 ml of distilled water with the precipitate still attached to the filter paper inside the filter funnel. The filtration process is rather slow and will last ~30 min.[20] Carefully remove the filter paper (with the powder bed kept intact), and place it on a large watch-glass. Leave this to dry overnight at ambient temperature.

During the next day, transfer the dried precipitate into an appropriately labelled specimen jar and close the lid tightly. Test the reaction of this powder

---

[19] For example, AGF 138–200 mm, pore size = 15 μm, 90 g m⁻², Frisenette Aps. The use of a Buchner filter funnel was avoided here, since the suction might cause the fine violarite particles to be drawn through the filter paper. The method as described here was proven to work; although it is acknowledged that other methods of filtration may be more applicable.

[20] The filtrate normally has a pale orange colouration, but it should be a clear liquid devoid of particulate matter.

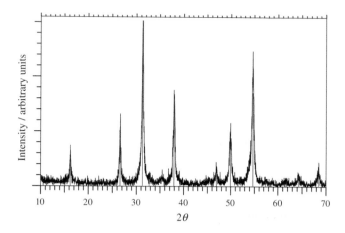

Figure 13.7   Powder XRD pattern (Cu-$K_{\alpha 1}$ radiation) of artificial polydymite, Ni$_3$S$_4$ precipitate, as prepared by the method described in this chapter. Cell constant, $a = 9.476$ Å. PDF 47-1739 for Ni$_3$S$_4$ is shown in red for comparison

(whilst inside the sample jar) towards a Nd$_2$Fe$_{14}$B (NEOMAX) magnet, in order to see if there are any ferrimagnetic phases present, such as, monoclinic pyrrhotite, Fe$_7$S$_8$, or, greigite, Fe$_3$S$_4$. Then submit a small sample ($\sim$100 mg) for analysis by powder X-ray diffraction (scan: 10–70° 2 θ). Compare your powder pattern with the PDF 47-1740 for FeNi$_2$S$_4$ (see Figure 13.6) noting particularly, the value of the cell constant. The presence of X-ray fluorescence in the powder pattern is incidental evidence for the incorporation of iron into the product material; although one can argue that the iron might be incorporated solely in a separate amorphous phase. It is informative to note that the fluorescence is absent in the powder pattern of artificial polydymite, Ni$_3$S$_4$ (cf. Figure 13.7).

The relatively low temperature for the decomposition of FeNi$_2$S$_4$ imposes an upper limit of about 300 °C for annealing powder compacts of this material. Unfortunately, the monoliths prepared at this temperature are poorly sintered and therefore friable. It is very likely that the solid-state diffusion rates are too low within FeNi$_2$S$_4$ at 300 °C to promote the sintering processes.

REFERENCES

1. E. H. Nickel, *The Australian Institute of Mining and Metallurgy. Conference, held at Perth*, May 1973, 111–116.
2. B. A. Wills and T. Napier-Munn, *Wills' Mineral Processing Technology: An Introduction to the Practical Aspects of Ore Treatment and Mineral Recovery*, 7th edn, Butterworth-Heinemann (Elsevier), Burlington, Massachusetts, 2006.

3. A. R. Burkin (ed.), *Extractive Metallurgy of Nickel*, Wiley, Chichester and New York, 1987, 160 pp.
4. J. R. Boldt and P. Queneau (Tech ed.), *The Winning of Nickel - Its Geology, Mining and Extractive Metallurgy*, 1967, Methuen, London.
5. W. Lindgren and W. M. Davy, *Economic Geology* **19** (1942) 309–319.
6. M. N. Short and E. V. Shannon, *American Mineralogist* **15** (1975) 1–22.
7. G. Harcourt, *American Mineralogist* **27** (1942) 63–113.
8. J. R. Craig, *American Mineralogist* **56** (1971) 1303–1311.
9. C. Tenailleau, B. Etschmann, R. M. Ibberson and A. Pring, *American Mineralogist* **91** (2006)1442–1447.
10. D. Lundqvist, *Arkiv för Kemi, Mineralogi och Geologi*, Band 24A No. 22 (1947) 1–12.
11. W. Hagedorn Jørgensen, *Preparation of ceramic iron–nickel thiospinel (Fe,Ni)$_3$S$_4$ by combined co-precipitation and sintering methods*, MSc Thesis, University of Southern Denmark, 2008.
12. D. H. Rousell, H. L. Gibson and I. P. Jonasson, *Exploration and Mining Geology* **6** (1997) 1–22.
13. E. H. Nickel, J. R. Ross and M. R. Thornber, *Economic Geology* **69** (1974) 93–107.
14. K. C. Misra and M. E. Fleet, *Economic Geology* **69** (1974) 391–403.
15. M. G. Townsend, J. R. Gosselin, J. L. Horwood, L.G. Ripley and R. J. Tremblay, *Physica Status Solidi* A **40** (1977) K25–K29.
16. T. E. Warner, N. M. Rice and N. Taylor, *Hydrometallurgy* **31** (1992) 55–90.
17. D. J. Vaughan and J. R. Craig, *American Mineralogist* **70** (1985) 1036–1043.
18. D. J. Vaughan, R. G. Burns and V. M. Burns, *Geochimica et Cosmochimica Acta* **35** (1971) 365–381.
19. D. J. Vaughan and J. A. Tossell, *American Mineralogist* **66** (1981) 1250–1253.
20. A. Manthiram and Y. U. Jeong, *Journal of Solid State Chemistry* **147** (1999) 679–681.
21. Y. U. Jeong and A. Manthiram, *Inorganic Chemistry* **40** (2001) 73–77.
22. W. H. Bird, *Economic Geology* **64** (1969) 91–94.
23. B. J. Skinner, R. C. Erd and F. S. Grimaldi, *American Mineralogist* **49** (1964) 543–555.
24. D. J. Vaughan and M. S. Ridout, *Journal of Inorganic Nuclear Chemistry* **33** (1971) 741–746.
25. L. Chang, A. P. Roberts, Y. Tang and B. D. Rainford, *Journal of Geophysical Research* **113** (2008) B06104/1–16.
26. S. Yamaguchi, H. Wada and H. Hiroshi, *Messtechnik (Braunschweig)* **78** (1970) 83–84.
27. M. Uda, *American Mineralogist* **50** (1965) 1487–1489.
28. X. Chen, X. Zhang, J. Wan, Z. Wang and Y. Qian, *Chemical Physics Letters* **403** (2005) 396–399.
29. D. R. Hudson and D. I. Groves, *Economic Geology* **69** (1974) 1335–1340.
30. C. E. Michener and A. B. Yates, *Economic Geology* **39** (1944) 506–514.
31. M. R. Thornber, *Chemical Geology* **15** (1975) 1–14.
32. J. R. Craig and J. B. Higgins, *American Mineralogist* **60** (1975) 36–38.
33. C. Tenailleau, A. Pring, B. Etschmann, J. Brugger, B. Grguric and A. Putnis, *American Mineralogist* **91** (2006) 706–709.
34. A. R. Burkin, *The Chemistry of Hydrometallurgical Processes*, 1966, Spon, London.
35. A. R. Burkin, *Chemical Hydrometallurgy – Theory and Principles*, 2001, Imperial College Press, London.

36. M. R. Thornber, *The Australian Institute of Mining and Metallurgy. Conference, held at Newcastle*, May–June 1972, 51–58.
37. M. R. Thornber, *Journal of Applied Electrochemistry* **13** (1983) 253–267.
38. D. Lundqvist, *Arkiv för Kemi, Mineralogi och Geologi*, Band 24A No. 21 (1947) 1–12.
39. G. Kullerud and R. A. Yund, *Journal of Petrology* **3** (1962) 126–175.
40. J. R. Craig, *Carnegie Institution of Washington Year Book* **66** (1968) 434–436.
41. D. Lundqvist, *Arkiv för Kemi, Mineralogi och Geologi* Band 24A No. 23 (1947) 1–7.
42. Y. Shimizu and T. Yano, *Chemistry Letters* **10** (2001) 1028–1029.
43. D. E. Furst, *Postgraduate Medicine* **87** (1990) 79–92.
44. G. B. Sanders, *Journal of the American Medical Association* **189** (1964) 874–878.
45. G. J. Brewer, *Expert Opinion on Pharmacotherapy* **7** (2006) 317–324.
46. J. Becher, H. Toftlund, P. H. Olesen and H. Nissen, *Inorganica Chimica Acta* **103** (1985) 167–171.
47. A. D. Pearson and M. J. Buerger, *American Mineralogist* **41** (1956) 804–805.
48. T. E. Warner, N. M. Rice and N. Taylor, *Hydrometallurgy* **41** (1996) 107–118.
49. C. G. Bergeron and S. H. Risbud, *Introduction to Phase Equilibria in Ceramics*, The American Ceramic Society, Westerville, Ohio, 1997.

PROBLEMS

1. According to Craig [8]: '*FeNi$_2$S$_4$ is unstable at 400 °C in the Fe–Ni–S ternary system, with respect to a mechanical mixture of pyrite, FeS$_2$; monosulfide solid solution, (Ni, Fe)$_{1-x}$S and nickel-rich violarite, Fe$_{1-x}$Ni$_{2+x}$S$_4$*'. Discuss whether or not this statement violates the phase rule; and if so, state the name of any phase(s) that you consider to be missing. (The reader is referred to Bergeron and Risbud [49] for an introduction to the phase rule).

2. Calculate the approximate temperature of a system that has a hydrostatic pressure of 15 bar.

3. Calculate the volume corresponding to 1 g of FeNi$_2$S$_4$.

4. Briefly explain the mechanism by which Cu-$K_{\alpha1}$ radiation induces X-ray fluorescence in iron-containing substances such as violarite.

14

Artificial Willemite $Zn_{1.96}Mn_{0.04}SiO_4$ by a Hybrid Coprecipitation and Sol-Gel Method

Willemite, $(Zn,Mn)_2SiO_4$, has been exploited for its bright green luminescent properties for over a century. It is a well-documented example of how the discovery of a mineral together with the subsequent discovery of its luminescence, has given chemists over several generations the inspiration to prepare artificial willemite with a desire to control the chemical composition and phase purity in order to understand its optical properties, and thereby develop the material for use as a phosphor in luminescent lamps, cathode ray tubes (oscilloscopes and radar screens) and most recently in plasma display panels.

Zn_2SiO_4 is an orthosilicate (or in mineralogical terminology; a nesosilicate) that crystallises with the phenakite (Be_2SiO_4) structure as shown in Figure 14.1. The crystal structure was first proposed during the early years of X-ray diffractometry by Zachariasen in 1926 [1] and Bragg and Zachariasen in 1930 [2]; then refined by Simonov *et al.* in 1977 [3]. An essential feature of this structure is that both the zinc and the silicon cations are tetrahedrally coordinated to oxygen. This makes Zn_2SiO_4 structurally different from olivine (Mg_2SiO_4) and spinel $(MgAl_2O_4)$ that incorporate both tetrahedral and octahedral cation sites within their structures. In Zn_2SiO_4, the SiO_4^{4-} tetrahedra are isolated from one another and are linked to a pair of corner-sharing ZnO_4^{6-} tetrahedra in an alternating fashion to form

Synthesis, Properties and Mineralogy of Important Inorganic Materials   By Terence E. Warner
© 2011 John Wiley & Sons, Ltd

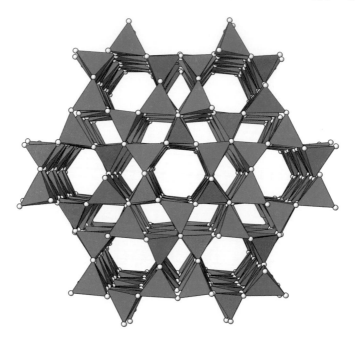

Figure 14.1   A perspective view for the crystal structure of willemite, Zn_2SiO_4 as projected down the c-axis. The ZnO_4^{6-} tetrahedra are shown green, the smaller SiO_4^{4-} tetrahedra are shown orange, and the oxide ions are depicted as small white spheres. Its honeycomb-like structure with a 3-fold axis is evident from the alternating SiO_4–ZnO_4–SiO_4–ZnO_4–SiO_4–ZnO_4 cyclic sequence around the central ring. This figure was drawn using data from Simonov *et al.* [3], cf. ICSD 20093

helical-type chains, such that the common oxygen is 3-fold coordinated (see Figure 14.2). These chains interconnect so as to yield infinite hexagonal tubes that exhibit trigonal symmetry around the c-axis. Each hexagonal tube is surrounded by six others, so as to form an ordered honeycomb-like structure belonging to the trigonal/rhombohedral crystal system and space group $R\bar{3}$.

An equilibrium phase diagram of the ZnO-SiO_2 binary system was constructed by Bunting [4] in 1930; see Figure 14.3. This is a classic piece of analytical work. Bunting encapsulated small ($\sim$15 mg) samples of a binary mixture of ZnO and SiO_2 inside individual platinum capsules for a total of 30 different $ZnO:SiO_2$ compositions. These were annealed at constant temperature and ambient pressure in order to attain a presumed state of equilibrium; then quenched, so as to avoid any significant displacement from the high temperature equilibrium. The samples therefore represent 'frozen' metastable material of high-temperature equilibrium phase assemblages, such that the occurrence of a glass phase is interpreted as evidence of a high-temperature liquid, and a mixture of two glass phases as evidence of two immiscible high-temperature liquids. 24 of these compositions were annealed at two different temperatures, thus generating a set of 54 specimens in all. These

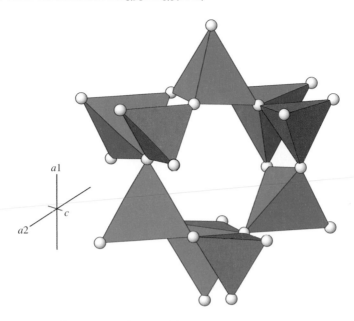

Figure 14.2 An oblique view of part of the willemite structure showing the isolated SiO$_4^{4-}$ tetrahedra (orange) linked to a pair of corner sharing ZnO$_4^{6-}$ tetrahedra (green). Note that the Zn^{2+} cations are in close proximity to each other through a common oxygen ion. This figure was drawn using data from Simonov *et al.* [3], cf. ICSD 20093

materials were examined under a petrographic microscope[1] for phase identification. From this work, Bunting showed that Zn$_2$SiO$_4$ is the only ternary phase that crystallises in the ZnO-SiO$_2$ system at ambient pressure and has a congruent melting point of 1512 ±3 °C.

The presence of a eutectic at 1432 °C, with the composition ZnO · SiO$_2$, coincides with what was once considered to be the metasilicate, ZnSiO$_3$. However, ZnSiO$_3$ is not found in nature and can not be made artificially at ambient pressure, but exists as a synthetic high-pressure phase that crystallises with the monoclinic clinopyroxene structure at > 30 kbar. This phase in turn transforms to the more densely packed hexagonal ilmenite structure, at the very much higher pressure of ~110 kbar and is a rare example of SiO$_6^{8-}$ octahedral coordination [5].

In the ZnO-SiO$_2$-H$_2$O ternary system, there exists an hydrated zinc silicate hydroxide Zn$_4$Si$_2$O$_7$(OH)$_2$ · H$_2$O, known naturally as the mineral hemimorphite (or in the earlier literature as, calamine), which has an identical ZnO:SiO$_2$ ratio (namely 2:1) to that in Zn$_2$SiO$_4$. Inconsistencies regarding its 'H$_2$O' content may have hindered the recognition of Zn$_2$SiO$_4$ and Zn$_4$Si$_2$O$_7$(OH)$_2$ · H$_2$O as distinct mineral species in former times. For

---

[1] NB powder X-ray diffractometry cannot resolve the presence of glass phases on account of their amorphous nature.

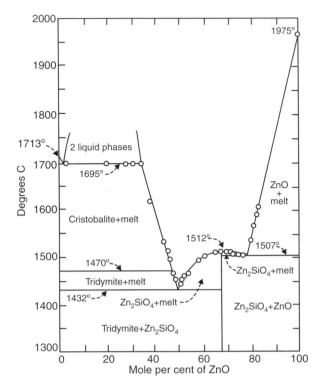

Figure 14.3 Phase relationships within the ZnO-SiO_2 binary system at ambient pressure (Reproduced with permission from J. Am. Ceram. Soc, Phase equilibria in the system: SiO_2–ZnO by E. N. Bunting, 13, 5–10 Copyright (1930) American Ceramic Society)

instance, the semianhydrous specimens of Hungarian calamine analysed[2] by James Smithson, certainly aroused some perplexity over this matter in 1803 [6]. Nowadays, hemimorphite is known to accommodate a loss of up to 50 mol.% 'H$_2$O' below 500 °C without undergoing any significant structural rearrangement [7].

Zn_2SiO_4 occurs in nature as the uncommon, though economically important mineral, willemite. The events surrounding its discovery give rise to a long, but curious, tale. The mineral was first discovered in New Jersey, USA, in an unusually large deposit located at Franklin Furnace, and Sterling Hill, Ogdensburg, which had been mined since the days of British colonialism [8]. The early mining ventures at Franklin failed because the zinc ore was mistaken for copper ore; so the new owner Samuel Fowler who acquired the mines in 1810, welcomed a number of eminent mineralogists from New York and Philadelphia to visit the mines in order to seek their advice [9].

---

[2] The results of the analysis performed on crystals of calamine from Regbania, Hungary: 25.0 SiO_2; 68.3 ZnO; 4.4 H_2O; 2.3 loss [6].

As a consequence of these visits, the mineral was first reported as specimen No. 6 "*siliceous oxide of zinc*" in a list of mineral species found in the vicinity of Franklin, New Jersey, by the mineralogists Lardner Vanuxem and William Keating in August, 1822, but no other details about the mineral were given [10]. As a sequel to this work, Vanuxem and Keating published the first accurate description and chemical analysis[3] of the mineral in 1824 [11]; so why they failed to give it a mineralogical name at that time is a mystery.

About the same time, in September, 1822, Torrey also reported a mineral specimen from Franklin Furnace, New Jersey – brought to him by his friend and mineral collector Mr Nuttell – as being a "*siliceous oxyd of zinc*" [12, 13]. Torrey described its appearance when pure as resembling granular quartz and that it occurs in irregular masses disseminated in franklinite and mixed with zincite, and also noted that the mineral dissolves in nitric acid such that the solution gelatinises strongly; observations that are indeed consistent with contemporary knowledge. Torrey goes on to mention quite courteously, that after examining this mineral, a certain Dr William Langstaff had informed him that *he* had discovered it many years earlier in the same place found by Nuttell; a claim that remains difficult to verify today. In order to obtain knowledge of the chemical composition, Torrey presented specimens of Nuttell's mineral to the Scottish chemist, Thomas Thomson, for chemical analysis.[4] In 1828, Thomson described the mineral as a "*ferruginous silicate of manganese*" [14]. This description is consistent with the mineral tephroite, $(Mn,Fe)_2SiO_4$, discovered at Stirling Hill, Ogdensburg in 1823, and therefore Thomson concluded that this mineral was nothing new [15]. Several years later, Thomson's chemical analysis turned out to be inaccurate, but this analytical blunder had by now prevented Nuttell's mineral from being given a name.

Meanwhile in Europe, the name *willemite* was given by the Frenchman Lévy in 1830, to a mineral exhibiting small prismatic crystals found at Altenburg, Moresnet, Liége, Belgium – in what was then part of the Netherlands – and named it in honour of King Willem I of the Netherlands (1772–1844) [16, 17]. Other variants of the name have been adopted in the early literature and include; villemite, wilhelmite, williamsite and hebetine. Not perturbed by this matter, Shepard surreptitiously adopted the erroneous analysis by Thomson, in ascribing the name *troostite* in 1832 – in honour of the Dutch-born American mineralogist Gerard Troost (1776–1850) – to the mineral listed in his 'Treatise on Mineralogy' as: "Ferruginous Silicate of Manganese. *Thomson*. Silicate of Zinc. *Vanuxem* and *Keating*." [18, 19].

---

[3] Vanuxem and Keating's results of the analysis performed on crystals of a light flesh colour: 25.00 silex (SiO_2); 71.33 oxide of zinc (ZnO); 2.66 oxide of manganese (MnO); 0.67 oxide of iron (FeO), 0.34 loss [11].
[4] Thomson's results of the analysis of Nuttell's mineral: 30.650 silica, 46.215 protoxide of manganese (MnO), 15.450 peroxide of iron (Fe_2O_3), 7.300 moisture and carbonic acid [14].

In 1846, Dellese and Descloizeaux [20] proved that the mineral willemite discovered by Lévy was actually the same mineral as the *"siliceous oxide of zinc"* described earlier by Vanuxem and Keating. Then shortly afterwards in 1849, Hermann [21] revealed the true chemical composition[5] of troostite (i.e., a predominance of zinc oxide over manganese oxide) and thereby exposed the errors in Thomson's chemical analysis. Hermann concluded therefore that willemite and troostite were equivalent mineral species and that troostite is a manganiferous variety of willemite; a varietal name that is still applied today [7]. In conclusion, it is interesting to note that if William Langstaff's claim to discovering the mineral prior to 1822 was indeed true, then the mineral is, unwittingly, well named.

Concerning the paragenesis of willemite, there is no geological evidence of this mineral crystallising directly, or indirectly, from molten silicate rock. There are only a few examples of willemite occurring within igneous rocks, and in all of these, willemite is considered to be a secondary mineral of rather exotic occurrence. For instance, through the alteration of primary sphalerite, $(Zn,Fe)S$, within a nepheline syenite from the Igaliko complex, Greenland [22]; through the metasomatic alteration of genthelvite, $Zn_4Be_3Si_3O_{12}S$, within an albitite from Ilímaussaq, Greenland [23]; and through the hydrothermal precipitation of willemite conglomerates (5–10 mm) within cavities in a nepheline syenite pegmatite from the Khibiny Massif, Kola Peninsula, Russia [24]. In hindsight, these observations would have appealed most favourably to the Neptunists![6]

The apparent absence of primary magmatic (i.e., plutonic and volcanic) willemite is not at all surprising given that zinc is present in trace amounts (93 ppm) in the earth's crust. As such, it readily substitutes for iron in biotites, particularly in more mafic hornblende granites [26]. When zinc is released into the environment through the processes of weathering, it is known to form complexes that are soluble, and so can readily migrate and become concentrated in hot brines, until precipitated as ZnS; for instance, on the sea-floor in black smokers. These are later consolidated into volcanogenic massive sulfide deposits. Basinal brines also concentrate zinc and precipitate it as the sulfide.[7]

Willemite is formed typically from the supergene and hypogene[8] alteration of pre-existing zinc sulfide deposits and other zinc-containing phases.

---

[5] Hermann's results of the analysis performed on specimens of troostite from the proximity of Sparta and Sterling in New Jersey: $1.00\ H_2O$; $26.80\ SiO_2$; $60.07\ ZnO$; $9.22\ MnO$; $2.91\ MgO$; trace FeO.

[6] The Neptunists were a school of geologists led by Professor A. G. Werner (1749–1817), Mining Academy in Freiberg, Saxony, who proclaimed that all rocks including basalt were aqueous precipitates and thus originated from the sea [25].

[7] The author is grateful to Dr Chris J. Stanley, Natural History Museum, London, for providing references to and information on the geochemisty of zinc.

[8] Supergene processes involving water percolating down from the earth's surface. Hypogene processes involving fluids ascending from within the earth.

Supergene deposits of willemite are the more widespread of the two classifications, but in these deposits, the major zinc minerals are commonly, hemimorphite $Zn_4Si_2O_7(OH)_2 \cdot H_2O$, smithsonite $ZnCO_3$ and hydrozincite $Zn_5(CO_3)_2(OH)_6$, with only minor amounts of willemite. Notable European occurrences are; La Calamine, Belgium; Tynagn, Ireland; Upper Silesia District, Poland; and Reocin, Spain. Hypogene deposits of willemite occur in Vazante, Brazil; Beltana, Australia; Kabwe, Zambia; Berg Aukas and Abenab West, Namibia. In these, willemite is the major zinc mineral, and there is currently an economic interest in their zinc values. The reader is referred to the recent reviews of Hitzman et al. [27] and Large [28] for further details. Interestingly, all of these hypogene deposits are associated with Neoproterozoic to Lower Cambrian carbonate rocks within the Pan-African orogenic belts of the southern hemisphere, and appear to have formed as a result of the mixing of a reduced zinc-rich/sulfur-poor fluid with an oxidised sulfur-poor fluid within a temperature range of 80–200 °C. The possibility of primary precipitation from hydrothermal fluids under certain conditions of low sulfur and high oxygen fugacities has been proposed recently as an alternative origin for these deposits based on the results of geochemical modelling [28, 29]. In summary, willemite is found only as a secondary mineral and as a hydrothermal precipitate, but nonetheless, this demonstrates that nature has found a way to form willemite through the interaction and precipitation of chemical species held in aqueous media.

In contrast to the above hypogene deposits, the Franklin and Sterling Hill hypogene deposits in New Jersey, USA, are a unique case in that they are associated with high-grade regional metamorphism, hosted in Mesoproterozoic carbonate rock. Another distinctive feature of willemite from Franklin and Sterling Hill is that most specimens exhibit luminescence (see Figure 14.4). Indeed, the walls of these mines glow with a spectacular green luminescence when exposed to ultraviolet radiation and have become a local tourist attraction. This luminescent property was first reported in 1904 by Baskerville and Kunz [30] on mineral specimens of willemite exposed to radium (an intense source of ionising radiation), X-ray and ultraviolet radiation. This property lent itself to an early application in controlling the loss of willemite in the tailings (discarded mineral waste) during the processing of zinc ore at New Jersey [31]. An electric arc was sprung across the air between iron poles situated over the tailings discharge so as illuminate them with ultraviolet light. This Frankensteinian contraption would then allow the presence of willemite to be vividly exposed through its bright green luminescence. It is the presence of manganese(II) within the willemite phase that is responsible for the luminescence; the substitution of this element being facilitated naturally by the intimate association of willemite with the manganese-bearing minerals, rhodonite $MnSiO_3$, zincite $(Zn,Mn)O$ and franklinite $(Zn,Mn,Fe)_3O_4$ in the Franklin and Sterling Hill deposits [32, 33]. A mechanism to account for the luminescence in willemite is as follows.

Figure 14.4 Specimen of willemite from Trotter's Dump, Franklin, New Jersey, U.S.A. The left photograph taken under normal light show pink-brown willimite crystals (the largest at the upper-right quadrant is $27 \times 24 \times 17$ mm). The right photograph taken under short-wavelength ultraviolet radiation shows brilliant green luminescence (Image used with permission courtesy Amethyst Galleries, Inc. http://www.galleries.com/minerals/silicate/willemit/willemit.htm Copyright (1998) Amethyst Galleries, Inc. Commerical rights reserved)

The cations Mn^{2+} and Zn^{2+} have similar ionic radii and carry the same charge, such that Mn^{2+} can substitute partially for Zn^{2+} in the willemite structure, (Zn,Mn)$_2$SiO$_4$. This forces the Mn^{2+} ($3d^5$) ions to adopt tetrahedral symmetry, and for a weak crystal field, as is the case here, have a ground-state electron configuration $e_g^2 t_{2g}^3$ with a high-spin state $S = 5/2$. Because the optical transitions for this particular electron configuration are both spin and parity forbidden,[9] this permits only very weak transitions in absorption (and in emission) to occur. However, Zn$_2$SiO$_4$ is a wide bandgap (~ 5 eV) semiconductor with a strong absorption band in the ultraviolet (250 nm) region, which is probably due to an electron being promoted from the valence band to the conduction band within the Zn–O substructure [35], and the creation of a hole defect localised on an oxygen ion.

$$O_O{}^x + h\nu_{\text{ultraviolet } (\sim 254 \text{ nm})} \rightarrow O_O{}^{\bullet} + e'$$

In the manganese-doped material, (Zn,Mn)$_2$SiO$_4$, the situation is similar. But, as a result of charge transfer, the hole is now located on a Mn^{2+} ion (present as a defect species within the Zn–O substructure); which is in effect a Mn^{3+} ion raised to a higher energy level (see Figure 14.5).

$$O_O{}^x + h\nu_{\text{ultraviolet } (\sim 254 \text{ nm})} \rightarrow O_O{}^{\bullet} + e'$$

$$O_O{}^{\bullet} + Mn_{Zn}{}^x \rightarrow O_O{}^x + Mn_{Zn}{}^{\bullet}$$

---

[9] The spin selection rule forbids electronic transitions between levels with different spin states ($\Delta S \neq 0$). The parity selection rule forbids electronic transitions between levels with the same parity; for example, within the same d-shell or, f-shell [34].

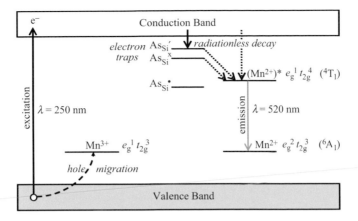

Figure 14.5 Schematic energy diagram for the excitation and emission mechanisms in manganese- and arsenic-doped willemite. The expressions are consistent with those described in the text

During the subsequent process of hole–electron recombination, the electron returning from the conduction band back to the valence band reduces the Mn^{3+} ion to Mn^{2+}. But most importantly, this Mn^{2+} defect ion now has an excited electron configuration, $e_g^1 t_{2g}^4$ with a low-spin state $S = 3/2$ (marked here as *).

$$Mn_{Zn}^\bullet + e' \rightarrow (Mn_{Zn}^{\ x})^*$$

This excited spin-state $(e_g^1 t_{2g}^4)$ then relaxes to the ground state $(e_g^2 t_{2g}^3)$ via the transition $^4T_1 \rightarrow {}^6A_1$ with a subsequent emission in the visible region.

$$(Mn_{Zn}^{\ x})^* \rightarrow Mn_{Zn}^{\ x} + h\nu_{\text{yellowish-green}}$$

This d–d transition is relatively sharp, and so the emission spectrum is relatively narrow (490–610 nm) peaking at $\sim$520 nm, thus appearing yellowish-green. Since this emission is due to what is essentially a forbidden transition ($^4T_1 \rightarrow {}^6A_1$), the luminescence has a comparatively long decay time of 25 ms, and by definition is termed *phosphorescence* [34].

Increasing the manganese concentration within the willemite causes an increase in the intensity of the phosphorescence, up to a critical concentration corresponding to approximately $Zn_{1.92}Mn_{0.08}SiO_4$. Beyond this level, the material experiences *concentration quenching*, with a rapid diminishment and loss of phosphorescence. This is a feature typical of many phosphors based on doped materials.

In 1918, Nichols and Howes noted that willemite is found occasionally in nature that exhibits a persistent phosphorescence (more correctly known as *afterglow*) under both photo- and cathode- excitation [36]. In 1940 Froelich patented the addition of arsenic (preferably as As_2O_3) in order to produce afterglow in various luminescent silicates, including artificial willemite [37]; followed in 1942 with an article describing the effects of different manganese and arsenic doping levels on the optical properties of artificial willemite [38]. Investigations on the nature of the electron traps in doped willemite in relation to the afterglow were first reported in 1945 by Randall and Wilkins [39, 40], then again in the 1980s through a series of articles by workers from the IBM Corporation [35], [41–44]. The oxidation state of the arsenic and its site occupancy within the willemite structure are not known for certain; and neither is the nature of the compensatory defects (e.g. cation vacancies, etc.) necessary to maintain charge and mass balance. However, it is generally accepted that As^{5+} acts as an electron trap. During the process of hole–electron recombination, the electron first becomes trapped at an arsenic site, presumably, through the consecutive reactions:

$$As_{Si}^{\bullet} + e' \rightarrow As_{Si}^{x}$$

and

$$As_{Si}^{x} + e' \rightarrow As_{Si}'$$

and is held temporarily at several metastable energy levels, before eventually reducing the metastable Mn^{3+} ion to yield Mn^{2+} with an excited spin-state, etc., as in the above (arsenic-free) case for phosphorescence (see Figure 14.5).

$$Mn_{Zn}^{\bullet} + As_{Si}' \rightarrow (Mn_{Zn}^{x})^* + As_{Si}^{x}$$

and subsequently,

$$Mn_{Zn}^{\bullet} + As_{Si}^{x} \rightarrow (Mn_{Zn}^{x})^* + As_{Si}^{\bullet}$$

The reader is referred to the above articles for further details, and to a comprehensive description of luminescence as given in the authoritative work on the subject by Blasse and Grabmaier [34].

Regarding the synthesis of artificial willemite, the earliest documented account is that given by Jacques Joseph Ebelman (1814–1852) in 1851, shortly before he died. Ebelman produced artificial willemite by heating boric acid, silica and zinc oxide in platinum capsules inside a muffle furnace [45, 46]. No details regarding the temperature and annealing time are given. The crystalline product was compared with natural specimens of willemite from Moresnet, Belgium. A decade later in 1861, Henri Etienne Sainte-Claire Deville (1818–1881) produced willemite through the more metallurgical approach of passing silicon fluoride over melted zinc and red hot zinc oxide [47]. In 1887, Alex Gorgeu [48] prepared willemite by heating an intimate mixture of approximately equivalent amounts of zinc sulfate and alkaline sulfate with one-thirtieth of their mass of hydrated silica for one hour at a temperature sufficiently high to evolve sulfur dioxide. The fused product was treated with water so as to leave hexagonal prismatic crystals of willemite amongst the residue. The alternative use of calcined silica in place of hydrated silica was noted to result in slower reaction rates.

The first published account of preparing willemite doped with manganese for the purpose of producing a luminescent artificial willemite, is that given by Andrews in 1922 [49]; i.e., 18 years after the discovery of luminescence in natural willemite. Andrews' account was based on the unpublished preparative description given to him by Mr W. L. Lemcke, of Franklin, Pennsylvania, and is summarised as follows. A mixture of 100 g of zinc oxide, 50 g of silica and 0.5 g of manganese(IV) oxide, were ground finely together so as to pass through a 200 mesh sieve ($\leq 75\,\mu m$). This mixture was then placed in a porcelain crucible and heated to $\sim 1200\,^{\circ}C$ for at least 30 min before cooling to room temperature. The product was reputed to be pure white with a bright green fluorescence under ultraviolet light. This work was conducted at the General Electric Company, Schenectady, New York, and led eventually to $Zn_2SiO_4{:}Mn^{2+}$ being commercialised in 1937 as a lamp phosphor (known as P1), and was quickly followed in 1938 by the Mn-doped beryllium-containing analogue $(Zn, Be)_2SiO_4{:}Mn^{2+}$ on account of its broader emission spectrum; shifted so as to include the red [34]. Due to the acidic behaviour of $(Zn,Be)_2SiO_4{:}Mn^{2+}$ towards the mercury vapour contained within the luminescent lamps, this phosphor had a relatively short life span. So, with the toxicity of beryllium as an additional problem, these lamps were replaced commercially in 1948 by lamps containing halophosphates such as $Ca_5(PO_4)_3(Cl,F){:}Mn^{2+}$, Sb^{3+} [34].

Willemite has also been exploited as a phosphor in cathode ray tubes, as used in oscilloscopes, radar screens and low-flicker terminal displays, where its phosphorescence and afterglow is particularly beneficial [34]. However, these properties make it unsuitable for use in television tubes, where the need for rapidly changing images is paramount. $Zn_2SiO_4{:}Mn^{2+}$ is of renewed interest today as a green phosphor in plasma display panels [50]. The reader is referred to the review article on inorganic luminescent materials by Feldman *et al.* [50] for a summary of commercial phosphors and applications, both past and present.

Preparative Procedure

Doped materials normally require more sophisticated methods of preparation than that described by Andrews [49], since the dopant species needs to be incorporated into the structure homogenously and at a controlled level of concentration. The preparative procedure described here, involves a hybrid coprecipitation and sol-gel method for the synthesis of manganese and arsenic coactivated zinc silicate $(Zn_2SiO_4{:}Mn^{2+},As^{5+})$. The principle aspects of this method were adopted from the pioneering work of Froelich and Fonda in

Figure 14.6 Ceramic monolith of arsenic-doped $Zn_{1.96}Mn_{0.04}SiO_4$ as prepared by the method described in this chapter under short-wavelength ultraviolet radiation. Width of field = 12 mm. (Photograph exposure time = 10 s)

1942 [38], together with certain preparative aspects by Yang *et al.* [51] (see Figure 14.6).

Previous attempts by the present author to prepare single-phase material have often resulted in a two-phase mixture; $(Zn,Mn)_2SiO_4$ (pale pink-brown) as the major phase, together with minor amounts of $(Zn,Mn)O$ (reddish-orange). This may be a consequence of incomplete chemical reaction in which a part of the silica content is unaccounted for; silica may be present as a glass phase, and as such is undetected by powder X-ray diffractometry. Alternatively, the system may be inadvertently deficient in silica. This problem of inhomogeneity often arises in solid-state synthesis, because it is difficult, in a practical sense, to prepare a reaction mixture that contains the precise quantities of the various chemical reagents required, so as to correspond exactly with the desired product. The reader should be aware that the willemite phase with the composition $Zn_{1.96}Mn_{0.04}SiO_4$ is represented graphically as a sharp point in the ZnO-MnO-SiO_2 ternary phase diagram. During the preparation, errors arise from the uncertainties regarding the purity or concentration of the chemical reagents used (for example, the moisture content is often unknown), inaccuracies during weighing, as well as, spillage and loss of material during handling. It might be prudent therefore, to add a slight excess of tetraethyl orthosilicate ($\sim$5%) during the initial stage of the synthesis and so hopefully avoid the presence of $(Zn,Mn)O$ in the product material. The presence of a small amount of free silica should not obscure the luminescence of the willemite since silica is photoinactive and transparent to ultraviolet and visible light. An excess of silica is apparently quite normal [38].

The reader has the possibility here of preparing specimens of the commercial-type phosphors P1 ($Zn_2SiO_4:Mn^{2+}$) and P39 ($Zn_2SiO_4:Mn^{2+},As^{5+}$). A detailed preparative method for making P39 that exhibits an afterglow lasting several seconds after exposure to short-wavelength ultraviolet radiation (254 nm) is given below. The value for the manganese concentration (1% (wt/wt) Mn) is adopted from Fonda [52] and Cho and Chang [53] where it was shown that manganese doping within 0.5–2% (wt/wt) Mn corresponds to an optimum and reproducible level of phosphorescence in arsenic-free willemite at 25 °C. The value for the arsenic concentration (0.05% (wt/wt) As_2O_5) is adopted from Froelich and Fonda [38] who showed that this level of arsenic doping corresponds to an optimal level of afterglow when codoped with 0.5% (wt/wt) Mn; the reader is advised to not exceed 0.1% (wt/wt) As_2O_5. The reader shall have noticed that these two manganese concentrations do not coincide with each other (1% vis-à-vis 0.5%). This is true; but by preparing P1 and P39 so as to contain an equal amount of manganese, allows for a direct comparison of their luminescent intensities. The nature of the system leads to a compromise between the relative intensities of the phosphorescence and the afterglow as a function of the arsenic concentration. If the reader wishes to prepare the phosphor P1, with bright phosphorescence at the expense of losing the afterglow, then follow the same preparative procedure as described below, but omit the arsenic component.

Prepare 20 g of $Zn_{1.96}Mn_{0.04}SiO_4$ doped with 0.05% (wt/wt) As_2O_5, in order to produce a sample of synthetic willemite containing 1% (wt/wt) Mn and 326 ppm As by the following method: The preparation of the precursor material (i.e., the silicon alkoxide sol-gel and zinc–manganese hydroxide/arsenate coprecipitated suspension) involves using the following reagents:

Zinc(II) acetate dihydrate [$Zn(CH_3CO_2)_2 \cdot 2H_2O$]	99.0%	FW = 219.50 g
Manganese(II) acetate tetrahydrate [$Mn(CH_3CO_2)_2 \cdot 4H_2O$]	99.99%	FW = 245.09 g
Tetraethyl orthosilicate (TEOS) [$Si(OC_2H_5)_4$]	98% (wt/wt)	FW = 208.33 g
Arsenic acid solution [H_3AsO_4]	80% (wt/wt)	FW = 141.91 g

Calculate the mass required for each of the above chemicals in order to produce 20 g of $Zn_{1.96}Mn_{0.04}SiO_4$ doped with 0.05% (wt/wt) As_2O_5. Weigh the appropriate amount of TEOS (including the 5% excess) by slowly pouring a sufficient amount of this liquid from the bottle, directly into a 500-ml glass beaker; making fine adjustments using a disposable plastic teat-pipette. Place this beaker on a magnetic stirrer together with a magnetic bead. Add 10 ml of ethanol (96% technical grade) and stir gently to homogenise the two liquids.

In a separate 100-ml glass beaker, place 20 ml of distilled water and add ~6 drops of concentrated hydrochloric acid solution, and stir gently with a glass rod. Pour this acidified water, slowly, into the TEOS/ethanol liquid, whilst stirring continuously with the magnetic stirrer. Cover the top of the glass beaker with a stretched sheet of plastic film, for example, parafilm™, and leave this mixture to stir magnetically for ~24 h. This will gradually hydrolyse the TEOS and thereby form the silicon alkoxide sol (with the appearance of a transparent and colourless homogenous liquid). The reader is referred to Schubert and Hüsing [54] for further information concerning the hydrolysis of TEOS and, sol-gels.

On the next day, weigh the required amount of arsenic acid 80% (wt/wt) aqueous solution (which should be about one drop) on a small clean watch glass using a 1-ml disposable plastic teat-pipette to a precision of ±0.5 mg using an analytical balance.[10] Rinse the watch-glass very carefully with 150 ml of distilled water so as to ensure that the entire measure of the arsenic acid (and the 150 ml of distilled water) enters a clean 500-ml glass beaker. Place this beaker on a magnetic stirrer together with a magnetic bead. Add the appropriate amounts of $Mn(CH_3CO_2)_2 \cdot 4H_2O$ and $Zn(CH_3CO_2)_2 \cdot 2H_2O$ in that order and stir for approximately 15 min until these salts are completely dissolved. Note that a very fine white cloudy precipitate of manganese–zinc arsenate will form at this stage. Then add drop-wise, using a glass burette, 40 ml of 24%(wt/wt) aqueous ammonia solution into the Zn^{2+}(aq) and Mn^{2+}(aq) solution and stir continuously. This will form a white $Zn(OH)_2$ and $Mn(OH)_2$ coprecipitate held in aqueous suspension. Then slowly add all of this zinc–manganese hydroxide/arsenate suspension to the silicon alkoxide sol (prepared the previous day) and continue to stir magnetically. You may need to rinse the beaker with a small amount (~20 ml or less) of distilled water in order to transfer all of the zinc–manganese hydroxide/arsenate suspension.[11] Cover the top of the glass beaker with a stretched sheet of plastic film, for example, parafilm™, and leave this mixture to stir magnetically for ~24 h.

On the next day, remove the magnetic bead, and then leave this sol-gel/coprecipitate mixture to dry in its glass beaker at ~70 °C in a drying oven for ~4 days in order to form a gel/coprecipitate intimate mixture. After this time, the contents of the beaker should have dried into a white to pale brown shrunken conglomerated lump that might exhibit a small degree of plasticity. Transfer this precursor material into a large alumina crucible (CC62 Almath Ltd) together with its matching lid, and place in a chamber (or pottery) furnace and heat to 1100 °C with a heating rate of 60 °C/h. After 10 hours at

---

[10] The reader is advised to handle arsenic acid with due care and attention on account of its high toxicity.

[11] Alternatively, the reader may wish to experiment by adding this the other way around; so as to add the silicon alkoxide sol to the zinc–manganese hydroxide/arsenate suspension.

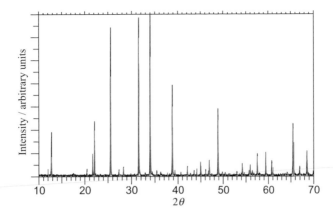

Figure 14.7 Powder XRD pattern (Cu-$K_{\alpha 1}$ radiation) of Zn_2SiO_4 as prepared by the method described in this chapter. This powder pattern corresponds closely to the PDF 37-1485 for Zn_2SiO_4 (shown in red)

1100 °C cool to room temperature (20 °C) with a cooling rate of 300 °C/h. This heating cycle will last ~32 h.

Remove the product powder from the crucible and place it in a sample bottle. Determine the percentage yield of the product and note its colour under normal white light. Observe its optical behaviour under ultraviolet radiation (365 and 254 nm) whilst in a dark room. Submit a small powdered sample (~3 g) for analysis by powder XRD (scan: 10–70° 2θ). Compare your powder XRD pattern with the PDF 37-1485 (cf. ICSD 2425) for Zn_2SiO_4 and, with Figure 14.7. Check any unaccounted peaks with the powder pattern for zinc oxide (ZnO) PDF 36-1451 and various relevant silica (SiO_2) phases, such; as quartz, tridymite and cristobalite.

A ceramic monolith of the product material can be made as follows. Press 9 g of the powdered product into a compacted disc using a 32-mm die and a uniaxial press under a load ~5 tonne. Place some of the remaining powdered product back into the original alumina crucible used in the synthesis, so as to form a sacrificial powder bed. Gently place the compacted disc on top of this powder bed and then place this in the chamber furnace. Heat at 200 °C/h to 1200 °C. After 10 h at 1200 °C cool to room temperature (20 °C) with a cooling rate of 200 °C/h. This heating cycle will last ~22 h. Observe the optical behaviour of the ceramic willemite under ultraviolet radiation (365 and 254 nm). The two faces of the ceramic monolith can be ground flat and polished, if required, using a grinding/polishing machine (for example, Struers RotoPol-11).

The reader may wish to observe the effect of temperature on the luminescence. The willemite monolith can be cooled by partial emersion in liquid nitrogen, and then exposed to ultraviolet light (254 nm). This should result in diminished phosphorescence, and the apparent loss of afterglow in P39. Upon warming to room temperature – with the aid of a hot-plate, and with the ultraviolet light now switched off – P39 may display thermoluminescence in a dark-room.

Finally, the preparative procedure described above for willemite is in principle, applicable for preparing other anhydrous mixed-metal silicate (and germanate) phases, with or without doping. For an example, the artificial clinopyroxene, $MgCuSi_2O_6$, has been prepared successfully using this method [55].

REFERENCES

1. W. H. Zachariasen, *Norsk Geologisk Tidsskrift* **9** (1926) 65–73.
2. W. L. Bragg and W. H. Zachariasen, *Zeitschrift für Kristallographie, Kristallgeometrie, Kristallphysik, Kristallchemie* **72** (1930) 518–528.
3. M. A. Simonov, P. A. Sandomirskii, Yu. K. Egorov-Tismenko and N. V. Belov, *Doklady Akademii Nauk SSSR*, **237** (1977) 581–584.
4. E. N. Bunting, *Journal of the American Ceramic Society* **13** (1930) 5–10.
5. M. Akaogi, H. Yusa, E. Ito, T. Yagi, K. Suito and J. T. Iiyama, *Physics and Chemistry of Minerals* 17 (1990) 17–23.
6. J. Smithson, *Philosophical Transactions of the Royal Society of London* **93** (1803) 12–28.
7. R. V. Gaines, H. C. W. Skinner, E. E. Foord, B. Mason and A. Rosenzweig, *Dana's New Mineralogy: The System of Mineralogy of James Dwight Dana and Edward Salisbury Dana*, 8th edn, 1997, Wiley-Blackwell, New York.
8. L. J. Spencer, *Journal of the Mineralogical Society* **21** (1927) 388–396.
9. J. C. Greene and J. G. Burke, 'The Science of Minerals in the Age of Jefferson', *Transactions of the American Philosophical Society*, New Series, **68** (1978) 1–113
10. L. Vanuxem and W. H. Keating, *On the geology and mineralogy of Franklin, in Sussex County, New Jersey, Journal of the Academy of Natural Sciences of Philadelphia* **2** (1822) 277–288.
11. L. Vanuxem and W. H. Keating, *Observations upon some of the minerals discovered at Franklin, Sussex County, New Jersey, Journal of the Academy of Natural Sciences of Philadelphia* **4** (1824) 3–11.
12. John Torrey, Mineralogical Notices, by Dr. Torrey: *American Journal of Science and Arts*, 1st Series, 5 (1822) 399–402.
13. P. J. Dunn, *Franklin and Stirling Hill, New Jersey: The world's most magnificent mineral deposits*, 1995, The Franklin-Ogdensburg Mineralogical Society.
14. T. Thomson, *Outlines of Mineralogy, Geology, and Mineral Analysis*, 1836, Baldwin & Cradock, London.
15. T. Thomson, *Annals of the Lyceum of Natural History of New York* 3 (1828) 9–86.
16. A. Lévy, *Neues Jahrbuch für Mineralogie, Geognosie, Geologie und Petrefaktenkunde*, 1 (1830) 71.
17. A. Lévy, *Annales des Mines (Paris)* 4th Series, 4 (1843) 507–520.
18. C. U. Shepard, *Treatise on Mineralogy*, 1st edn, volume 1, part. 1, (1832) page 154, Hezekiah Howe, New Haven.

19. C. Palache, *United States Geological Survey (USGS) Professional Paper 180* (1935).
20. A. Dellese and A. Descloizeaux, *Annales des Mines*, 4th series, **10** (1846) 211–214.
21. R. Hermann, *Journal für Praktische Chemie* **47** (1849) 1–15.
22. J. Metcalf-Johansen, *Mineralogical Magazine* **41** (1977) 71–75.
23. A. A. Finch, *Mineralogical Magazine* **54** (1990) 407–412.
24. Z. V. Shlyukova, E. V. Vlasova and A. I. Tsepin, *Mineralogicheskii Zhurnal* **2** (1980) 100–102.
25. A. Holmes, *Principles of Physical Geology*, Thomas Nelson and Sons Ltd, 2nd edn, (1965), London pp 1288.
26. A. B. Blaxland, *Mineralium Deposita* **6** (1971) 313–320.
27. M. W. Hitzman, N. A. Reynolds, D. F. Sangster, C. R. Allen and C. E. Carman, *Economic Geology*, **98** (2003) 685–714.
28. D. Large, *Erzmetall*, **54** (2001) 264–274.
29. J. Brugger, D. C. McPhail, M. Wallace, J. Waters, *Economic Geology*, **98** (2003) 819–835.
30. G. F. Kunz and C. Baskerville, *Chemical News and Journal of Industrial Sciences* **89** (1904) 1–6.
31. E. K. Judd, *Engineering and Mining Journal* **83** (1907) 803.
32. S. F. Squiller and C. B. Sclar, *Proceedings of the Quadrennial IAGOD Symp.*, 5th (1980), Meeting Date 1978, 759–66. ed. J. D. Ridge, John Drew; Schweizerbart, Stuttgart.
33. C. Frondel and J. L. Baum, *Economic Geology and the Bulletin of the Society of Economic Geologists* **69** (1974) 157–80.
34. G. Blasse and B. C. Grabmaier, *Luminescent Materials*, Spinger-Verlag, Berlin (1994) pp 232.
35. P. Avouris, I. F. Chang, D. Dove, T. N. Morgan and Y. Thefaine, *Journal of Electronic Materials* **10** (1981) 887–899.
36. E. L. Nichols and H. L. Howes, *Proceedings of the National Academy of Sciences of the United States of America* **4** (1918) 305–312.
37. H. C. Froelich, *Manufacture of Luminescent Materials*, United States Patent, 2,206,280, 2 July 1940 pp3.
38. H. C. Froelich and G. R. Fonda, *Journal of Physical Chemistry* **46** (1942) 878–885.
39. J. T. Randall and M. H. F. Wilkins, *Proceedings of the Royal Society of London. Series A, Mathematical and Physical Sciences* **184** (1945) 347–364.
40. J. T. Randall and M. H. F. Wilkins, *Proceedings of the Royal Society of London. Series A, Mathematical and Physical Sciences* **184** (1945) 365–389.
41. I. F. Chang and G. A. Sai-Halasz, *Journal of the Electrochemical Society* **127** (1980) 2458–2464.
42. I. F. Chang, P. Thioulouse, E. E. Mendez, E. A. Giess, D. B. Dove and T. Takamori, *Journal of Luminescence* **24/25** (1981) 313–316.
43. D. B. Dove, T. Takamori, I. F. Chang, P. Thioulouse, E. E. Mendez and E. A. Giess, *Journal of Luminescence* **24/25** (1981) 317–320.
44. D. J. Robbins, N. S. Caswell, P. Avouris, E. A. Giess, I. F. Chang and D. B. Dove, *Journal of the Electrochemical Society: Solid-State Science and Technology* **132** (1985) 2784–2793.
45. J. J. Ebelmen, *Annales des Chimie et des Physique* 3e series XXXIII (1851) 34–74.
46. J. J. Ebelmen, M. Salvétat, E. Chevreul, *Chimie, Céramique, Géologie, Métallurgie*, Mallet-Bachelier, Paris, 1861, pp 628.

47. H. S. C. Deville, *Comptes Rendus*, **52** (1861) 1304–1308.
48. A. Gorgeu, *Comptes Rendus Hebdomadaires des Seances de l'Academie des Sciences* **104** (1887) 120–123.
49. W. S. Andrews, *American Mineralogist* **7** (1922) 19–23.
50. C. Feldmann, T. Jüstel, C. R. Ronda and P. J. Schmidt, *Advanced Functional Materials* **13** (2003) 511–516.
51. P. Yang, M. K. Lü, C. F. Song, S. W. Liu, D. R. Yuan, D. Xu, F. Gu, D. X. Cao and D. H. Chen, *Inorganic Chemistry Communications* **5** (2002) 482–486.
52. G. R. Fonda, *Journal of Physical Chemistry* **43** (1939) 561–577.
53. T. H. Cho and H. J. Chang, *Ceramics International* **29** (2003) 611–618.
54. U. Schubert and N. Hüsing, *Synthesis of Inorganic Materials*, 2nd edn, Wiley-VCH, Weinheim, 2005, pp409.
55. F. Adam, *A Study of the synthesis of the alkaline earth copper silicates $MgCuSi_2O_6$ and $BaCuSi_2O_6$ and their phase relationships*, MSc Thesis, 2007, The University of Southern Denmark.

PROBLEMS

1. Write appropriate chemical equations to describe the acid-catalysed hydrolysis of TEOS.

2. State the solubility (in units of $g\ dm^{-3}$) for the compounds; $Zn(CH_3CO_2)_2 \cdot 2H_2O$ and $Mn(CH_3CO_2)_2 \cdot 4H_2O$, in water at 25 °C. It may help to consult various chemical data-books in order to answer this Problem.

3. In this work, an excess of aqueous ammonia solution was added with respect to the amount of metal cations present in aqueous solution. Calculate the minimum volume of 24% aqueous ammonia solution required in order to precipitate these cations as $Zn(OH)_2$ and $Mn(OH)_2$.

4. Manganese(II) hydroxide is precipitated from the solution as a white solid. If this precipitate is exposed to air, it can rapidly darken because of oxidation by atmospheric oxygen. Write a chemical equation to describe this phenomenon and comment on the oxidation states of manganese in the various phases concerned.

5. Use Outokumpu HSC Chemistry® software (or appropriate thermodynamic data from elsewhere) in order to compare and contrast the changes in standard Gibbs free energy (ΔG°_R) for the following chemical reactions as a function of temperature over the range 0–1500 °C. It is informative to display these data in the form of a plot.
   (a) $Mn_2O_3(s) \rightarrow 2MnO(s) + 0.5O_2(g)$
   (b) $Mn_2O_3(s) + SiO_2(s) \rightarrow Mn_2SiO_4(s) + 0.5O_2(g)$

6. If, at some stage during the preparation of the precursor material, the oxidation as described in *Problem 4* was to occur, do you think this would affect the outcome of the synthesis? Comment on how the presence of silica in the system may influence matters here.

7. Suggest an alternative chemical reagent for adding the 'As_2O_5' component to the reaction mixture in place of the arsenic acid as used here. Can you foresee any advantages, or disadvantages, in using your chosen chemical reagent compared to that of arsenic acid?

8. State the coordination and symmetry of the manganese(II) ions in $Zn_{1.96}Mn_{0.04}SiO_4$ and in Mn_2SiO_4.

9. What would you expect to produce if you attempted to synthesise '$ZnMnSiO_4$'?

10. The halophosphates with the apatite structure are an important class of commercial phosphors. Use your knowledge and experience acquired during the synthesis of $Zn_2SiO_4:Mn^{2+},As^{5+}$ and $CuTiZr(PO_4)_3$ (see Chapter 7) to propose a method for preparing the artificial luminescent apatite, $Ca_5(PO_4)_3F:Mn^{2+},Sb^{3+}$. The reader is referred to Blasse and Grabmaier [34] for a description of the defect chemistry of $Ca_5(PO_4)_3F:Mn^{2+},Sb^{3+}$.

15

Artificial Scheelite CaWO$_4$ by a Microwave-Assisted Solid-State Metathetic Reaction

Calcium tungstate, Ca^{2+}WO$_4^{2-}$ is a salt that is formed, in principle, through the reaction of calcium hydroxide, Ca(OH)$_2$ with tungstic acid, H$_2$WO$_4$. It exhibits a bluish-white luminescence upon exposure to X-rays and short-wavelength (254 nm) ultraviolet radiation.[1] CaWO$_4$ is used commercially as a blue X-ray phosphor in X-ray intensifying screens for general application in radiography,[2] a use that originated shortly after the discovery of X-rays by Wilhelm Röntgen in 1895. For this particular application it is desirable to prepare CaWO$_4$ as a powder with a high degree of purity and small crystallite size, suitable for depositing as a thin film. Single crystals of CaWO$_4$ are used as an intrinsic scintillator[3] for γ-rays, and in the search for dark matter [1]. Single crystals of CaWO$_4$ doped with various lanthanides have potential electro-optic applications, for example; in laser technology. CaWO$_4$ is also of interest as a ceramic material in high-frequency microwave technology, on account of its dielectric properties. This chapter describes the preparation of CaWO$_4$ as a powder with particles of micrometre-sized dimensions, through

---

[1] Throughout this chapter, *X-ray luminescence*, refers to luminescence that is excited by X-rays; whereas, *photoluminescence*, refers to luminescence that is excited by ultraviolet radiation.
[2] The term *radiography* describes the process of using X-rays (or other ionising radiation) to make a *radiograph*. A radiograph is an image of an object, for example, human bones (cf. the analogous expression: photograph).
[3] A *scintillator* is a material that emits a minute flash of visible light when struck by a charged particle or high-energy photon.

Synthesis, Properties and Mineralogy of Important Inorganic Materials   By Terence E. Warner
© 2011 John Wiley & Sons, Ltd

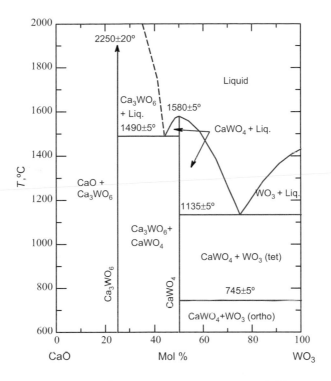

Figure 15.1 An equilibrium phase diagram of the CaO–WO₃ binary system. After Chang *et al.* [2]; ACerS-NIST Phase Equilibria Diagram 2306. tet = tetragonal phase; ortho = orthorhombic phase (Reproduced with permission from Journal of the American Ceramic Society, Alkaline-Earth Tungstates: Equilibrium and Stability in the M-W-O Systems by L. L. Y. Chang, M. G. Scroger and B. Phillips, 49, 385–390 Copyright (1966) American Ceramic Society

a microwave-assisted solid-state metathetic[4] reaction between the hydrated salts, $CaCl_2·2H_2O$ and $Na_2WO_4·2H_2O$. This method can be exploited for the preparation of many other inorganic materials.

An equilibrium phase diagram of the $CaO-WO_3$ binary system was constructed by Chang *et al.* [2] in 1966, through a study of quenched samples encapsulated in platinum, as analysed by reflected light microscopy and X-ray diffractometry; see Figure 15.1. From this diagram it can be seen that $CaWO_4$ has a congruent melting point at 1580 °C; and in this sense, the material can be considered as a refractory ceramic. $CaWO_4$ forms a binary eutectic with WO_3 at 1135 °C, and likewise with Ca_3WO_6 at 1490 °C.

$CaWO_4$ crystallises with the prototype structure as shown in Figure 15.2. It belongs to the tetragonal crystal system and space group $I4_1$, with cell constants $a = 5.243$ Å and $c = 11.376$ Å [3]. The crystal structure was first

---

[4] A metathetic reaction involves the exchange of anions (X and Y) between two different cations (A and B), for instance: AX + BY → AY + BX.

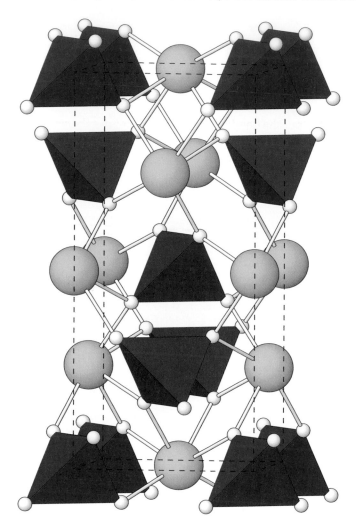

Figure 15.2   Crystal structure of scheelite, CaWO₄ showing the tetragonal unit cell. The Ca^{2+} ions are shown yellow, and the tungstate, WO_4^{2-} tetrahedra are shown blue. The oxide ions are shown white with reduced radii in order that the structure becomes more 'transparent'. This figure was drawn using data from Zalkin and Templeton [3], cf. ICSD 15586

determined in 1920 by Dickinson using single-crystal X-ray diffractometry [4], and refined by Zalkin and Templeton in 1964 using the same technique [3]. The scheelite structure can be visualised as comprising two pseudocubes in rotary mirror image to one another. The bottom face of the lower pseudocube contains four tungsten ions located at the four corners, with one calcium ion at the centre. Two tungsten ions are centred at two opposing sides; and likewise, two calcium ions are centred at the other two opposing sides. The top face of the lower pseudocube contains four calcium ions located at the four corners, with one tungsten ion at the centre; this top face also constitutes the bottom face of the upper pseudocube. The upper pseudocube is a mirror image of the lower pseudocube that has undergone, in addition, a 90° rotation around the c-axis. The tungsten ions are tetrahedrally coordinated to four oxide ions, forming isolated WO_4^{2-} tetrahedra that are flattened in the c-direction [5]. The Ca^{2+} ions are 8-fold coordinated to oxide ions from eight separate WO_4^{2-} tetrahedra. The bonding sequence throughout the crystal is therefore: W–O–Ca–O–W–... etc.

$Ca^{2+}WO_4^{2-}$ is essentially an ionic compound, although the W O bonds have a significant degree of covalency. Both the excitation and the emission processes in $CaWO_4$ are considered to be localised at the WO_4^{2-} complex anion [6], despite the short distances that isolate these tetrahedra within the crystal structure. It is generally agreed that the highest occupied bonding orbital in the WO_4^{2-} complex anion is dominated by the oxygen $2p$ (π) state; whilst the lowest unoccupied anti-bonding orbital is dominated by the tungsten $5d$ (e) state [7]. The absorption spectra for $CaWO_4$ at room temperature (and also at cryogenic temperatures) show a strong and broad absorption band in the ultraviolet, with an absorption edge at 280 nm, and a peak near 250 nm. This is interpreted as a charge transfer from O^{2-} to W^{6+} within an individual WO_4^{2-} complex anion, whereupon the electron is trapped momentarily at the d_{z^2} orbital on the W^{5+} ion [8]. The corresponding hole is localised at the oxygen $2p$ (π) orbital, and is considered by Herget et al. [9] to be accompanied by a trigonal Jahn–Teller distortion of the WO_4 tetrahedron. When the electron returns to the ground state, it produces a broad band emission in the visible and near-ultraviolet region of the electromagnetic spectrum, peaking at 420 nm, which appears as blue light (see Figures 15.3 and 15.4). $CaWO_4$ is a reasonably efficient photoluminescent material at 298 K, although above 250 K, the emission intensity decreases (when used as a scintillator) due to thermal quenching [10].

The discovery of X-rays by Wilhelm Röntgen in November 1895, created an urgent need for a scintillation material that was sensitive to X-rays. Photographic film is rather insensitive to X-rays, therefore long exposure times (~1 h) were necessary in order to obtain a radiograph; thus endangering the health of the persons exposed to the radiation. In March 1896, Thomas Edison [11], [12] reported that $CaWO_4$ is an effective blue X-ray phosphor, that can be used as a powder film in X-ray intensifying screens (see Figure 15.5).⁵ X-rays are absorbed by the $CaWO_4$ lattice resulting in the

---

⁵ During the early part of 1896, Thomas A. Edison sent a CaWO₄ fluorescent screen to Mihajlo I. Pupin (a Serbian immigrant in the United States of America) who placed it upon a photographic plate and obtained a clear image of the lead pellets and bones of the hand relating to a prominent lawyer who had recently received a blast from a shotgun to the hand [16]. Later, in 1907, Edison patented the use of CaWO₄ as an X-ray phosphor in a fluorescent lamp; but there is no evidence that this particular invention was commercialised [17]. Contrary to popular belief, Edison did not patent CaWO₄ for use as a photoluminescent material as excited by ultraviolet radiation, but only as an X-ray phosphor.

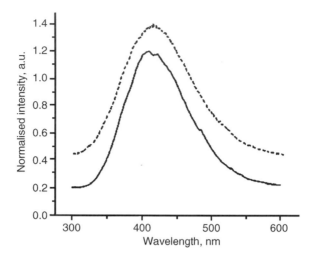

Figure 15.3   Luminescence spectra of CaWO₄ at 296 K as excited by X-rays (in dash) and short-wavelength (260 nm) ultraviolet radiation (full line) (Reproduced with permission from Radiation Measurements, Two-photon excitation and luminescence of a CaWO₄ scintillator by V. B. Mikhailik, I. K. Bailiff, H. Kraus, P. A. Rodnyi, and J. Nikovic, 38, 585–588 Copyright (2004) Elsevier Ltd)

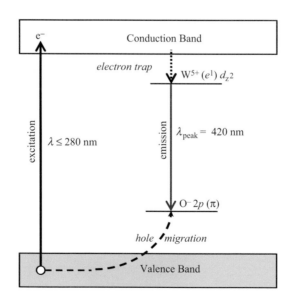

Figure 15.4   Schematic energy diagram for the excitation and emission mechanisms in CaWO₄. The expressions are consistent with those described in the text. The broad conduction and valance bands are shown here in order to emphasise the effects of neighbouring interactions between the WO₄²⁻ complex anions within the solid state

Figure 15.5 An engraving (1896) of the Edison Fluoroscope in use [31]; 'By placing the object to be observed, such as the hand, between the vacuum tube and the fluorescent screen, the shadow is formed on the latter and can be observed at leisure' (Reproduced from Electrical Engineer, T. A. Edison, 21, p 340 Copyright (1896) Copyright Holder unknown)

formation of electron–hole pair defects (see above). Upon thermalisation, these pairs recombine, resulting in the emission of blue light [13]. The blue X-ray luminescence darkens the photographic film, which is then developed to produce a radiograph (see Figure 15.6). By using an X-ray intensifying screen, the exposure time to the X-rays is reduced by almost a thousand-fold,

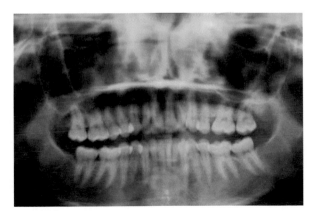

Figure 15.6 An orthopantomograph (i.e. a particular type of radiograph that gives a panoramic image) of human teeth and bone, in which the jaw appears flat and not curved (Reproduced with permission from N. Webb)

Figure 15.7   Scheelite: Xinjiang, China. Field of view: 28 × 42 mm. (Photograph reused with permission from Ole Johnsen. Copyright (2010) Ole Johnsen)

to the order of a few seconds. To be most effective in creating a sharp image, the CaWO$_4$ should be deposited as a thin and coherent film on the photographic film; necessitating the production of powders with small crystallites, ideally within the range, 5–10 μm. Particles < 5 μm, have a diminished emission intensity as a consequence of internal scattering, whilst particles > 10 μm, can cause mechanical difficulties in forming thin films [14]. Likewise, the presence of afterglow in the X-ray phosphor should be avoided since this creates undesirable 'ghosting' in the radiograph. Unfortunately, CaWO$_4$ is rather inefficient as an X-ray phosphor, with a conversion efficiency of only 5% [15], which has stimulated a search for alterative scintillation materials. However, CaWO$_4$ is an inexpensive and stable material with a short decay time of 6 μs (at room temperature) and a very low afterglow, and so it is still used in general radiography.[6]

   CaWO$_4$ occurs naturally as the mineral scheelite (see Figure 15.7). The mineral was named in 1821 for the Swedish chemist Carl W. Scheele (1742–1786) who proved in 1781 that it contains tungsten trioxide, WO$_3$ [5]. Scheelite is found typically in skarns[7] and high-temperature hydrothermal veins. In 1930, van Horn [18] described one of the earliest observations of blue photoluminescence in scheelite under ultraviolet radiation, regarding a specimen from the East Pool Mine at Camborne, Cornwall.[8] A few years later, in 1936, Claude [19] described the use of CaWO$_4$ as a blue phosphor for fluorescent lighting (excited by ultraviolet radiation from a mercury vapour

---

[6] At the time of writing CaWO$_4$ is still manufactured and marketed as a blue-emitting X-ray phosphor for X-ray intensifying screens by; Mitsubishi Chemical Corporation and Nichia Corporation, both located in Japan.

[7] Skarns are metamorphic deposits formed by the intrusion of granitic magma into calcareous rocks, such as, limestone.

[8] van Horn [18] referred to certain earlier observations of fluorescence in scheelite under ultraviolet radiation, as conveyed to him in a letter from Mr W. L. Lemcke.

lamp), which led eventually to the production of 'insect lamps'.[9] However, the main commercial interests have focused on the use of $CaWO_4$ as an X-ray phosphor and scintillator whilst its photoluminescent properties remain largely unexploited.

Scheelite, $CaWO_4$ and wolframite, $(Fe,Mn)WO_4$ are the two principle ore minerals for the production of tungsten metal and its alloys. In 1847, Robert Oxland [20] patented an extractive metallurgical process for obtaining sodium tungstate, Na_2WO_4 which is used to produce high-purity tungsten metal.[10] Coincidently, this compound in its hydrated form, $Na_2WO_4 \cdot 2H_2O$ is a convenient chemical reagent for the WO_3 component required for the preparation of high purity $CaWO_4$ as described in this chapter. There is some irony in the fact that scheelite is mined as an ore mineral, from which its tungsten value is extracted as sodium tungstate, Na_2WO_4 and then used to reform the original phase, $CaWO_4$.[11]

Artificial scheelite, $CaWO_4$ has been prepared by numerous workers using a wide variety of methods and preparative routes. The method described here utilises the ability of microwave radiation to activate a metathetic reaction between the hydrated salts: $CaCl_2 \cdot 2H_2O$ and $Na_2WO_4 \cdot 2H_2O$. This innovative approach to its synthesis began in 1988, when Baghurst and Mingos [21] reported that tungsten oxide, WO_3, is a strong absorber of microwave radiation. It was noted that ~ 5 g of WO_3 reached a temperature of $532\,^\circ C$ after just 30 s inside a commercially available 500-W microwave oven operating at 2.45 GHz [22]. Baghurst and Mingos [21] then exploited this feature to show that barium tungstate, $BaWO_4$ can be synthesised successfully from a mixture of BaO and WO_3 in a 500-W microwave oven (2.45 GHz) within 30 min. In 2001, Cirakoglu et al. [23] adopted this method in an attempt to prepare the calcium analogue, $CaWO_4$ from a compacted mixture of CaO (-20 mesh) and WO_3 ($10-20\,\mu m$) in the molar ratio 1:1. This was placed in a 500-W microwave oven (2.45 GHz) and irradiated for 15 min. The product, unfortunately, had a greenish-yellow coloration, and comprised, $CaWO_4$ together with substantial amounts of unreacted WO_3 and $Ca(OH)_2$ (the latter forming upon exposure to the atmosphere).

In 2004, Kloprogge et al. [24] prepared $CaWO_4$ by a microwave assisted hydrothermal synthesis. A dilute equimolar aqueous solution of calcium nitrate and sodium tungstate was held inside a Teflon®-lined beaker within a

---

[9] The excitation source was a mercury vapour discharge lamp. The emission spectrum from insect lamps lures insects towards an electrified grill.

[10] Tungsten is an important refractory metal and has the highest melting point of all known metals at 3422 °C.

[11] High-technological applications involving $CaWO_4$ as an X-ray phosphor demand a high degree of purity and a controlled particle size. Naturally occurring scheelite is commonly contaminated with other metal cations (e.g., molybdenum) that alter its luminescent properties. Naturally occurring material is therefore unsuitable for these applications; hence the need to prepare $CaWO_4$ powders artificially.

high-pressure vessel (made of high-density polypropylene) and placed inside a laboratory microwave oven. This was heated to $150\,^{\circ}C$ under an autogeneous water vapour pressure for $1\,h$. Powder X-ray diffractometry indicated that the product was single-phase $CaWO_4$. Phuruangrat *et al.* [25] carried out a similar reaction by dissolving equimolar amounts of $Ca(NO_3)_2$ and Na_2WO_4 in propylene glycol, with the intention that this liquid will serve as a medium for microwave absorption. This mixture was exposed to $600\,W$ microwave radiation, intermittently, for $30\,min$; so as to raise the temperature monotonically to $\sim130\,^{\circ}C$. The product material comprised, $CaWO_4$ with a crystallite size of nanometre dimensions, and therefore, unsuitable for the production of X-ray intensifying screens because the grain size is too small.

In contrast to the above methods, Thangadurai *et al.* [26] exploited a metathetic reaction (performed originally by de Schulten [27] a century earlier[12]) to prepare $CaWO_4$ at room temperature by mixing together aqueous solutions of $CaCl_2$ and Na_2WO_4 without recourse to microwave radiation. Their product formed spontaneously as a white precipitate and comprised spherical particles of $CaWO_4$ of micrometer dimensions. In 2008, Parhi *et al.* [28] devised a novel method that combined a solid-state version of this metathetic reaction with the assistance of microwave radiation in order to activate an exothermic reaction between $CaCl_2$ and Na_2WO_4. This reaction is driven, in part, by the relatively large lattice energy $(-778\,kJ\,mol^{-1})$ for NaCl [29], and is described in terms of the anhydrous salts as follows:[13]

$$\Delta H_R^{\circ}(298\,K) = -127\,kJ\,mol^{-1}$$

$$CaCl_2(s) + Na_2WO_4(s) \rightarrow 2NaCl(s) + CaWO_4(s)$$

Parhi *et al.* [28] reported that $CaWO_4$ can be prepared by reacting a well-ground mixture of $CaCl_2$ and Na_2WO_4 in a 1:1 molar ratio.[14] This mixture was placed in a crucible (of unspecifed material) and exposed to microwave radiation in a 1100-W domestic microwave oven ($2.45\,GHz$) for a duration of $10\,min$. The microwave radiation is absorbed strongly by the water

---

[12] In 1903 Schulten [27] prepared artificial scheelite in the form of square-based pyramidal colourless crystals of $\sim60\,\mu m$ diameter by the following method (as reported originally in French). A dilute aqueous solution of sodium tungstate ($3.5\,g$ of 'ordinary crystalline' sodium tungstate per litre) was added drop-wise into a weakly acidified aqueous solution of calcium chloride ($10\,g$ of $CaCl_2$ per 3 litre) containing 1 ml concentrated hydrochloroic acid.

[13] Lattice energy is defined here as the heat released during the course of the following condensation reaction: $Na^+(g) + Cl^-(g) \rightarrow NaCl(s)$, and therefore has a negative value. Certain authors prefer to consider this reaction from the reverse direction and ascribe a positive sign to the lattice energy.

[14] The present author is grateful to Professor V. Manivannan of Colorado State University for the advice to use the dihydrates: $CaCl_2 \cdot 2H_2O$ and $Na_2WO_4 \cdot 2H_2O$ for the synthesis of $CaWO_4$; as opposed to the *anhydrous* salts as reported for this purpose, however, by Parhi *et al.* [28].

molecules of crystallisation within the hydrated salts: CaCl$_2$·2H$_2$O and Na$_2$WO$_4$·2H$_2$O, resulting in their rapid dehydration. This creates hot particles of the corresponding anhydrous salts with highly reactive surfaces that facilitate the subsequent and rapid metathetic reaction. The product material, CaWO$_4$ is a poor absorber of microwave radiation, and so the amount of heat generated by this irradiation process is self-limited. The product material was washed with distilled water to remove the sodium chloride and then dried at 80 °C. The phase purity of the final product was confirmed by powder X-ray diffractometry.

Preparative Procedure

Prepare 10 g of CaWO$_4$ by the following method; as adopted from the work of Parhi *et al.* [28]. With a precision of ±0.001 g weigh appropriate amounts of calcium chloride dihydrate, CaCl$_2$·2H$_2$O and sodium tungstate dihydrate, Na$_2$WO$_4$·2H$_2$O (Sigma-Aldrich, ACS reagent grade, 99%) in a 1:1 molar ratio. Grind these two dry powders together thoroughly with a porcelain pestle and mortar. It is important to avoid spillage of material at all stages of the preparation since this will effect the composition of the product material. Compact the entire powder into a cylindrical monolith using a 32-mm die and a uniaxial press under a load of ~10 tonne.

Place the compacted monolith in a Pyrex™ glass Petri dish. Since the microwave radiation is absorbed strongly by the reactants and the reaction is exothermic, the Petri dish should be supported on a thick mat of glass wool, in order to protect the glass base-plate inside the domestic microwave oven from overheating.[15] Introduce these into the microwave oven (for example, 2.45 GHz and 700 W) and expose the reaction mixture to microwave radiation for the duration of 10 min. Allow the product material to cool to room temperature, and make a note of its appearance. Then expose the product material to short-wavelength ultraviolet radiation (254 nm) in a dark room as a test for the bluish-white photoluminescence of artificial scheelite.

The sodium chloride byproduct can be removed by the following method. Crush and grind the product material with a porcelain pestle and mortar. Transfer the powdered material onto a wetted sheet of medium speed filter paper (e.g., *Whatman* No. 2) in a Buchner filter funnel and wash it with distilled water. Finally, wash the product with a small amount of acetone (in order to accelerate the drying process), and leave it to dry for ~10 min whilst in the Buchner filter funnel under vacuum. Carefully place the product material on a watch-glass and leave it to dry for a further 10 min. Once the powder is dry, expose it once again under short-wavelength ultraviolet radiation (254 nm) to observe the photoluminescence. Submit ~3 g sample for analysis by powder X-ray diffractometry (10–70° 2θ). Compare the

---

[15] Glass wool has a low thermal conductivity and is microwave transparent.

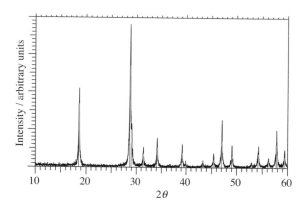

Figure 15.8   Powder XRD pattern (Cu-$K_{\alpha 1}$ radiation) of artificial scheelite, CaWO₄, as prepared by the method described in this chapter. This powder pattern corresponds closely to the PDF 77-2233 for the mineral scheelite, CaWO₄ (shown in red)

powder pattern with the PDF 77-2233 for the mineral scheelite, CaWO₄ and with Figure 15.8. Keep the remaining material in an appropriately labelled specimen jar.

REFERENCES

1. R. H. Gillette, *The Review of Scientific Instruments* **21** (1950) 294–301.
2. L. L. Y. Chang, M. G. Scroger and B. Phillips, *Journal of the American Ceramic Society* **49** (1966) 385–390.
3. A. Zalkin and D. H. Templeton, *Journal of Chemical Physics* **40** (1964) 501–504.
4. R. G. Dickinson, *Journal of the American Chemical Society* **42** (1920) 85–93.
5. R. V. Gaines, H. C. W. Skinner, E. E. Foord, B. Mason and A. Rosenzweig, *Dana's New Mineralogy: The System of Mineralogy of James Dwight Dana and Edward Salisbury Dana*, 8th edn, 1997, Wiley-Blackwell, New York, 1819 pp.
6. C. Feldmann, T. Jüstel, C. R. Ronda and P. J. Schmidt, *Advanced Ceramic Materials* **13** (2003) 511–516.
7. Y. Zhang, N. A. W. Holzwarth and R. T. Williams, *Physical Review B* **57** (1998) 12738–12750.
8. M. Nikl, V. V. Laguta and A. Vedda, *Physica Status Solidi B* **245** (2008) 1701–1722.
9. M. Herget, A. Hofstaetter, T. Nickel and A. Scharmann, *Physica Status Solidi B* **141** (1987) 523–528.
10. V. B. Mikhailik and H. Kraus, *Physica Status Solidi B* **247** (2010) 1583–1599.
11. T. A. Edison, *Nature* **53** (1896) 470.
12. T. A. Edison, *Electrical Engineer* **21** (1896) 305.
13. S. V. Moharil, *Bulletin of Materials Science* **17** (1994) 25–33.
14. L. H. Brixner, *Materials Chemistry and Physics*, **16** (1987) 253–281.
15. M. Nikl, Measurement Science and Technology **17** (2006) 37–54.
16. M. Pupin, *From Immigrant to Inventor*, Scribners, 1923, New York, 396 pp.

17. T. A. Edison, *Fluorescent electric lamp*, 1907, US Patent 865367.
18. F. R. van Horn, *American Mineralogist* **15** (1930) 461–469.
19. G. Claude, *Comptes Rendus Hebdomadaires des Séances de l'Academie des Sciences* **203** (1936) 1203–1206.
20. R. Oxland, British Patent 11848, 1847.
21. D. R. Baghurst and D. M. P. Mingos, *Chemical Communications* **12** (1988) 829–830.
22. D. M. P. Mingos and D. R. Baghurst, *Applications of Microwave Dielectric Heating Effects to Synthetic Problems in Chemistry*, Chapter 1, 3–53, in *Microwave-Enhanced Chemistry: Fundamentals, Sample Preparation, and Applications*, eds. H. M. Kingston and S. J. Haswell, American Chemical Society, 1997, Washington, DC.
23. M. Cirakoglu, W. A. Prisbrey, J. R. Jokisaari, S. Bhaduri and S. B. Bhaduri, *Ceramic Transactions* **111** (2001) 173–180.
24. J. T. Kloprogge, M. L. Weier, L. V. Duong and R. L. Frost, *Materials Chemistry and Physics* **88** (2004) 438–443.
25. A. Phuruangrat, T. Thongtem and S. Thongtem, *Journal of the Ceramic Society of Japan* **116** (2008) 605–609.
26. V. Thangadurai, C. Knittlmayer and W. Weppner, *Materials Science and Engineering B* **106** (2004) 228–233.
27. A. de Schulten, *Bulletin de la Societe Francaise de Mineralogie* **26** (1903) 112–113.
28. P. Parhi, T. N. Karthik and V. Manivannan, *Journal of Alloys and Compounds* **465** (2008) 380–386.
29. D. Cubicciotti, *Journal of Chemical Physics* **31** (1959) 1646–1651.
30. V. B. Mikhailik, I. K. Bailiff, H. Kraus, P. A. Rodnyi and J. Nikovic, *Radiation Measurements* **38** (2004) 585–588.
31. T. A. Edison, *Electrical Engineer* **21** (1896) 340.

PROBLEMS

1. Calculate the standard enthalpy change and standard entropy change for the following reaction at 298 K: $CaBr_2 + Na_2WO_4 \rightarrow 2NaBr + CaWO_4$. Compare and contrast these values with those for the corresponding reaction involving calcium chloride in place of calcium bromide.

2. Compare and contrast the crystal structures of scheelite and chalcopyrite (cf. Chapter 11).

3. Explain briefly why $CaWO_4$ does not normally exhibit photoluminescence under long-wavelength ultraviolet light.

16

Artificial Hackmanite $Na_8[Al_6Si_6O_{24}]Cl_{1.8}S_{0.1}$ by a Structure-Conversion Method with Annealing Under a Reducing Atmosphere

Hackmanite is a rare variety of the common rock-forming mineral sodalite, $Na_8[Al_6Si_6O_{24}]Cl_2$ and has a composition approximating to $Na_8[Al_6Si_6O_{24}]$ $Cl_{1.8}S_{0.1}$. It is an unusual example of a mineral that exhibits tenebrescence.[1] Hackmanite changes from almost colourless to reddish purple on exposure to short- and long-wavelength ultraviolet radiation (including sunlight), and is bleached on exposure to white light; such that the process is cyclical.[2] Besides being a mineralogical curiosity, there has been an interest in preparing artificial hackmanite since the 1950s for potential applications in filter optics, display panels, digital storage information technology and more recently as a photochromic pigment. This chapter describes a high-temperature method for preparing a powder specimen of artificial hackmanite through a so-called, 'structure-conversion method', using a commercial source of zeolite 4A,

---

[1] *Tenebrescence* is the property of darkening, especially reversibly, in response to incidental radiation.
[2] The closely related mineral tugtupite, $Na_8[Be_2Al_2Si_8O_{24}](Cl,S)_{2-\delta}$ behaves similarly, although the coloration is a purplish pink, and is used as a gemstone.

Synthesis, Properties and Mineralogy of Important Inorganic Materials   By Terence E. Warner
© 2011 John Wiley & Sons, Ltd

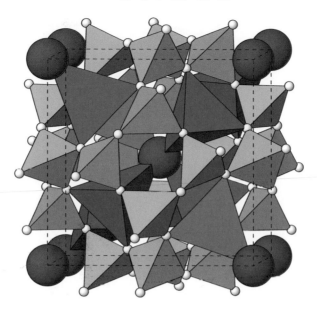

Figure 16.1 The sodalite Na$_8$[Al$_6$Si$_6$O$_{24}$]Cl$_2$ structure with the AlO$_4{}^{5-}$ and SiO$_4{}^{4-}$ tetrahedra (gray) the [NaO$_3$Cl] tetrahedra (orange) and the Cl$^-$ ions (purple-pink). The unit cell is shown by the dashed lines. This figure was drawn using data from Hassan and Grundy [2], cf. ICSD 29443

NaAlSiO$_4$·xH$_2$O, as the precursor compound for the sodium aluminosilicate component within the material.

Sodalite is a tectosilicate[3] and crystallises with the prototype structure as shown in Figure 16.1. It belongs to the cubic crystal system and space group $P\bar{4}3n$. The crystal structure was first determined by Linus Pauling in 1930 using single-crystal diffractometry [1], and refined by Hassan and Grundy in 1984 [2]. Sodalite has essentially an aluminosilicate framework formed by the alternating linkage of SiO$_4{}^{4-}$ and AlO$_4{}^{5-}$ tetrahedra, yielding four and six-membered rings, which link directly together to form large cubo-octahedral cavities (see Figures 16.2 and 16.3).[4] This negatively charged framework is counter balanced by the incorporation of extraframework sodium cations. The centre of each cubo-octahedral cavity is occupied by a chloride ion, which is tetrahedrally coordinated to four sodium ions, whose nuclei are also located within the cavity. These sodium ions are tetrahedrally coordinated to three oxide ions and one chloride ion. Thus, sodalite is characteristically,

---

[3] *Tectosilicates* are a class of silicates with three-dimensional tetrahedral frameworks in which Al may occupy some of the tetrahedral sites.

[4] Naturally occurring sodalite minerals typically exhibit ordering among the SiO$_4{}^{4-}$ and AlO$_4{}^{5-}$ tetrahedra; in synthetic analogues this sequence is typically disordered. These structural differences very likely reflect the differences in the crystallisation rate of the mineral under geological conditions compared with the preparative methods employed in producing artificial material.

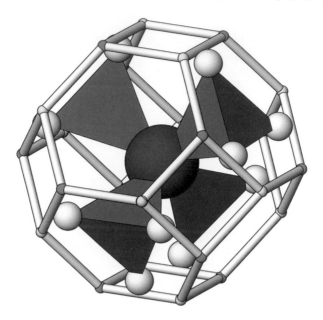

Figure 16.2 A detail of the aluminosilicate cubo-octahedral cage containing the central chloride ion (purple-pink) and the four [NaO$_3$Cl] tetrahedra (orange). The silicon ions are shown yellow, and the aluminium ions are shown blue

a tetrahedrally coordinated material, which partly accounts for its open structure and low density (2.30 g cm^{-3}). The chloride ions have a large ionic radius (180 pm), and this enables them to be substituted by relatively large complex ions, such as, sulfate ions and certain polysulfide species, as will be discussed below.

Sodalite $Na_8[Al_6Si_6O_{24}]Cl_2$ is one of the common rock-forming feldspathoid minerals [3]. It is most commonly encountered in alkaline plutonic rocks, such as, nepheline syenites, and also in silica-poor volcanic rocks, such as, phonolite and alkali basalts. Sodalite is also found in metasomatised calcareous rocks at the contact with alkaline igneous bodies [4]. An occurrence of sodalite from the Igaliko complex, Gardar Province, South Greenland, is considered to have formed from nepheline, NaAlSiO$_4$ through the action of metasomatic fluids containing NaCl at subsolidus temperatures [5]. This geological process is similar to the method adopted in this chapter for the preparation of artificial hackmanite.

$$6NaAlSiO_4 + 2NaCl \rightarrow Na_8[Al_6Si_6O_{24}]Cl_2$$

The sodalite structure can accommodate a variety of chemical substitutions involving both the anions and the cations. This is illustrated in nature by the fact that sodalite is the prototype for a group of minerals that include: nosean, $Na_8[Al_6Si_6O_{24}]SO_4$; and the rare minerals, haüyne,

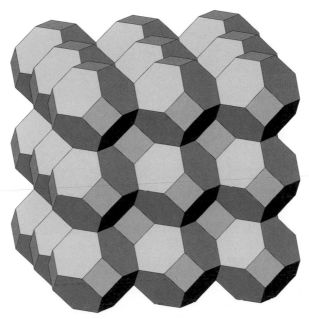

Figure 16.3 The aluminosilicate cubo-octahedral cages (as illustrated in Figure 16.2) pack together to form the above space-filling tetradecahedral structure, that describes the sodalite structure

(Na,Ca)$_8$[Al$_6$Si$_6$O$_{24}$](SO$_4$,S)$_{2-\delta}$; lazurite (a variety of haüyne), (Na,Ca)$_8$ [Al$_6$Si$_6$O$_{24}$](Cl,SO$_4$,S)$_{2-\delta}$; and tugtupite, Na$_8$[Be$_2$Al$_2$Si$_8$O$_{24}$](Cl,S)$_{2-\delta}$ ($0 \leq \delta < 1$).

The mineral, lazurite, (Na,Ca)$_8$[Al$_6$Si$_6$O$_{24}$](Cl,SO$_4$,S)$_{2-\delta}$ has been exploited as a blue pigment since antiquity, as mentioned briefly in Chapter 1. From 1828 onwards, lazurite was superseded by its artificial equivalent, which is known commercially as the pigment, ultramarine blue [6]. In sharp contrast to this, the closely related synthetic and stoichiometric compound, sulfosodalite, Na$_8$[Al$_6$Si$_6$O$_{24}$]S is almost colourless [7]. Whereas, the corresponding sulfur-rich analogue, Na$_8$[Al$_6$Si$_6$O$_{24}$]S$_{1+x}$ (prepared under a sulfur vapour pressure of 1 bar) has a deep blue coloration; though not as deep as ultramarine [8]. Likewise, reacting artificial hackmanite with an excess of elemental sulfur encapsulated in an evacuated quartz glass ampoule to 600 °C results in a pale blue coloration to the material; whilst conducting the reaction at 850 °C imparts a pale green coloration [9]. These synthetic materials mimic the blue and green varieties of naturally occurring sodalite; see Figure 16.4. The blue coloration of lazurite and ultramarine is attributed to the blue chromophore S$_3^-$ radical anion [10, 11]. Related materials with

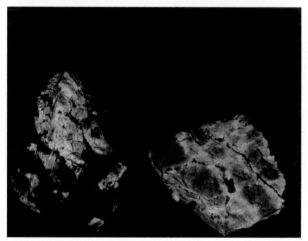

Figure 16.4　Top photograph: Natural specimens of sodalite showing variations in blue and green coloration. The top specimen is of unknown location and from the author's private collection. The bottom-left specimen is from Greenland and from the author's private collection. The bottom-right specimen is from Ilímaussaq-Alkaline-Complex, Greenland; courtesy of the Geological Museum at the University of Copenhagen. Width of field = 240 mm. Bottom photograph: Taken under long-wavelength (365 nm) ultraviolet radiation

green, or greener hues, are attributed to the additional presence of the yellow chromophore S_2^- radical anion (hence the impression of being green) [10, 11]. Polysulfur radical anions are highly unstable under atmospheric conditions; and their existence within these materials is a consequence of the protective environment offered by the sodalite cage.

Another example is the substitution of a sulfate ion for a pair of chloride ions. When taken to the extremity, this becomes known as the mineral nosean, $Na_8[Al_6Si_6O_{24}]SO_4\square$. The square symbol in the formula is used here to emphasise the anionic vacancy on the chloride site; whilst the SO_4^{2-} anion is also located randomly on the chloride site. Complete solid solution between nosean and sodalite occurs at high temperature, but the solid solution is discontinuous below 1000 °C. Below 700 °C, it is restricted to within $\sim$5 mol.% at either end of this pseudobinary system [12].

The mineral, hackmanite is a rare variety of sodalite and was known originally as 'pink sodalite'. In 1834, Allan [13] described the original 'pink sodalite' (as collected by Giesecke *ca.* 1807 in Greenland) as being normally green, but when freshly fractured, displays a bright pink tinge, which then fades under visible light within a few hours. In 1904, Vredenburg [14] described the fracturing of large metre-sized blocks of elaeolite[5] from Rajputana in India, as if they had become suffused with blood! The present author has observed this phenomenon on a more modest scale, during the fracturing of the green sodalite specimen as shown in Figure 16.4. The chemical composition of a 'pink sodalite' from the Kola Peninsula in Russia was determined by Borgström in 1901 [15] and was shown to contain 0.39 wt% sulfur. Borgström's analysis indicated that this sulfur is present as monosulfide; which is consistent with the composition, $Na_8[Al_6Si_6O_{24}]$ $Cl_{1.76}S_{0.12}$. Hackmanite can therefore be viewed as a reduced member of the sulfate-deficient sodalite–nosean solid solution. Its crystal structure has been confirmed by Hassan and Grundy [16].

Nowadays, it is generally considered that hackmanite has the property of changing from colourless, or pale pink, to a deep reddish purple upon exposure to ultraviolet radiation (including sunlight), as described originally by Lee in 1936 [17] in regards to a specimen from near Bancroft, Ontario, Canada; see Figure 16.5. Hackmanite is normally reported with the nominal composition; $Na_8[Al_6Si_6O_{24}]Cl_{2-2x}S_x\square_x$ (in which x is typically $\sim$ 0.1 and the square symbolises an anionic vacancy). This colour is bleached when exposed to light that is restricted to the visible region; although wavelengths longer that green (> 590 nm) are considered to be less effective for this purpose [17]. The depth of the coloration is dependent upon the intensity, wavelength and exposure time of the ultraviolet radiation, as well as, the grain size of the hackmanite. Therefore, single crystals of millimetre dimensions yield a much deeper coloration than powder specimens under the same exposure to ultraviolet radiation. It is speculated that the tenebrescence is related to trace amounts of both mono- and poly-sulfide residing in the chloride site of the sodalite structure [18]. One mechanism that is commonly postulated to account for the tenebrescence, is that a photoactivated

---

[5] *Elaeolite* is a former expression for nepheline syenite; i.e., an alkali-rich intermediate intrusive (plutonic) rock comprising essentially nepheline with sodalite (hackmanite) as an accessory mineral.

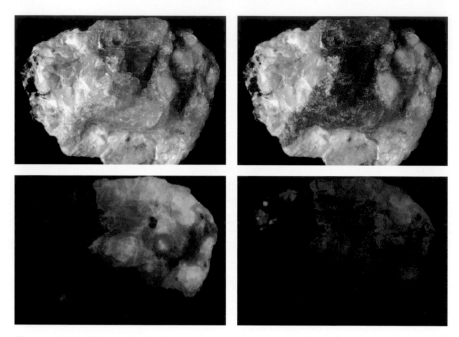

Figure 16.5 Mineralogical specimen containing hackmanite from Kvanefjeld, Greenland. The photograph at the top left was taken under white tungsten light and shows pale purplish pink to colourless transparent crystals of hackmanite, with minor amounts of albite (bottom centre; opaque and dirty white). The photograph at the top right was taken after a 2-min exposure to ultraviolet light (254 nm) and shows a deep reddish purple coloration of the hackmanite. The photograph at the bottom left was taken under ultraviolet light (365 nm). The photograph at the bottom right was taken under ultraviolet light (254 nm); the blue-violet coloration is essentially reflected unfiltered visible light from the UV source. Width of field = 90 mm. The specimen is from the author's private collection

electron-transfer process takes place between the sulfide species and the anionic vacancies ($V^{\bullet}_{Cl}$) resulting in the creation on an F-centre:[6]

$$\text{ultraviolet light} \rightarrow \text{reddish purple}$$

$$S'_{Cl} + V^{\bullet}_{Cl} \rightleftharpoons S^X_{Cl} + e^X_{Cl}$$

$$\text{colourless} \leftarrow \text{visible light}$$

Selenium and tellurium are known to substitute for sulfur in ultramarine-type sodalites in relatively high concentrations, resulting in a permanent bright red coloration in the case of selenium, and a bright greenish-blue

---

[6] An F-centre or *Farbzentrüm* (in English: colour centre) is an example of a crystal defect in which an electron is trapped at a negative ion vacancy; in this case a chloride ion vacancy.

coloration in the case of tellurium [11]. Likewise, selenium and tellurium have also been substituted for the much smaller sulfur content in hackmanite. These materials have been prepared through the incorporation of controlled amounts of Na$_2$SeO$_4$ and Na$_2$TeO$_4$ within the chlorosodalite phase, followed by annealing under a hydrogen atmosphere for about half an hour at 900 °C, to form the reduced chalcogenide species *in situ* [19]. These materials exhibit a very similar photochromism, in terms of coloration, intensity and activation, to that in normal sulfur-containing hackmanite. This is somewhat surprising since one might have expected the differences in the ionisation energies associated with these three chalcogenide species, to manifest as differences in the position of the ultraviolet absorption band, and thus lead to variations in the intensity of the photochromic coloration for a given dose of radiation. Consequently, Ballentyne and Bye [19] suggested that the ultraviolet absorption band may *not* be due to a *direct* electronic transition within a sulfur, selenium or tellurium ion.

In addition to exhibiting photochromism, natural specimens of hackmanite are often strongly photoluminescent, emitting orange-yellow luminescence whilst under long-wavelength (365 nm) ultraviolet radiation[7], see Figures 16.4 and 16.5. Interestingly, certain specimens also exhibit a prolonged after-glow of white light. This photoluminescence has been attributed to the presence of the disulfide, S$_2^{2-}$ anion [20, 21], and more recently, to the S$_2^-$ radical anion [22]. How far the mechanism for the photoluminescence is entangled with the photochromism, is unclear.[8] But Kirk [21] concluded that artificial hackmanite containing a high sulfide/sulfate ratio (as annealed in hydrogen) produced the best tenebrescence, at the expense of the luminescence. Conversely, similar material with a low sufide/sulfate ratio (prepared by a subsequent exposure to air at 900 °C for 15 min) produced the brightest luminescence, at the expense of the tenebrescence. Whereas, pure (sulfur-free) sodalite is colourless, nontenebrescent and nonluminescent; i.e., photoinactive.

There is a technological interest in hackmanite, particularly since the tenebrescence can be activated by long- (and short-) wavelength ultraviolet radiation.[9] Potential applications include, filter optics, information storage,

---

[7] The present author has examined several specimens of hackmanite originating from Greenland as held in the collection at the Geological Museum, Copenhagen, Denmark, in order to derive this conclusion. By comparison, artificial hackmanite as prepared according to this chapter does not exhibit luminescence under ultraviolet light. This aspect is in full agreement with the findings of Kirk [21], and it is consistent with the presence of monosulfide, S^{2-} ions only.

[8] Medved [24], however, proposed a model that incorporated both photoluminescence and photochromism.

[9] The expression *photochromic sodalite* is synonymous (in this context) with *tenebrescent sodalite*; and both expressions may imply the mineral, *hackmanite*. However, these two expressions can be used in a wider generic context to include artificially prepared materials with a chemistry extending well beyond that normally associated with naturally occurring hackmanite. Furthermore, this general class of photochromic materials is also known by the term, *scotophors*.

display panels, and pigments. One rather fascinating device that was suggested by Chang (IBM Corporation) [23] is to use hackmanite as the sensitive cathodochromic material in a storage cathode-ray tube (CRT); thus requiring no refreshing. Presumably, in such a device, when the electric power (and hence the source of cathode rays) is switched-off, the image is still retained on the screen; until bleached. With regards to technological use, artificial hackmanite is preferred to the natural mineral, since this offers more scope to control the photochromism and cathodochromism in a reproducible manner. Moreover, the natural occurrences of hackmanite are both rare and sporadic. Consequently, there is an interest in preparing artificial hackmanite. The pioneering work in this area has been reported by: Medved [24-26]; Kirk [20, 21]; and Chang [23].

Sodalite, $Na_8Al_6Si_6O_{24}Cl_2$ can be synthesised by a variety of methods that include hydrothermal, solid-state sintering, sol-gel, flux, and structure conversion. Preparation of hackmanite involves a similar approach but with further requirements for the controlled substitution of sulfide for chloride within the sodalite structure. This has been performed through the direct use of Na_2S as the sufiding reagent, and also through the indirect use of Na_2SO_4; with either an in situ or a postannealing stage in a reducing atmosphere.

When planning a synthesis route in solid-state chemistry it is prudent to consider the kinetic aspects, such as the chemical reactivity of the starting reagents. Certain oxides are notoriously sluggish in their reaction with one another, because of the slow rates of solid-state diffusion, and alumina and silica fall into this category. Therefore, it can often be beneficial to employ a precursor material that already comprises several of the components desired in the product material; especially if these components are available in the precise ratio required for the synthesis, and if the reagent is readily available. In this case, nepheline, $NaAlSiO_4$; zeolite 4A, $NaAlSiO_4 \cdot xH_2O$; and kaolinite (china clay), $Al_2Si_2O_5(OH)_4$ can all be considered as potential chemical reagents.

For an example, artificial sodalite can be prepared successfully, by reacting kaolinite,[10] $Al_2Si_2O_5(OH)_4$ (as a precursor material for the Al_2O_3 and SiO_2 components) with stoichiometric amounts of Na_2CO_3 and NaCl. An excess of these two salts are added so as to provide a flux in order to promote the reaction at 950 °C, with a reaction period of 36 h at this temperature. The excess salts are then removed by washing the product with water. Unfortunately, attempts at preparing artificial hackmanite by a similar method, with the addition of Na_2SO_4, and a subsequent annealing stage under hydrogen, yielded a sodalite phase, but this material did not display tenebrescence. The source of kaolinite as used in this preparation was found to contain a significant amount of copper, whose presence is known to inhibit tenebrescence in sodalite [26]; thereby highlighting the problems associated with chemical impurities.

---

[10] The sample of china clay (supreme grade) was kindly supplied by, Imerys Minerals Ltd., St. Austell, Cornwall, United Kingdom.

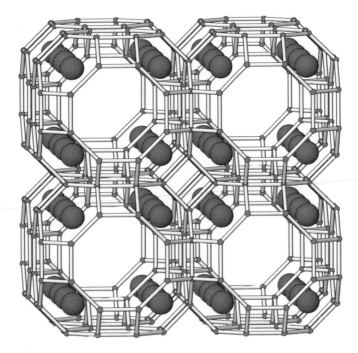

Figure 16.6 Crystal structure of zeolite 4A showing the aluminosilicate framework (green) together with the sodium ions (orange). This particular zeolite takes its name from the diameter of the channels that connect the cavities in the structure; in this case ~ 4 Å. This figure was drawn using data from Broussard and Shoemaker [29], cf. ICSD 43802

This narrows the choice in the present context to either, nepheline or, zeolite 4A. Both of these are attractive, since sodalite can be perceived as the pseudobinary compound, $3NaAlSiO_4 \cdot NaCl$. Natural sources of nepheline, however, invariably contain potassium substituting for part of the sodium content, and therefore are not ideal for this purpose. Whereas, zeolite 4A has the attractions in that the metal impurities are low, and that it has an open structure (see Figure 16.6) that should enhance its reactivity towards sodium chloride, sodium sulfate and sodium sulfide. Preparative routes involving zeolites as precursor compounds for this purpose have been attempted since the late 1960s by William *et al*. [27], Schipper *et al*. [28], Chang [23] and more recently by, Armstrong and Weller [18].

Preparative Procedure

This chapter describes the preparation of a powdered specimen of artificial hackmanite with the stoichiometry, $Na_8[Al_6Si_6O_{24}]Cl_{1.8}S_{0.1}$ by a so-called, 'structure-conversion method'. This preparative method is adopted, with certain modifications, from the original work by Chang [23]. The chemical reagents include a commercial source of zeolite 4A ($NaAlSiO_4 \cdot xH_2O$) together with the required stoichiometric amounts of NaCl and Na_2SO_4. Na_2SO_4 is added as the source for the 'Na_2S' component that is required in the final material. Commercially available zeolite 4A normally contains a certain amount of water of crystallisation as represented by, $NaAlSiO_4 \cdot xH_2O$ where $x \sim 0.35$.

Therefore, first determine the water content of the commercial zeolite 4A by the following method. Weigh a 10 g sample (to a precision of ± 0.001 g) and spread it evenly into a clean alumina boat (SRX110 Almath Ltd). Insert this into a tube furnace and heat under a flowing atmosphere of argon (100 ml/min at STP) to 600 °C for 6 h with heating and cooling rates of 200 °C/h. After dehydrating the zeolite, weigh it immediately, and thus determine the mole fraction of H_2O in, $NaAlSiO_4 \cdot xH_2O$.

Prepare 15 g of $Na_8[Al_6Si_6O_{24}]Cl_{1.8}S_{0.1}$ by the following method: Weigh the appropriate amounts of the commercial zeolite 4A ($NaAlSiO_4 \cdot xH_2O$), sodium chloride (NaCl) and anhydrous sodium sulfate (Na_2SO_4) as dry powders, directly into a 500-ml polyethylene bottle with a screw lid. Add about a dozen zirconia grinding balls and sufficient acetone to produce a consistency similar to that of normal paint, then screw the lid on very tightly and ball mill these powders together overnight.

The next day, transfer the resultant slurry into a round-bottomed flask. Use a pair of forceps to remove the zirconia grinding balls whilst washing them gently with a small amount of cyclohexane, so as to avoid an unnecessary loss of material. Then attach the flask to a rotary evaporator and a condenser in order to recover the cyclohexane. Retain the zirconia balls for further use.

Next day, place the dry powdered mixture in a large alumina boat (SRX110 Almath Ltd.) and then insert this in the central zone of the tube furnace. Attach the silicone rubber bungs to both ends of the work-tube and ensure that the exhaust gases from the tube furnace are vented appropriately.[11] Heat at 200 °C/h to 900 °C under a flowing atmosphere of 7% (vol./vol.) hydrogen in argon (a commonly available welding grade) at 500 ml/min. After 8 h at 900 °C, cool to ambient temperature (20 °C) with a cooling rate of 300 °C/h. This heating cycle will last ~ 15 h.

Observe and note the optical behaviour of your product material under ultraviolet light (365 and 254 nm). Submit a small powdered sample (~ 3 g) for analysis by powder X-ray diffractometry (scan: 5–70° 2θ). Compare the powder pattern with PDF 72-29 for sodalite and Figure 16.7.

---

[11] The reader is advised to conform to the legal requirements and regulations that may exist in the country in which the work is being conducted.

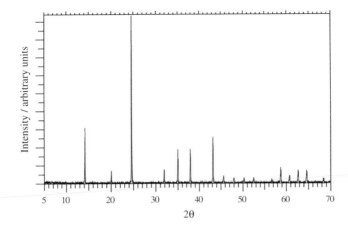

Figure 16.7 Powder XRD pattern (Cu-$K_{\alpha 1}$ radiation) of hackmanite, Na$_8$[Al$_6$Si$_6$O$_{24}$]Cl$_{1.8}$S$_{0.1}$ as prepared by the method described in this chapter. PDF 72–29 for chlorosodalite, Na$_8$[Al$_6$Si$_6$O$_{24}$]Cl$_2$ is shown in red for comparison

The reader may wish to prepare a ceramic monolith of this material. If so, then place a sufficient quantity of this powder ($\sim$1.2 g) into a silicone rubber mould (14 mm ID), and introduce this into a condom, eliminate the air and seal with a tight knot. Place this in an oil press under an isostatic pressure of 15 kbar for 10 min. Spread the remaining powdered product in the original alumina boat (SRX110 Almath Ltd.) so as to form a sacrificial powder bed.

Figure 16.8 The polished surface of a ceramic monolith of artificial hackmanite, Na$_8$[Al$_6$Si$_6$O$_{24}$]Cl$_{1.8}$S$_{0.1}$ (as prepared by the method described in this chapter) after exposure to short-wavelength (254 nm) ultraviolet radiation under a mask; creating a Mark Rothko like impression of St. George's Cross. Width of field = 8 mm

Place the *green* pellet on top of this powder bed and then place this in the central zone of the tube furnace. Repeat the annealing procedure at 900 °C as described above, yet with a slower flow rate of gas; 100 ml/min 7% (vol/vol) hydrogen in argon. Finally, observe and note the optical behaviour of your ceramic monolith under ultraviolet radiation (365 and 254 nm); cf. Figure 16.8.

REFERENCES

1. L. Pauling, *Zeitschrift für Kristallographie, Kristallgeometrie, Kristallphysik, Kristallchemie* **74** (1930) 213–225.
2. I. Hassan and H. D. Grundy, *Acta Crystallographica B* **40** (1984) 6–13.
3. W. A. Deer, R. A. Howie and J. Zussman, *An Introduction to the Rock-Forming Minerals*, 2nd edn, 1992, Longman, Harlow, Essex.
4. R. V. Gaines, H. C. W. Skinner, E. E. Foord, B. Mason and A. Rosenzweig, *Dana's New Mineralogy: The System of Mineralogy of James Dwight Dana and Edward Salisbury Dana*, 8th edn, 1997, Wiley-Blackwell, New York.
5. A. A. Finch, *Mineralogical Magazine* **55** (1991) 459–463.
6. R. Ashok, *Artists' Pigments*, in *Handbook of their History and Characteristics*, Vol. 2, 1997, Oxford University Press, New York.
7. G. Yamaguchi and Y. Kubo, *Bulletin of The Chemical Society of Japan* **41** (1968) 2641–2645.
8. G. Yamaguchi and Y. Kubo, *Bulletin of The Chemical Society of Japan* **41** (1968) 2645–2650.
9. J. Hutzen Andersen, 'The creation of tenebrescence (reversible photochromism) in synthetic sodalite by novel methods' Unpublished MSc Thesis, 2008, University of Southern Denmark.
10. D. Arieli, D. E. W. Vaughan, and D. Goldfarb, *Journal of the American Chemical Society* **126** (2004) 5776–5788.
11. D. Reinen and G.-G. Linder, *Chemical Society Reviews* **28** (1999) 75–84.
12. T. Romisaka and H. P. Eugster, *Mineralogical Journal* **5** (1968) 249–275.
13. R. Allan, *A Manual of Mineralogy*, 1834, A & C Black, Edinburgh.
14. E. Vredenburg, *Records of the Geological Survey of India* **31** (1904) 43.
15. L. H. Borgström, *Geologiska föreningens i Stockholm förhandlingar* **23** (1901) 557–566.
16. I. Hassan and H. D. Grundy, *Canadian Mineralogist* **21** (1983) 549–552.
17. O. I. Lee, *American Mineralogist* **21** (1936) 764–775.
18. J. A. Armstrong and M. T. Weller, *Chemical Communications* **10** (2006) 1094–1096.
19. D. W. G. Ballentyne and K. L. Bye, *Journal of Physics D: Applied Physics* **3** (1970) 1438–1443.
20. R. D. Kirk, *Journal of the Electrochemical Society* **101** (1954) 461–465.
21. R. D. Kirk, *American Mineralogist*, **40** (1955) 22–31.
22. A. Sidike, A. Sawuit, X-M. Wang, H-J. Zhu, S. Kobayashi, I. Kusachi and N. Yamashita, *Physics and Chemistry of Minerals* **34** (2007) 477–484.
23. I. F. Chang, *Journal of the Electrochemical Society* **121** (1974) 815–820.
24. D. B. Medved, *American Mineralogist* **39** (1954) 615–629.

25. D. B. Medved, *Journal of Chemical Physics*, **21** (1953) 1309–1310.
26. D. B. Medved, *Tenebrescent Sodalite* (1956) US Patent 2761846 19560904.
27. E. F. Williams, W. G. Hodgson and J. S. Brinen, *Journal of the American Ceramic Society* **52** (1969) 139–145.
28. D. J. Schipper, C. Z. van Doorn and P. T. Bolwijn, *Journal of the American Ceramic Society* **55** (1972) 256–259.
29. L. Broussard and D. P. Shoemaker, *Journal of the American Chemical Society* **82** (1960) 1041–1051.

PROBLEMS

1. Write a balanced chemical equation to describe the conversion of zeolite 4A to artificial hackmanite, $Na_8[Al_6Si_6O_{24}]Cl_{1.8}S_{0.1}$ by the preparative method as described in this chapter.

2. Write a balanced chemical equation to describe the conversion of kaolinite, $Al_2Si_2O_5(OH)_4$ to artificial hackmanite, $Na_8[Al_6Si_6O_{24}]Cl_{1.8}S_{0.1}$ by a preparative method similar to that as described in this chapter.

3. Describe what should happen if hackmanite, $Na_8[Al_6Si_6O_{24}]Cl_{1.8}S_{0.1}$ were accidentally washed with hydrochloric acid solution.

4. State the mineralogical name that is given to the mineral with the composition, $Na_4Al_3Si_9O_{24}Cl$.

5. Comment on the use of 7% hydrogen in argon instead of 100% hydrogen for the reduction of $Na_8Al_6Si_6O_{24}Cl_{1.8}(SO_4)_{0.1}$.

6. (a) Calculate the minimum volume of gas (7%(vol./vol.) hydrogen in argon) that is necessary for the complete reduction of Na_2SO_4 to Na_2S as used in your work.
   (b) From the flow rate of gas used in your work, calculate the time that corresponds to the volume of gas calculated in part (a).

17

Gold-Ruby Glass from a Potassium–Antimony–Borosilicate Melt with a Controlled Annealing

It is generally considered that the technology for the production of glass was originally developed in Mesopotamia ca. 1550 BC [1]. Today, over 90% of the world's glass production concerns the synthesis of soda-lime-silica glass [2]. This common type of glass is normally pale green and is used extensively in the manufacture of windows, bottles and vessels etc.[1] Silicate glasses can be made in a variety of colours, but glass that is truly red was extremely rare in former times. Indeed there are only two classical types of red glass prior to the 20th century, namely, *copper-ruby glass* and *gold-ruby glass*; the latter is reputed to be the finer of the two. Artefacts made of gold-ruby glass can be dated from the late Roman period (fourth century AD), a notable example being the gaily decorated 'Lycurgus Cup' now held in the collection of the British Museum [3].[2] This chapter describes a method for

---

[1] Soda-lime-silica glass is an amorphous phase consisting approximately; 73 wt% SiO_2; 14 wt% Na_2O; 11 wt% CaO; and 2 wt% Al_2O_3; with minor amounts of contaminant oxides, such as; K_2O; MgO; FeO; etc. The reader is referred to the monographs by; Doremus [5], Shelby [2] and Vogel [6] for information regarding the physicochemical properties of silicate glasses.

[2] The Lycurgus Cup contains ~40 ppm Au and ~300 ppm Ag [7]. Minute particles of a silver-gold-copper alloy, 50–100 nm in diameter, make the glass appear dichroic; red by transmitted light, and green by reflected light [3]. British Museum reg. no. 1958,12–2,1.

Synthesis, Properties and Mineralogy of Important Inorganic Materials   By Terence E. Warner
© 2011 John Wiley & Sons, Ltd

preparing an ingot of gold-ruby glass from a $K_2O–Sb_2O_3–B_2O_3–SiO_2$ melt and discusses the chemistry associated with its production.

The surviving documentation of the events that led to the development of gold-ruby glass are fragmentary; the following is a brief summery of a few prominent points.[3] In 1546, the German mineralogist and metallurgist Georgius Agricola (*alias* Georg Bauer) wrote a short comment on gold-ruby glass in his monograph, '*De Natura Fossilium*' [4]. He mentioned that during his lifetime, a famous variety of dyeing glass was made at Murano[4] from gold and used to tint glass a transparent ruby red.

Much of the early experimentation, however, was oriented towards imitating the gemstone, ruby; rather than producing a red glass per se. One notable example is the recipe[5] written by the French physician Antonius Mizaldus in 1566 [11]. Likewise, in 1597, the German chemist Andreas Libavius [12] believed that since gold and ruby occur together, in nature, then gold must surely be capable of degrading into ruby;[6] and that a ruby can be obtained by dissolving a red tincture of gold with crystal glass.[7] Libavius may have been on the right path for making a red glass, but his original intention of preparing an artificial ruby is, of course, based on the wrong assumptions. In 1612, the Florentine priest Antonio Neri [13] wrote a brief comment regarding the preparation of a transparent ruby-red glass, in which gold was apparently heated in aqua regia in order to form a red gold powder, and then added to crystal glass to impart a ruby-red coloration.[8]

Thereafter, it became apparent that the physical state of the gold reagent is extremely important in the preparation of gold-ruby glass. In this respect, the efforts of the German chemist Johann Glauber (1604–1670) are pre-eminent. Glauber added *liquor of flints* (potassium silicate) to a solution of gold in aqua regia in order to form a gold precipitate for use in his experiments in colouring

---

[3] An account of the history and early chemistry of gold-ruby glass is described in the monograph on coloured glasses by Weyl [8] and in the review articles by Hunt [9] and Frank [10].

[4] Murano is a small archipelago of islands situated 3 km north of Venice. Agricola also mentions that the finest glass at that time was produced at Murano and was made famous by the Venetians. The trade secrets for Murano glass were protected by not allowing anybody who worked on the process to leave the island.

[5] A ruby is imitated in the following manner: Take 4 ounces (113 g) of Sal alcali (crude alkali carbonate obtained from plant ashes), 3 ounces (85 g) of powdered crystal glass, and 6 grain (0.39 g) of gold leaves. Mix these and then place in a crucible, and melt in a reverberatory furnace, and finally leave to cool spontaneously [11].

[6] Indeed, both gold and ruby are known to occur in certain alluvial sands and gravels on account of their high density and durability. Bearing in mind that one of the major objectives of the alchemists was to transmute the base-metals into gold, it is an interesting notion that gold per se might be capable of being degraded into another substance.

[7] *Crystal glass*, is essentially, a transparent and colourless glass free from inclusions, and is of high optical quality.

[8] *Aqua regia* is a mixture of concentrated nitric and hydrochloric acids. Neri's description on this matter is too brief to convey any practical meaning.

glass [14]. Glauber also described the original method for the precipitation of gold from its solution in *aqua regia* with the aid of an aqueous solution of tin. The German physician Andrea Cassius the elder (1605–1673) performed similar practices to those of Glauber a few years later, which were recorded by his son, Andrea Cassius the younger, in 1685; but without any reference to his father [15]. Sadly, Glauber's achievements in this field were ignored as well, so that the name for the gold precipitate was ascribed to Cassius (namely 'Purple of Cassius' [16]); although it should, and more rightfully, have been ascribed to Glauber.

The first production of gold-ruby glass (as opposed to experimentation) is attributed to the German alchemist-*cum*-chemist Johann Kunckel (1630–1703). Kunckel was the son of a glassmaker and worked in the glass business at Potsdam, Brandenburg. He succeeded in producing a large number of objects of gold-ruby glass on a commercial basis at Potsdam between 1679 and 1689 [9], although his methods for doing this were unreliable. Due to the novelty of its bright red colour, and the difficulties and mysteries associated with in its production, Kunckel's gold-ruby glass was greatly desired by the nobility and therefore commanded a high price (see Figure 17.1). Quite naturally, Kunckel tried to keep his preparative methods secret, but nevertheless, this knowledge spread during his lifetime to Bavaria[9] and Bohemia [17]; and continued to trickle down in a rather haphazard way to the present day. After Kunckel died, certain details concerning gold-ruby glass became available in his book, *Laboratorium Chymicum* [18], published posthumously in 1716; some of these details are described below.[10]

Hunt [9] included a few passages from Kunckel's, *Laboratorium Chymicum* (transcribed into English) in his own article, written in 1976. From these, it is apparent that Kunckel obtained some of his ideas from Cassius; particularly the method for the precipitation of gold with tin, and of introducing the gold precipitate into glass.[11] Furthermore, Kunckel believed that Cassius may

---

[9] The German film director Werner Herzog made a dramatisation of these events, as set in a Bavarian village during the year 1700, in his film '*Herz aus Glas*' (1976). The film portrays, in a traumatic manner, the human consequences accompanying the loss of knowledge; in this case, the loss of the recipe for gold-ruby glass.

[10] Essential preparative details can be found on pages: 4; 382–383; and 650; of Kunckel's *Laboratorium Chymicum* [18].

[11] The following passage is transcribed by Hunt [9] from pages 382–383 of Kunckel's *Laboratorium Chymicum* [18], and describes Kunckel's recipe for the gold precipitate as used in making his gold-ruby glass. '*I take two parts of good aqua fortis (nitric acid) and one part of spirits of salt (hydrochloric acid), or failing this some strong salt water. When these are mixed throw in from time to time a little of the purist tin so that it does not heat up but dissolves slowly.... With this solution the gold can be precipitated in such a beautiful colour that it cannot be more beautiful, and thereby can crystal glass be given the finest ruby red colour if the gold has previously been dissolved in three parts of aqua regia and one part of spirits of salt*'.

Figure 17.1 Tankard, gold ruby glass, gold-plated silver mounts; Brandenburg (Germany), beginning of the 18th century. Stiftung museum kunst palast, Glasmuseum Hentrich, Düsseldorf, Germany, inv. no. P 1940-135. (Photograph used with kind permission from Dr Dedo von Kerssenbrock-Krosigk Copyright (2010) Glasmuseum Hentrich)

have acquired this knowledge, in turn, from Glauber. Kunckel also emphasised the trouble that he, personally, had to go through to find the right composition and obtain a durable red coloration.

Zschimmer [19] included in his article, written in 1930, several quotations (in German[12]) from Kunckel's, *Laboratorium Chymicum*. From these, it can be read that Kunckel's recipe for making a beautiful gold-ruby glass, involved adding 1 part of finely divided gold to 1280 parts of the glass batch (i.e., 781 ppm Au). Kunckel described that after the first stage of the glass-making process, the glass is colourless and appears similar to crystal glass. Kunckel was very curious as to what had happened to the gold within this *colourless* glass; he believed that the gold had changed into something else, and suggested that this may be a vitrified state of gold. Kunckel described most informatively, that

---

[12] The present author is grateful to Dr M. Dörr for translating and commenting on the articles by Zschimmer.

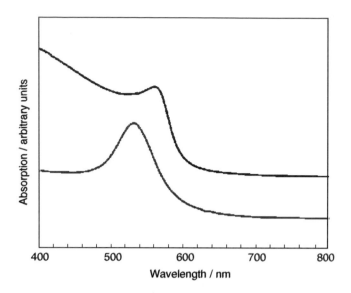

Figure 17.2   The red line is the absorption spectrum of the gold-ruby glass ingot annealed at 650 °C, corresponding to the red specimen as shown in Figure 17.3. The black line is for a specimen of cadmium chalcogenide glass (These measurements were performed on a Shimadzu UV-1601PL spectrometer; Courtesy of M. Dörr.)

upon subsequent heating at a milder temperature, this colourless glass becomes completely red.[13]

Many years passed by before it was realised, by Michael Faraday in 1857, that the red coloration was due to the presence of metallic gold particles within the glass; rather than the occurrence of gold in solid solution [20]. Gold-ruby glass is indeed, a two-phase system comprising discrete particles of metallic gold (ideally, 5–20 nm) dispersed in a silicate glass matrix [21]. The depth of the colour is proportional to the gold content; typically, 200–500 ppm Au. Whilst the hue is strongly influenced by the size of the gold particles, such that the presence of relatively large particles imparts a purple tint to the red, and can turn the glass blue. The actual colour of gold-ruby glass is due to an intense plasma absorption band that peaks near 530 nm (corresponding to the green region of the visible spectrum; see Figure 17.2). Doremus [22] suggested that this plasma absorption band results from the collective oscillations of free electrons within the small gold particles (typically, 5–60 nm diameter [23]) being excited by light of a particular

---

[13] Zschimmer [19] also reported that certain notes containing a recipe for gold-ruby glass as written by Kunckel survived his death and were discovered by Metzger in Kunckel's old workshop at Zechlin (50 km north of Berlin). Apparently, Metzger published these notes in *Polytechnischen Centralblatt* 2 (1836) I, 385; wherein, SiO_2, $NaNO_3$, Na_2CO_3, borax, mennige (Pb_3O_4), As_2O_3 and weinstein ($C_4H_4O_6KH$) are listed in Kunckel's formulation.

wavelength. For very small particles this absorption band broadens and ultimately disappears for particles ≤ 1 nm in diameter. The scattering of light is believed to play only a minor contribution to the coloration [6].

The chemistry of gold-ruby glass is complex, as is the thermal treatment required to form a glass with a deep red colour. The nucleation of the gold particles is favoured kinetically in glasses of a relatively low viscosity, such that a slight variation in the composition of a potassium borosilicate base-glass, for example, affects the viscosity of the glass, which in turn affects the kinetics of the formation of the gold particles, which influences their eventual size, and thus, the colour of the glass [24]. This is consistent with the use of lead oxide (PbO) as a traditional component in gold-ruby glasses; being added unwittingly perhaps, to lower the viscosity of the glass for this very reason. The function of lead oxide is generally ascribed to increasing the solubility of gold within the glass melt; as is the addition of tin oxide (SnO_2), especially in the case of lead-free glasses [8]. It is, of course, quite plausible that PbO plays a dual role here. However, gold-ruby glasses formulated with a high content of PbO, comparable to traditional English lead crystal at ≥ 30 wt% PbO, have a tendency to yield a strong purple tint to the red colourant.

It is usually necessary to quench the glass melt, in order to create a high degree of supersaturation of the dissolved gold within the glass prior to 'striking'; i.e., the annealing process.[14] The annealing temperature and time are of paramount importance to the development of the colour; see Figure 17.3. If the annealing temperature is too low, or the annealing period is too short, then the glass forms a pale pinkish coloration that is comparable to a feeble-coloured 'Cranberry glass'.[15] Conversely, if the annealing temperature is too high, then the glass acquires a bluish-purple coloration. An excessive annealing period at the correct temperature does not appear to have any detrimental effect on the colour. The *Holy Grail* in the glass business is to find a glass composition that will afford a self-striking glass; i.e., a glass that will turn a deep red spontaneously upon cooling from the melt, without the need for annealing.

In the fabrication of fine or delicate artefacts, rapid cooling is not always feasible, and so tin oxide is considered an essential component in these circumstances [8]. It has been suggested recently, by Haslbeck *et al.* [25] that small additions of SnO_2 (20–200 ppm Sn) accelerate the creation of embryonic gold clusters, with the effect that this increases the number of gold particles, and thus restricts their eventual size for a given amount of gold; resulting in the desired purplish-red coloration. But excessive amounts of tin (≥ 2 wt% Sn) results in an undesirable brown coloration [24], sometimes referred to as a 'liverish' colour.

---

[14] *Striking the glass* is a traditional term to describe the annealing process that is responsible for inducing the coloration.

[15] 'Cranberry glass' is an American expression for gold-ruby glass; which includes a pale pinkish-red variety.

Figure 17.3 A series of glass ingots as prepared by the method described in this chapter, annealed for ~ 60 h at different temperatures. From left to right: unannealed, colourless; 600 °C results in a pale pink coloration; 650 °C results in a deep red coloration (the ideal gold-ruby glass); 700 °C results in a bluish purple coloration

Potash glasses (i.e., those containing K_2O) are reputed to give a brighter colour than soda ash glasses (i.e., those containing Na_2O). Therefore, K_2CO_3 is included in the formulation of the glass as prepared in this chapter for this very reason. Whereas, B_2O_3 is included, primarily, to yield a glass with enhanced resistance to thermal shock (cf. Pyrex™ glass) and also for traditional reasons.

The role of small additions of antimony(III) oxide is somewhat ambiguous. Sb_2O_3 is commonly added during glass making procedures as a so-called 'finer' for the specific function of removing gas bubbles from the glass melt [2]. Vogel [6] stated that a high content of Sb_2O_3 in the glass melt controls the quantitative reduction of gold ions to the elemental state. Whereas, Weyl [8] suggested that Sb_2O_3 might also act as a scavenger for the removal of chloride ions within the glass melt through the formation of volatile $SbCl_3$; chloride ions being considered as detrimental to the coloration process. If so, these ideas might explain the benefit in maintaining the glass melt at a high temperature (~ 1400 °C) for several hours before casting the glass. The present author's experience is that a small addition (2–3 wt%) of Sb_2O_3 reduces the viscosity of the glass melt and that at least one of the following p-block metal oxides, PbO, SnO_2, Sb_2O_3 and Bi_2O_3, needs to be included within the formulation, to obtain a satisfactory red colour in the gold-ruby glass.

In summary, there have been numerous methods spanning many generations that have utilised different base-glass compositions with various quantities of chemical additives, and with different heating and annealing procedures, in attempts to make gold-ruby glasses. There have also been many theories and hypotheses devoted to the subject; some of which appear almost alchemical in their approach. But one of the outstanding features regarding these *p*-block metal oxide additives (PbO, SnO_2, Sb_2O_3 and Bi_2O_3), is that they all contain *soft* metal cations that display mixed valency. It is very likely that these soft cations exert a polarisation on the oxide environment within the silicate melt, which affects the gold chemistry; favouring the stability and solubility of atomic gold within the melt.

Indeed, one important aspect that still remains a controversial issue is the question regarding the oxidation state of the gold throughout the various preparative stages, which include the glass melt and the glass prior to, and during, the annealing process. The conclusion derived from a recent study by Wagner *et al.* [23] involving ^{197}Au Mössbauer spectroscopy on quenched samples of gold-ruby glass before and after annealing, suggests that gold is present as a dispersed monovalent gold(I) species within the glass matrix prior to 'striking'. But these spectra were recorded at room temperature, and therefore do not necessarily reflect the state of the gold in a high-temperature melt, nor within a glass at the annealing temperature. On the contrary, it is quite plausible that gold dissolves within the silicate melt as a neutral atomic species (as indeed presumed by Weyl [8], and contemplated by Zschimmer [26]), and remains as such in the quenched glass, albeit in a state of metastability with respect to gold clusters. The oxidation state of the gold within these substances at high temperature must surely be left unresolved, at present.

Besides its ornamental use, gold-ruby glass has been exploited in certain early technological applications, for example, in the red signals on the railway and for maritime navigation. The American glass company, Corning Inc. imported a glass, made in France, for one of its red filters (Gold Ruby No. 2270), but this trade came to an end with the outbreak of the Second World War.[16] However, the need for gold-ruby glass was largely superseded by the discovery of glasses that contain cadmium chalcogenides as the colourant. The discovery that minute crystallites of cadmium sulfoselenide, $CdS_{1-x}Se_x$ can act as a red colourant in silicate glasses was made during the early 1930s, and this enabled red glass to become popular.[17] By comparison

---

[16] The present author is grateful to Kristine Gable, Research Consultant, Corning Incorporated Archives, for sharing information concerning gold-ruby glass.

[17] Glass containing $CdS_{1-x}Se_x$ was used universally in red traffic lights, prior to its replacement by light emitting diodes (LEDs) at the turn of the millennium. $CdS_{1-x}Se_x$ is a direct-bandgap semiconductor with a sharp cut-off in transmission in the visible or near-infrared (depending on the composition). This means that all wavelengths of electromagnetic radiation shorter than the corresponding energy of the direct bandgap are absorbed. The energy of the direct bandgap, and thus the cutoff wavelength, is a function of the chemical composition. Therefore, a progressive and controlled variation in its composition results in colours ranging from yellow to orange to red, and eventually to total absorption of visible light; black [2].

with the cadmium chalcogenide glasses, gold-ruby glass has a substantial transmission in the blue-violet region of the visible spectrum (see Figure 17.2), and this gives a blue or purplish tint to the red colour of the glass, which limits its technological use [8]. Nevertheless, as an ornamental red glass, gold-ruby glass remains par excellence.

Preparative Procedure

This chapter describes a method for preparing a gold-ruby glass based on a potassium–antimony–borosilicate glass (K_2O–Sb_2O_3–B_2O_3–SiO_2) with a small addition of $H_3O[AuCl_4]$ as the gold reagent. $H_3O[AuCl_4]$ was chosen for the simple reason that it is commercially available as a chemical reagent. Alternatively, according to Weyl [8], a solution of gold in aqua regia (a mixture of concentrated nitric and hydrochloric acids) can be sprinkled over the glass batch in place of $H_3O[AuCl_4]$. After reading this chapter, the reader may wish to explore this possibility, and perhaps compare this with using 'Purple of Cassius' as the gold reagent.[18]

Prepare 100 g of gold-ruby glass by the following method: The chemical formulation for the gold-ruby glass is as follows:

Component	Composition (wt%)
SiO_2	60
K_2O	24
B_2O_3	14
Sb_2O_3	2
Au	0.05 ($\equiv$ 500 ppm as elemental gold)

The components, K_2O and Au are added in the form of the compounds K_2CO_3 and $H_3O[AuCl_4]$, respectively; therefore, first calculate the mass of K_2CO_3 and $H_3O[AuCl_4]$ required. Take care when handling $H_3O[AuCl_4]$; this gold compound is toxic, hygroscopic and expensive, and should be stored in an evacuated desiccator on account of its hygroscopicity. Weigh the oxides on weighing paper to a precision of (±0.001 g) and pour them into a 500-ml polyethylene bottle with a screw lid. The gold compound $H_3O[AuCl_4]$ should be weighed last of all, and to a precision of ±0.0005 g. Due to its hygroscopic nature, $H_3O[AuCl_4]$ should be weighed on a small watch-glass (without weighing paper), before placing into the polyethylene bottle, using a small spatula. Wash the watch-glass carefully with a small quantity of cyclohexane, using a disposable teat-pipette, so as to ensure that the entire measure of H_3O $[AuCl_4]$ enters the polyethylene bottle. Add about a dozen zirconia grinding balls ($\sim$ 12 mm diameter) and sufficient cyclohexane ($\sim$ 50 ml) to produce a consistency similar to that of normal paint, then screw the lid on very tightly

---

[18] A modern preparative method for making 'Purple of Cassius' is given by Bishop and Sutton [16].

and ball-mill these powders together overnight. This will result in a light brown-coloured slurry.

After milling, transfer the resultant slurry into a round-bottomed flask. Use a pair of forceps to remove the zirconia grinding balls whilst washing them gently with a small amount of cyclohexane, so as to avoid an unnecessary loss of material. Then attach the flask to a rotary evaporator and a condenser in order to recover the cyclohexane. Retain the zirconia balls for further use.

Upon evaporation, there may be some segregation of the constituents within the powder mixture, therefore carefully stir the dry powders together using a spatula and place them in a large alumina crucible (CC62 Almath Ltd). Place this crucible inside an alumina dish (LR94 Almath Ltd), which will act as a shield and thereby prevent accidental spillage[19] of the contents (particularly the molten glass) onto the furnace floor. Then place this in the chamber furnace and heat at a rate of 86 °C/h to 1400 °C in order to form the molten glass. Programme the chamber furnace in order to hold the temperature at 1400 °C for 5 h before cooling to ambient temperature (20 °C) with a cooling rate of 200 °C/h. It is important to start the heating profile of the furnace at 3 p.m., so as to provide a sensible schedule for the next day.

On the following morning, at 10 a.m. (i.e., after a period of 3 h at 1400 °C), remove the crucible from the furnace and pour the molten glass into the graphite moulds (Ingot IM-05T Graphitestore Inc.), so as to form a set of quenched glass ingots (see Figure 17.4). It is important that the operator for this part of the procedure, wears a face shield, foundry suit and foundry boots, together with good thermally insulated gloves (e.g., aluminium-rayon *värmeskyddshand*; Procurator AB, Sweden), and uses a pair of steel tongs (~0.5 m long) for retrieving the hot crucible. The molten glass can be poured into three or four moulds as required. The furnace (now empty, besides the alumina dish) should remain programmed so as to cool to ambient temperature (20 °C) with a cooling rate of 200 °C/h. Once the glass ingots have cooled (~30 min), observe and note the colour and transparency of the glass at this stage in the preparation.

The glass must now be annealed by a process that is known traditionally as 'striking' the glass, see Figure 17.3. Remove the glass ingots (at ambient temperature) and place them in a large alumina boat (preferably SRX150 or SRX110 Almath Ltd). Place this into the centre of the chamber furnace. Heat in air to 650 °C with a heating rate of 200 °C/h. After 60 h at 650 °C, cool to

---

[19] Due to the effects of surface tension and low viscosity, certain molten glasses (particularly those with high lithium contents) have a strong tendency to flow along ceramic surfaces, which can result in material flowing up and out of the alumina crucible. This feature is exploited during the glazing of pottery.

Figure 17.4 The casting of the 'white-hot' gold-ruby glass melt from the alumina crucible into the graphite mould (Photograph used with kind permission from Eivind Skou. Copyright (2010) Eivind Skou)

ambient temperature (20 °C) with a cooling rate of 200 °C/h. This heating cycle will last ∼72 h (3 days).

Once at ambient temperature, observe and note the colour and transparency of the glass. If the glass has acquired a deep red colour, then retrieve the glass ingot. If the glass is colourless or has only a pale pink coloration, then leave it in the crucible so that the glass can undergo a reannealing at the slightly higher temperature of 675 °C or, for a longer annealing, at 650 °C.[20] The faces of the red coloured glass block can be ground flat and polished, if required, using a grinding/polishing machine (e.g., Struers RotoPol-11). It should be interesting to record the absorption spectrum for the glass, as prepared, and compare it with those reported in the literature, for example, by Badger *et al.* [27], and with Figure 17.2.

The production of gold-ruby glass continues on a modest scale at provincial glass works. Therefore, it should be a worthy pursuit to devise a method to

---

[20] Note that, Badger *et al.* [27] regarded 650 °C as an optimal annealing temperature for *their* soda-lime-silica glass, requiring only 4.5 h in order to produce a deep ruby coloration. The same glass annealed at 675 °C was more bluish. Whereas, the slightly higher temperature of 700 °C resulted in a spoilt glass; blue in transmitted light and brown in reflected light.

produce a genuine self-striking gold-ruby glass. As a final comment, a recent innovation by Eichelbaum *et al.* [28] in the controlled growth of *embryonic* gold particles within silicate glasses, by synchrotron X-ray lithography, resulted in gold enhanced photoluminescent microstructures, within the otherwise, transparent and colourless glass. These microstructures were invisible under normal light, but appeared yellowish green under long-wavelength ultraviolet radiation. In the future, these materials could serve as interesting optical storage media, with a good prospect for longevity.

REFERENCES

1. A. J. Shortland, *Social Influence on the Development and Spread of Glass Technology* Chapter 11, 212–222, in *The Social Context of Technological Change: Egypt and the Near East, 1650–1550 BC*, A. J. Shortland (ed.), Oxbow Books, Oxford, (2001) 288 pp.
2. J. E. Shelby, *Introduction to Glass Science and Technology*, Royal Society of Chemistry, Cambridge, 1997.
3. I. Freestone, N. Meeks, M. Sax and C. Higgitt, *Gold Bulletin* **40** (2007) 270–277.
4. G. Agricola, *De Natura Fossilium*, 1546. Translated by M. C. Bandy and J. A. Bandy, The Geological Society of America Special Paper 63, 1955, pp 240.
5. R. H. Doremus, *Glass Science*, 2nd edn, Wiley-Interscience, New York, (1994) 352 pp.
6. W. Vogel, *Glass Chemistry*, 2nd edn, 1992, Springer-Verlag, Berlin and Heidelberg.
7. R. H. Brill, *Proceedings of the 7th International Congress on Glass*, Brussels, 1965, Volume 2, Paper 223, 1-13.
8. W. A. Weyl, *Coloured Glasses*, Dawson's of Pall Mall, London, 1959.
9. L. B. Hunt, *Gold Bulletin* **9** (1976) 134–139.
10. S. Frank, *Glass Technology* **25** (1984) 47–50.
11. A. Mizaldus, *Memorabilium, utilium ac jucundorum Centuriae IX in awhorismos arcanorum omnium digestae*, Francfurt, 1566.
12. A. Libavius, *Alchymia Andreae Libavii*, 1597.
13. A. Neri, *L'arte Vetraria*, 1612, German translation by J. Kunckel, *Ars Vitraria Experimentalis; oder, vollkommene Glasmacker-Kunst*, 1679.
14. J. R. Glauber, *Des Teutschlandts Wohlfahrt*, Amsterdam, 1659, Part IV, 35–36.
15. A. Cassius (junior), *De Auro*, Hamburg, 1685.
16. P. T. Bishop and P. A. Sutton, *Decorative Gold Materials*, Chapter 15, 317–368 in, *Gold: Science and Applications*, eds. C. Corti and R. Holliday, 2009, CRC Press, Boca Raton, Florida, 444 pp.
17. O. Drahotová, *Glass Review* **28** (1973) 8–11.
18. J. Kunckel, *Collegium Physico-Chymicum Experimentale, oder Laboratorium Chymicum*, ed. J. C. Engelleder, Hamburg, 1716.
19. E. Zschimmer, *Sprechsaal* **34** (1930) 642–644.
20. M. Faraday, *Philosophical Transactions of the Royal Society* **147** (1857) 145–181.
21. F. M. Veazie and W. A. Weyl, *Journal of the American Ceramic Society* **25** (1942) 280–281.
22. R. H. Doremus, *Langmuir* **18** (2002), 2436–2437.

23. F. E. Wagner, S. Haslbeck, L. Stievano, S. Calogero, Q. A. Pankhurst and K-P. Martinek, *Nature* **407** (2000) 691–692.
24. J. Vosburgh and R. H. Doremus, *Journal of Non-Crystalline Solids* **349** (2004) 309–314.
25. S. Haslbech, K.-P. Martinem, L. Stievano and F. E. Wagner, *Hyperfine Interactions* **165** (2005) 89–94.
26. E. Zschimmer, *Sprechsaal* **63** (1930) 832–835 and 852–854.
27. A. E. Badger, W. A. Weyl and H. Rudow, *Glass Industry* **20** (1939) 407–414.
28. M. Eichelbaum, K. Rademann, W. Weigel, B. Löchel, M. Radtke and R. Müller, *Gold Bulletin* **40** (2007) 278–282.

PROBLEMS

1. Why are $H_3O[AuCl_4]$ and K_2CO_3 used as reagents instead of elemental gold and K_2O?

2. Consult the Alfa Aesar Online Catalogue to find the current price for 10 g of hydrogen tetrachloroaurate (III), 99.9% (metals basis), and thereby calculate the cost of the amount of gold compound used in your work.

3. Give an example of another gold compound that could be used to make gold-ruby glass.

4. At what temperature does K_2CO_3 decompose in air? Use the Outokumpu HSC Chemistry® software (or similar), to construct an appropriate predominance diagram for the K_2CO_3 system; such as, P_{CO_2} as a function of T, in order to help answer this question.

5. The maximum temperature during the preparation of the glass was 1400 °C. Although this temperature is very high, it is nevertheless significantly lower than the temperature obtained in *Problem 4* above; and yet it is evident that the K_2CO_3 has decomposed in forming the glass phase. Explain why this is.

6. State the empirical formula for the ternary crystalline phases that exist within the $K_2O–B_2O_3–SiO_2$ ternary system.

7. State a definition of the term, 'base glass'.

8. Categorise the following oxides; K_2O; B_2O_3; SiO_2; PbO; according to the terms; glass formers; glass intermediates; or, glass modifiers. It may help to consult Shelby [2] or Doremus [5] before attempting this question.

9. It would be interesting to know whether the annealing temperature for 'striking' the glass (650 °C) is above or below the glass transition temperature, T_g. Therefore, search the literature to find a value for the glass transition temperature, T_g for:
(a) Pyrex™ glass (borosilicate glass).
(b) A potassium borosilicate glass with a base-glass composition somewhat similar to that as prepared here.

10. State the name of another metallic element that is known to form colloidal particles within a silicate glass matrix and likewise forms a reddish coloured glass?

Index

Page numbers for figures are given in *italic* type.

Synthesis, Properties and Mineralogy of Important Inorganic Materials   By Terence E. Warner
© 2011 John Wiley & Sons, Ltd